Instabilities in space and laboratory plasmas

INSTABILITIES IN SPACE AND LABORATORY PLASMAS

D. B. MELROSE

School of Physics, University of Sydney

CAMBRIDGE UNIVERSITY PRESS

Cambridge
New York   New Rochelle
Melbourne   Sydney

CAMBRIDGE UNIVERSITY PRESS
Cambridge, New York, Melbourne, Madrid, Cape Town, Singapore,
São Paulo, Delhi, Dubai, Tokyo, Mexico City

Cambridge University Press
The Edinburgh Building, Cambridge CB2 8RU, UK

Published in the United States of America by
Cambridge University Press, New York

www.cambridge.org
Information on this title: www.cambridge.org/9780521379625

First published 1986
First paperback edition 1989

A catalogue record for this publication is available from the British Library

Library of Congress Cataloguing in Publication Data
Melrose, D.B., 1940–
Instabilities in space and laboratory plasmas.
Bibliography: p.
Includes index.
1. Plasma instabilities.   2. Space plasmas.
I. Title.
QC718.5.S7M45   1986   530.4´4   86–2233

ISBN 978-0-521-30541-1 Hardback
ISBN 978-0-521-37962-5 Paperback

Contents

Preface

This book is intended as an introduction to the theory of plasma instabilities. It is directed at graduate students, advanced undergraduate students with some background knowledge of plasma physics, and to researchers seeking to become more familiar with the field.

In most applications of plasma physics, plasma instabilities of various kinds play important roles. Some laboratory examples include instabilities limiting inertial or magnetic confinement of fusion plasmas, instabilities which produce enhanced radiation and anomalous transport coefficients in current-carrying plasmas, and instabilities which provide coherent sources of radiations in gyrotrons and free electron masers. Some examples from space plasmas include instabilities which produce nonthermal wave and particle distributions in the magnetosphere and the interplanetary medium and nonthermal radiation from the planets and the solar corona, instabilities involved in nonlinear propagation effects in the ionosphere, and instabilities leading to scattering and acceleration of fast particles in astrophysical plasmas. The richness and variety of the plasma instabilities and the diversity of their applications preclude any thorough treatment in a single book. In this book I have attempted to be thorough only in the coverage of the qualitative kinds of plasma instability which are possible. The main emphasis is on instabilities at moderate to high frequencies, that is frequencies from about the ion gyrofrequency to above all the natural frequencies. However, I felt it important to include at least brief discussions of the low frequency instabilities (the MHD instabilities, tearing instabilities, fluid instabilities and drift instabilities). No attempt has been made to be thorough in the discussion of the applications. The choices made reflect my research interests an those of graduate students who attended my lectures at the University of Sydney. These interests include magnetospheric physics, solar radiophysics, plasma astrophysics, the gyrotron and RF heating of plasmas. Also I have not attempted to be thorough in reviewing the literature. References are given only sparingly in the text, and 'Bibliography notes' are included to direct the interested reader to sources where the literature is cited more extensively.

The historical development of plasma physics in general and of the piecemeal development of the theory of plasma instabilities in particular has left us with some unfortunate legacies in terminology and nomenclature. Names used to describe plasma instabilities include some which are historical (e.g. the Buneman and Weibel instabilities), some which are descriptive of the source of free energy (e.g. bump-in-tail instability, loss-cone instability), some which are descriptive of the growth mechanism (e.g. maser instability, resistive instability), some which are descriptive of the waves involved (e.g. ion sound instability, drift instability), and some which are descriptive of the theory used (e.g. MHD instability, parametric instability). Needless to say, the lack of a systematic nomenclature makes it difficult for those trying to become familiar with the field. However, it has a more serious effect in that it tends to obscure some actual taxonomic problems in the classification of various kinds of instability. A relevant example is the distinction made here between 'reactive' and 'resistive' instabilities, cf. Chapters 3 and 4. Although this distinction is widely recognized there is no widely accepted nomenclature for the two types of instability and there is no widely adopted classification scheme which distinguishes clearly between them. The student or reader cannot avoid these difficulties with nomenclature, which simply has to be mastered in order to understand the literature.

It was my intention to write this book in a leisurely fashion starting in September 1984, based on lecture notes prepared in 1983. However, when Ray Grimm, who had just started presenting a lecture course to Honours Year students, died suddenly in August 1984, I took over the course and prepared and used the first nine chapters of this book as lecture notes for the course. The remaining chapters were written in November 1984 and I lectured using them to most of these (newly graduated) and some other graduate students early in 1985. These students helped me with corrections and other criticisms, with the assistance provided by Lewis Ball being worthy of special mention. Nearly all the manuscript was typed by Pat Moroney, who maintained her very high standards despite the difficulties involved.

School of Physics   *D. B. Melrose*
The University of Sydney

Plasma formulary

Physical constants

Physical quantity	Symbol	Value	Units
electron mass	m_e	9.11×10^{-31}	kg
proton mass	m_p	1.67×10^{-27}	kg
elementary charge	e	1.60×10^{-19}	C
speed of light	c	3.00×10^8	$\mathrm{m\,s^{-1}}$
Plank constant	$\hbar = h/2\pi$	1.05×10^{-34}	J s
Boltzmann constant*		1.3807×10^{-23}	$\mathrm{J\,K^{-1}}$
		$1/1.1605 \times 10^4$	$\mathrm{eV\,K^{-1}}$
classical electron radius	r_0	2.82×10^{-15}	m
Thomson cross section	$(8\pi/3)r_0^2$	6.65×10^{-29}	$\mathrm{m^2}$
energy of 1 eV		1.60×10^{-19}	J
permittivity of free space	ε_0	8.85×10^{-12}	$\mathrm{F\,m^{-1}}$
permeability of free space	μ_0	$4\pi \times 10^{-7}$	$\mathrm{H\,m^{-1}}$

*Throughout this book Boltzmann's constant is set equal to unity, implying that temperatures are expressed in energy units. The results quoted are conversion factors to kelvins.

Plasma constants (SI units)

Physical quantity	Formula	Value	Units
electron plasma frequency	$\omega_p = \left(\dfrac{n_e e^2}{\varepsilon_0 m_e}\right)^{1/2}$	$56.4 n_e^{1/2}$	$\mathrm{s^{-1}}$
ion plasma frequency	$\omega_{pi} = \left(\dfrac{n_i q_i^2}{\varepsilon_0 m_i}\right)^{1/2}$	$1.32 Z_i n_i^{1/2} A_i^{-1/2}$	$\mathrm{s^{-1}}$
electron gyrofrequency	$\Omega_e = \dfrac{eB}{m_e}$	$1.76 \times 10^{11} B$	$\mathrm{s^{-1}}$
ion gyrofrequency	$\Omega_i = \dfrac{q_i B}{m_i}$	$9.85 \times 10^7 Z_i B A_i^{-1}$	$\mathrm{s^{-1}}$

Plasma constants (SI units)

Physical quantity	Formula	Value	Units
Alfvén speed	$v_A = \dfrac{B}{(\mu_0 \eta)^{1/2}}$	$2.18 \times 10^{16} \, B n_i^{-1/2} A_i^{-1/2}$	$\mathrm{m\,s^{-1}}$
ion sound speed	$v_s = \left(\dfrac{T_e}{m_i}\right)^{1/2}$	$90.9 T_e^{1/2} A_i^{-1/2}$	$\mathrm{m\,s^{-1}}$
electron thermal speed	$V_e = \left(\dfrac{T_e}{m_e}\right)^{1/2}$	$3.89 \times 10^3 T_e^{1/2}$	$\mathrm{m\,s^{-1}}$
ion thermal speed	$V_i = \left(\dfrac{T_i}{m_i}\right)^{1/2}$	$90.9 T_i^{1/2} A_i^{-1/2}$	$\mathrm{m\,s^{-1}}$
electron Debye length	$\lambda_{De} = \dfrac{V_e}{\omega_p}$	$69 T_e^{1/2} n_e^{-1/2}$	m
plasma beta	$\beta = \dfrac{n_e T_e}{B^2/2\mu_0}$	$3.47 \times 10^{-28} n_e T_e B^{-2}$	—
collision frequency	$\nu_0 = \dfrac{\omega_p \ln \Lambda}{4\pi n_e \lambda_{De}^3}$	$13.7(\ln \Lambda) n_e T_e^{-3/2}$	$\mathrm{s^{-1}}$

All quantities are in SI units, i.e. number densities are per cubic metre, temperatures are in kelvins, B is in tesla, and A_i is the mass of the ion in units of the proton mass.

Conversion factors (Gaussian units/SI units)

Physical Quanitity	Symbol	Factor	Units
charge	q	3×10^9	statcoulomb/coulomb
electrical potential	Φ	$1/(3 \times 10^2)$	statvolt/volt
electric field	E	$1/(3 \times 10^4)$	statvolt cm^{-1}/volt m^{-1}
magnetic field	B	10^4	gauss/tesla
electrical conductivity	σ	9×10^9	s^{-1}/mho m^{-1}

PART I
Introduction to plasma theory

1
Introduction

1.1 Preliminary remarks

One of the most obvious features of the plasma state is the rich variety of wave motions which plasmas can support. Waves of a particular kind are said to be in a particular wave mode. The idea of a wave mode is familiar from other contexts. For example in a compressible gas there are *sound waves* and, if there is a gravitational field present, there are also *internal gravity waves*. These waves have specific dispersion relations, which relate the frequency ω to the wave vector $\mathbf{k}$. For sound waves and internal gravity waves the dispersion relations are $\omega = kc_s$ and $\omega = (gk)^{1/2}$ respectively, where c_s is the sound speed and g is the gravitational acceleration. One could cite numerous examples of wave modes in other media, e.g. spin waves in ferromagnetic media and seismic waves in the solid Earth, each of these is characterized by its dispersion relation and other properties which determine the nature of the wave motion. The wave modes of a plasma depend on the plasma properties and these are described in terms of various plasma parameters. Of particular importance are the natural frequencies of the plasma: the electron plasma frequency ω_p, the electron cyclotron frequency Ω_e, the corresponding ion frequencies ω_{pi} and Ω_i, and various collision frequencies ($\nu_{ei}, \nu_{ee}, \nu_{ii}$) between electrons (e) and ions (i). In most plasmas these frequencies are ordered from highest to lowest as follows: $\omega_p, \Omega_e, \omega_{pi}, \Omega_i, \nu_{ei}, \nu_{ee}, \nu_{ii}$. The properties of the wave modes are strong functions of frequency from the high-frequency modes ($\omega \gtrsim \omega_p, \Omega_e$) to low-frequency modes ($\omega \ll$ all natural frequencies).

Unfortunately the nomenclature for the various wave modes of plasmas is far from systematic, being a mixture of historical names (e.g. Langmuir, Alfvén, Bernstein), of names descriptive of the wave motion (e.g. ion acoustic waves, electron cyclotron waves, transverse waves), and of names characteristic of the theory used to derive the wave properties (e.g. magnetoionic waves, MHD waves, drift waves).

A *plasma instability* involves waves in some mode growing exponentially or faster. The nomenclature for plasma instabilities is even more cumbersome than for the wave modes themselves. First, the growing mode needs to be identified. In addition one would like to indicate both the nature of the wave growth and also the source of free energy driving the instability. Classification schemes have been presented by Mikhailovskii (1974a, b) and by Cap (1976), the latter being attributed to B. Lehnert. One important feature of the classification schemes is the distinction between macroinstabilities and microinstabilities.

A *macroinstability* is driven by the structure of the medium in configuration space. A familiar example of a macroinstability is for a convectively unstable system: when the temperature gradient is superadiabatic internal gravity waves grow to large amplitude and cause a large-scale convection of the fluid, which tends to reduce the temperature gradient. Other familiar examples are the Rayleigh-Jeans instability when a denser fluid is supported by a less dense fluid, and the Kelvin–Helmholtz instability when one fluid flows over another fluid, e.g. wind over water, causing surface waves to grow. In plasmas the macroinstabilities occur in the low-frequency regime and usually involve the magnetic field. Examples include flute (or interchange) and ballooning instabilities.

Microinstabilities are usually driven by a velocity space anisotropy in the plasma. A consequence of a microinstability is a greatly enhanced level of fluctuations in the plasma associated with the unstable mode. These fluctuations are called *microturbulence*. Microturbulence can lead to enhanced radiation from the plasma and to enhanced scattering of particles resulting in 'anomalous' transport coefficients, e.g. anomalous electric and thermal conductivities.

The simplest example of a microinstability is a beam-driven instability in an unmagnetized plasma. The existence of such an instability was indicated by Langmuir in his pioneering investigation of arc discharges in the mid 1920s; however the earliest treatments of such instabilities were not given until 1948, cf. the detailed discussion by Bohm and Gross (1949a, b). In introducing plasma instabilities here, we start with the case of an unmagnetized plasma. In the remainder of this Chapter the properties of the waves in such a plasma are outlined, partly in a historical context, and then the response tensor for the plasma is defined and the causal condition is discussed. More detailed discussions of the response tensor and of the wave properties are given in Chapter 2, and specific instabilities are discussed in Chapters 3 and 4.

Before proceeding it is appropriate to explain some terminology. A plasma is said to be *unmagnetized* if the spiralling motion of particles in any ambient magnetic field is ignored. A given plasma may be regarded as unmagnetized for some purposes, e.g. in treating Langmuir waves, and yet the inclusion of the magnetic field may be important in another connection, e.g. in discussing the polarization of escaping radiation. Similarly a plasma, or a particle species in the plasma, is said to be *cold* if the thermal motions are neglected. Again a given plasma may be regarded as cold for some purposes and yet inclusion of the thermal motions might be essential for other purposes. Another important example is that a *collisionless* plasma means a plasma in which interparticle collisions are neglected. All plasmas are collision dominated at sufficiently low frequencies, and most plasmas may be regarded as collisionless, at least to a first approximation, at frequencies of order ω_p and Ω_e.

Initially we concentrate here on unmagnetized collisionless plasmas. Collision-dominated and magnetized plasmas are discussed in Chapters 8 & 9 and 10 to 13 respectively.

1.2 Langmuir waves and ion sound waves

The name 'plasma' was coined by Langmuir in the mid 1920's to mean an ionized gas. Langmuir investigated arc discharges and he identified certain phenomena with wave motions in the arc plasma. The theory for these was presented by Tonks & Langmuir (1929); in modern terminology their theory is for a thermal (i.e. Maxwellian velocity distributions) unmagnetized plasma. The only independent plasma parameters are the electron plasma frequency ω_p, the ion plasma frequency ω_{pi}, the electron thermal speed V_e and the ion thermal speed V_i. Let n_e and n_i be the number densities, $-e$ and $Z_i e$ be the charges, m_e and m_i be the masses and T_e and T_i be the temperatures of the electrons and ions respectively. Then we have

$$\omega_p = \left(\frac{n_e e^2}{\varepsilon_0 m_e}\right)^{1/2}, \qquad \omega_{pi} = \left(\frac{Z_i^2 n_i e^2}{\varepsilon_0 m_i}\right)^{1/2} \tag{1.1a, b}$$

and

$$V_e = \left(\frac{T_e}{m_e}\right)^{1/2}, \qquad V_i = \left(\frac{T_i}{m_i}\right)^{1/2}. \tag{1.2a, b}$$

Note that we set Boltzmann's constant equal to unity and measure temperatures in units of energy (1 eV corresponds to 1.16×10^4 K).

There are three wave modes in a thermal unmagnetized plasma. One is well known: transverse waves in a plasma have refractive index $N = (1 - \omega_p^2/\omega^2)^{1/2}$. The two additional modes discussed by Tonks & Langmuir are both *longitudinal* (also called electrostatic), i.e. the electric vector is parallel to the wave vector **k** and there is no wave magnetic field. One of these is now called the *Langmuir mode*. It has dispersion relation $\omega = \omega_L(\mathbf{k})$ with

$$\omega_L(\mathbf{k}) \approx (\omega_p^2 + 3k^2 V_e^2)^{1/2} \tag{1.3}$$

for $k^2 \lambda_{De}^2 \ll 1$ where

$$\lambda_{De} = V_e/\omega_p \tag{1.4}$$

is the electron Debye length. These waves have frequency close to the plasma frequency and in some of the older literature they are called electron plasma oscillations. The other mode identified by Tonks & Langmuir is now called either the *ion sound mode* or the *ion acoustic mode*. The dispersion relation is $\omega = \omega_s(\mathbf{k})$ with

$$\omega_s(\mathbf{k}) \approx \frac{k v_s}{(1 + k^2 \lambda_{De}^2)^{1/2}} \tag{1.5}$$

where

$$v_s = \omega_{pi} \lambda_{De} \tag{1.6}$$

is the *ion sound speed*. For $k^2 \lambda_{De}^2 \ll 1$ the dispersion relation reduces to

$\omega \approx kv_s$, which is of the form of that for sound waves. For $k^2\lambda_{De}^2 \gtrsim 1$ the waves have frequency close to ω_{pi} and are called *ion plasma oscillations*. Note that the same mode (1.5) is given two different names in different regimes ($k^2\lambda_{De}^2 \ll 1$ and $k^2\lambda_{De}^2 \gtrsim 1$). This is often the case and reflects the fact that the identification of particular modes cannot always be made in a unique way and is often model dependent. Further discussion of these wave properties is given in §§2.4 and 2.5 below.

At this stage let us digress briefly to remark on the history of the development of the theory of waves in plasmas and of plasma theory itself. Just after the work of Langmuir and his colleagues on arc discharges, the study of the propagation of radio waves in the ionosphere led to the magnetoionic theory (Hartree 1931 and Appleton 1932), which is the theory for waves in a cold magnetized plasma (§10.3). The inclusion of the motion of the ions, to give what is now called cold plasma theory, was not made for nearly two decades (Åström 1950). In the meantime, motivated by primarily astrophysical applications, Alfvén (1942) had investigated the properties of a plasma from a fluid viewpoint; assuming the fluid to be highly conducting, magnetized and incompressible, Alfvén predicted the wave motions which now bear his name. The generalization to the compressible case was made by Herlofsen (1950); the waves of a magnetized fluid are called collectively the MHD waves (§8.2). The general theory of waves, based on the kinetic theory of plasmas, with both the magnetic field and thermal motions included was developed in the mid 1950's, e.g. Sitenko & Stepanov (1957). The history of the rapid development of plasma physics in the early 1950's is obscured by the fact that much of the important work was classified.

The kinetic theory of plasmas was developed through the late 1930's and 1940's. One of the important aspects of kinetic theory, especially from the viewpoint of instability theory, is the treatment of collisionless or Landau damping (Landau 1946). Landau damping may be interpreted as the absorption process corresponding to Cerenkov emission. This is discussed further in the next section.

1.3 Cerenkov emission and Landau damping

Cerenkov emission was discovered experimentally by Cerenkov in 1934 as a bluish glow from water near some radioactive material emitting β-rays. The theory for Cerenkov emission, first presented by Tamm & Frank in 1937, is based on the fact that when a particle moves at faster than the phase speed ($v_\phi = c/N = \omega/k$) of waves in any specific wave mode it radiates waves in that mode. Note that the statement that only accelerated particles radiate is not correct when there are waves with $N > 1$. A familiar example of a process analogous to Cerenkov emission is the emission of sound waves by a supersonic aircraft.

Langmuir waves have phase speeds which can be as low as several times V_e, and thus fast electrons or ions with $v \gg V_e$ can emit Langmuir waves with $\omega/k \lesssim v$ through Cerenkov emission. Even electrons in the tail of the Maxwellian distribu-

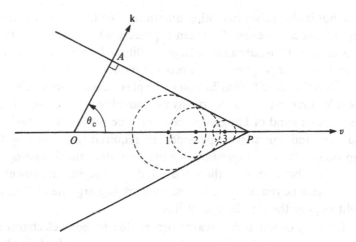

Fig. 1.1 The geometry for Cerenkov emission: a particle at O at $t = 0$ emits waves with phase speed $v_\phi = \omega/k$. At t the particle is at P a distance $OP = vt$ from O and the wavefront reaches A a distance $OA = v_\phi t$ from O. The wavefronts emitted at all times between 0 and t form a bowfront PA, as indicated for fronts emitted at $t = 1, 2, 3$. The Cerenkov angle θ_c is also indicated.

tion ($v \gtrsim$ several V_e) can emit Langmuir waves. The condition for emission, called the *Cerenkov condition*, is

$$\omega - \mathbf{k}\cdot\mathbf{v} = 0. \tag{1.7}$$

The angle $\theta_c = \arccos(\omega/kv)$ is called the Cerenkov angle. The waves form a front, as for a supersonic aircraft or for the bowfront of a ship moving through water, which is at an angle $\pi/2 - \theta_c$ to the direction $\mathbf{v}$, cf. Figure 1.1.

The damping of Langmuir waves includes a contribution from collisions, but this is weak for $v_{ei} \ll \omega_p$. The most important damping mechanism was first pointed out by Landau (1946). The physical nature of *Landau damping* is not apparent from its derivation: in the kinetic theory the response tensor involves a denominator $\omega - \mathbf{k}\cdot\mathbf{v}$ inside an integral over velocity space, and the *Landau prescription* is a procedure for integrating around the pole at $\omega - \mathbf{k}\cdot\mathbf{v} = 0$. This prescription is based on a causal argument (cf. §1.5 below), and it is not obvious why the imposition of causality should imply the existence of collisionless damping. However, one can derive Landau damping in a different way. To every emission process there must be a corresponding absorption process, because otherwise the second law of thermodynamics would be violated. It follows that Langmuir waves with given ω/k can be absorbed by plasma particles with $v > \omega/k$ by the inverse of Cerenkov emission. When one calculates this absorption for thermal electrons one finds that the absorption coefficient is that for Landau damping. Hence Landau damping is equivalent to Cerenkov absorption. (In the presence of a beam of fast particles Landau damping can be negative, leading to growth of Langmuir waves; this so-called bump-in-the-tail instability is discussed in §4.1).

The foregoing argument provides a physical interpretation of Landau damping on one level, but it also raises two other questions. The first concerns the fact that by a 'damping' process we usually mean a process which creates entropy, i.e. a randomizing process. Consider a single large amplitude Langmuir wave in a plasma. It damps and its energy goes into electrons with $v \approx \omega/k$ which are slightly accelerated. Thus the velocity distribution in the plasma is distorted. This distortion would persist for ever in the absence of any randomizing or 'phase-mixing' process. In practice a broad band of Langmuir waves may be regarded as having random phases and this randomness itself provides the required phase mixing for a true damping to occur. The other question is more profound: the foregoing argument implies a relation between causality (the Landau damping argument) and the second law of thermodynamics (the Cerenkov damping argument). The interested reader might ponder the significance of this.

Landau damping of ion sound waves occurs due to both electrons and ions. Almost all electrons contribute to the damping because of the low phase speed $\omega/k \approx v_s$. This damping is relatively strong, with a damping rate of order v_s/V_e times the frequency, cf. (2.51) below. The damping by the ions is so strong that it prevents the waves from existing for $\omega/k \approx v_s \lesssim V_i$, when most of the thermal ions contribute. Ion sound waves exist only for $v_s \gg V_i$ which corresponds to $Z_i T_e \gg T_i$. Thus ion sound waves exist in a hydrogen plasma only if the electron temperature is much greater than the ion temperature.

Transverse waves are not subject to Landau damping because they have phase speeds greater than the speed of light: the condition (1.7) cannot be satisfied for $\omega/k > c$ and $v < c$. Collisionless damping of transverse waves is possible in a magnetized plasma. The Cerenkov condition (1.7) is then replaced by a resonance condition

$$\omega - s\Omega - k_\parallel v_\parallel = 0 \qquad (1.8)$$

where $\parallel$ denotes components parallel to the magnetic field, where $\Omega = |q|B/m\gamma$ is the gyrofrequency of the particle, and where $s = 0, \pm 1, \pm 2, \ldots$ is the harmonic number. For waves with refractive index less than unity (1.8) can be satisfied only for positive s.

When a magnetic field is included the only waves which can escape directly from an astrophysical plasma have refractive index less than unity. Such waves cannot be generated by Cerenkov emission or by an instability involving negative Landau damping. They can be generated by an electron cyclotron instability (§11.3) in which (1.8) is satisfied with $s = 1, 2 \ldots$. (An instability which can lead to direct emission of transverse waves in an unmagnetized plasma (or in vacuo) is the free-electron maser (§7.3).) Specifically the only waves which can escape directly from an astrophysical plasma are o-mode waves at $\omega > \omega_p$ and x-mode waves at $\omega > \omega_x$ (cf. §10.3). Other waves encounter a cutoff or a resonance along prospective escape paths and are either reflected or absorbed. This severe limitation on the waves which can escape applies to laboratory plasmas but is not so relevant because surface layers can couple energy from microturbulence inside the plasma into escaping radiation. In astrophysical plasmas surface layers are usually ineffective in leading to such so-called mode coupling, whereas in laboratory plasmas the gradients are large in the surface layers and such coupling can be quite efficient. Waves in the Earth's magnetosphere, such as whistler waves (§§13.2 & 13.3) and other low frequency waves (e.g. §12.3) can be detected on the ground again because of a boundary layer; in this case the oscillating currents set up by the waves in the ionosphere (which separates the magnetosphere from the non-conducting atmosphere) lead to radiation which can propagate in the non-conducting region between the ionosphere and the ground.

In astrophysical plasmas the term 'radiation' is used to refer to escaping radiation. The number of 'radiation processes' is actually quite small, including bremsstrahlung, gyromagnetic emission, 'plasma emission' (§6.6) and perhaps several others. The same applies to laboratory plasmas and to the magnetosphere, but in these contexts the term 'radiation' is not always used in such a strict sense.

1.4 The response tensor

The response of a medium in the older branch of electromagnetic theory, called the electrodynamics of continuous media or phenomenological electrodynamics, involves a dielectric tensor (or, equivalently, an electric susceptibility tensor) for the electric response, a magnetic permeability tensor (or a magnetic susceptibility tensor) for the magnetic response, and also magneto-electric response tensors for magneto-electric media. This description cannot be avoided when considering the static response of a medium, but there is a simpler description available for the response at non-zero frequency. Indeed this older description becomes ill-defined due to the fact that there is no unique separation of the electromagnetic field into electric and magnetic parts when the field is varying in both time and space.

The procedure used to describe the response of a plasma is based on Fourier transforming in both time and space and including the entire electromagnetic disturbance in one field vector and the entire electromagnetic response in one vectorial quantity. One then requires only one tensorial quantity, called the response tensor, which relates the response to the disturbance.

Consider a function $f(t, \mathbf{x})$. Its Fourier transform in time and space is defined by

$$\tilde{f}(\omega, \mathbf{k}) = \int dt d^3\mathbf{x} \, e^{i(\omega t - \mathbf{k} \cdot \mathbf{x})} f(t, \mathbf{x}) \tag{1.9}$$

and the inverse transform is

$$f(t, \mathbf{x}) = \int \frac{d\omega \, d^3\mathbf{k}}{(2\pi)^4} e^{-i(\omega t - \mathbf{k} \cdot \mathbf{x})} \tilde{f}(\omega, \mathbf{k}). \tag{1.10}$$

Where no confusion should result we shall omit the tilde on Fourier transformed quantities. Some properties of Fourier transforms are outlined in Exercise 1.4.

Maxwell's equations are

$$\operatorname{curl} \mathbf{E}(t, \mathbf{x}) = -\frac{\partial \mathbf{B}(t, \mathbf{x})}{\partial t}, \tag{1.11}$$

$$\operatorname{div} \mathbf{B}(t, \mathbf{x}) = 0, \tag{1.12}$$

$$\operatorname{curl} \mathbf{B}(t, \mathbf{x}) = \mu_0 \mathbf{J}(t, \mathbf{x}) + \frac{1}{c^2} \frac{\partial \mathbf{E}(t, \mathbf{x})}{\partial t}, \tag{1.13}$$

$$\operatorname{div} \mathbf{E}(t, \mathbf{x}) = \frac{\rho(t, \mathbf{x})}{\varepsilon_0}, \tag{1.14}$$

where $\mathbf{E}$ is the electric field strength, $\mathbf{B}$ is the magnetic induction and ρ and $\mathbf{J}$ are the charge and current densities. After Fourier transforming in time and space, (1.11) to (1.14) become

$$\mathbf{k} \times \mathbf{E}(\omega, \mathbf{k}) = \omega \mathbf{B}(\omega, \mathbf{k}), \tag{1.15}$$

$$\mathbf{k} \cdot \mathbf{B}(\omega, \mathbf{k}) = 0, \tag{1.16}$$

$$\mathbf{k} \times \mathbf{B}(\omega, \mathbf{k}) = -i\mu_0 \mathbf{J}(\omega, \mathbf{k}) - \frac{\omega}{c^2} \mathbf{E}(\omega, \mathbf{k}), \tag{1.17}$$

$$\mathbf{k} \cdot \mathbf{E}(\omega, \mathbf{k}) = -\frac{i\rho(\omega, \mathbf{k})}{\varepsilon_0}. \tag{1.18}$$

Let us also note (1.14) and (1.13) imply the equation of charge continuity

$$\frac{\partial}{\partial t} \rho(t, \mathbf{x}) + \operatorname{div} \mathbf{J}(t, \mathbf{x}) = 0 \tag{1.19}$$

whose Fourier transform is

$$\omega \rho(\omega, \mathbf{k}) - \mathbf{k} \cdot \mathbf{J}(\omega, \mathbf{k}) = 0. \tag{1.20}$$

We may replace (1.18) by (1.20) without altering the physical content of the set of Fourier transformed equations.

Equation (1.15) relates $\mathbf{B}(\omega, \mathbf{k})$ to $\mathbf{E}(\omega, \mathbf{k})$, and implies that $\mathbf{B}(\omega, \mathbf{k})$ contains no information which is not already implicit in $\mathbf{E}(\omega, \mathbf{k})$. We may therefore re-interpret (1.15) as a definition of a subsidiary vector $\mathbf{B}(\omega, \mathbf{k})$ and regard $\mathbf{E}(\omega, \mathbf{k})$ as containing all information on the (non-static) electromagnetic field. Equation (1.16) is redundant, being implied by the scalar product of (1.15) with $\mathbf{k}$. Equation (1.18) has been replaced by (1.20) which may be re-interpreted as defining a subsidiary scalar quantity $\rho(\omega, \mathbf{k})$ in terms of $\mathbf{J}(\omega, \mathbf{k})$. We may regard the total response of the medium as being included in $\mathbf{J}(\omega, \mathbf{k})$. Thus our set of Maxwell's equations reduces to a redundant equation, two subsidiary equations and a basic equation, called the *wave equation* for $\mathbf{E}(\omega, \mathbf{k})$ in terms of $\mathbf{J}(\omega, \mathbf{k})$:

$$\frac{c^2}{\omega^2} \mathbf{k} \times \{\mathbf{k} \times \mathbf{E}(\omega, \mathbf{k})\} + \mathbf{E}(\omega, \mathbf{k}) = -\frac{i\mu_0 c^2}{\omega} \mathbf{J}(\omega, \mathbf{k}). \tag{1.21}$$

In tensor notation (1.21) becomes

$$\left\{ \frac{c^2}{\omega^2} k_i k_j - \frac{c^2 k^2}{\omega^2} \delta_{ij} + \delta_{ij} \right\} E_j(\omega, \mathbf{k}) = -\frac{i\mu_0 c^2}{\omega} J_i(\omega, \mathbf{k}), \tag{1.22}$$

where i and j run over Cartesian components (x, y, z or $1, 2, 3$), and the sum over the repeated index j is implied. The Kronecker delta

$$\delta_{ij} = \begin{cases} 1 & \text{for } i = j \\ 0 & \text{for } i \neq j \end{cases} \tag{1.23}$$

corresponds to the unit tensor.

We now postulate that $\mathbf{J}(\omega, \mathbf{k})$ may be separated into an induced (ind) part and an extraneous (ext) part,

$$\mathbf{J}(\omega, \mathbf{k}) = \mathbf{J}^{\text{ind}}(\omega, \mathbf{k}) + \mathbf{J}^{\text{ext}}(\omega, \mathbf{k}), \tag{1.24}$$

and that the induced part may be expanded in powers of $\mathbf{E}(\omega, \mathbf{k})$:

$$\mathbf{J}^{\text{ind}}(\omega, \mathbf{k}) = \sum_{n=1}^{\infty} \mathbf{J}^{(n)}(\omega, \mathbf{k}). \tag{1.25}$$

By hypothesis $\mathbf{J}^{(1)}$ is a linear function of $\mathbf{E}$, and hence must be of the form

$$J_i^{(1)}(\omega, \mathbf{k}) = \sigma_{ij}(\omega, \mathbf{k}) E_j(\omega, \mathbf{k}). \tag{1.26}$$

The tensor σ_{ij} is the *conductivity tensor*.

The response may be described in a variety of different ways, cf. Exercise 1.2. The most widely used form of the response is in terms of the equivalent dielectric (or permittivity) tensor $\varepsilon_{ij}(\omega, \mathbf{k}) = \varepsilon_0 K_{ij}(\omega, \mathbf{k})$. The tensor $K_{ij}(\omega, \mathbf{k})$ is convenient because it is dimensionless. It may be defined by

$$K_{ij}(\omega, \mathbf{k}) = \delta_{ij} + \frac{i}{\varepsilon_0 \omega} \sigma_{ij}(\omega, \mathbf{k}). \tag{1.27}$$

If we make the separation (1.24) in the wave equation (1.22) and include $\mathbf{J}^{(1)}$ on the left hand side, then we obtain the alternative form for the wave equation

$$\Lambda_{ij}(\omega, \mathbf{k}) E_j(\omega, \mathbf{k}) = -\frac{i\mu_0 c^2}{\omega} J_i^{\text{ext}}(\omega, \mathbf{k}) \tag{1.28}$$

with

$$\Lambda_{ij}(\omega, \mathbf{k}) = \frac{c^2}{\omega^2}(k_i k_j - k^2 \delta_{ij}) + K_{ij}(\omega, \mathbf{k}), \tag{1.29}$$

and where the nonlinear responses ($n \gtrsim 2$ in (1.25)) have been included in $\mathbf{J}^{\text{ext}}$, i.e. as source terms.

The response tensor $K_{ij}(\omega, \mathbf{k})$ may be regarded as completely specifying the (linear) electromagnetic properties of a medium. Thus by a 'thermal unmagnetized plasma' one means that $K_{ij}(\omega, \mathbf{k})$ is taken to be the expression obtained under the assumption that the plasma is thermal and unmagnetized. The treatment of waves and instabilities in plasmas is thereby reduced (a) to calculating $K_{ij}(\omega, \mathbf{k})$ given a specific model for the plasma, and (b) to solving the wave equation (1.28).

It should be emphasized that 'dielectric tensor' as used in plasma physics has a different meaning from its use in the electrodynamics of continuous media, and 'equivalent dielectric tensor' is used when any confusion might result. There are two alternative descriptions, one involving both electric and magnetic responses (and maybe magneto-electric response) while the other involves a single electro-magnetic response. This distinction is illustrated by comparing the two descriptions for an isotropic medium in Exercise 1.1.

1.5 The Landau prescription

Landau (1946) imposed the causal condition on the response tensor by using a Laplace transform in time rather than a Fourier transform. Let $F(p)$ be the Laplace transform of $f(t)$

$$F(p) = \int_0^\infty dt\, e^{-pt} f(t). \tag{1.30}$$

The inverse transform is

$$f(t) = \int_{-i\infty+\Gamma}^{i\infty+\Gamma} dp\, e^{pt} F(p), \tag{1.31}$$

where Γ is chosen such that the contour of integration is to the right of all singularities of $F(p)$ in the complex p plane, cf. Figure 1.2a. We may deform the contour of integration so that it runs along the imaginary p axis and deviates to avoid poles on the imaginary p axis and in the right half plane as indicated in Figure 1.2b.

Simply by writing $p = -i\omega$ in (1.30) we may identify $F(-i\omega)$ as the (temporal) Fourier transform of $f(t)$ for $t > 0$. A *causal function* is one which vanishes for $t < 0$. For example, (1.26) must describe the response at time t to disturbances at earlier times $t' \leqslant t$, and hence $\sigma_{ij}(\omega, \mathbf{k})$ must be the Fourier transform of a function of $t - t'$ which vanishes for $t - t' < 0$. Hence for a causal function we must interpret the ω-integral in the inverse transform in the manner implied by the Laplace transform. This requires that the contour be in the upper half ω-plane above all singularities. We may deform this contour so that it is along the real ω-axis and deviates to avoid poles as illustrated in Figure 1.3.

Consider a pole on the real axis at $\omega = \omega_0$. The contour is deformed so that it involves a semicircle in the upper half plane centred on $\omega = \omega_0$, and the radius of this semicircle may be allowed to shrink to zero. Then the integral consists of parts along the real axis from $-\infty$ to $\omega_0 - \eta$ and from $\omega_0 + \eta$ to ∞ with η allowed to go to zero, plus the contribution from the semicircle. These two contributions correspond to the Cauchy principal value of the integral and to the 'semiresidue' ($-i\pi$ times the residue) at $\omega = \omega_0$. This result may be summarized in terms of the Plemelj formula

$$\frac{1}{\omega - \omega_0 + i0} = P\frac{1}{\omega - \omega_0} - i\pi\delta(\omega - \omega_0), \tag{1.32}$$

where P denotes the Cauchy principal value of the integral. The left hand side corresponds to placing the pole at $\omega = \omega_0 - i0$ infinitesimally below the contour, rather than leaving the pole at $\omega = \omega_0$ and deviating the contour

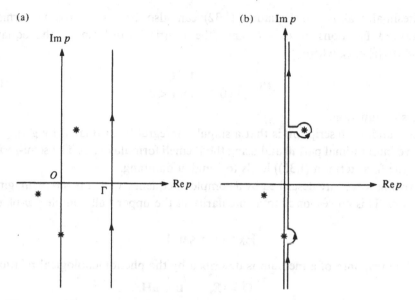

Fig. 1.2(a) The contour of integration in the complex p-plane is to the right of all
singularities; three singularities are denoted by *.
(b) The contour may be deformed so that it lies along the imaginary p-axis and
deviates around poles on the axis and to the right of it.

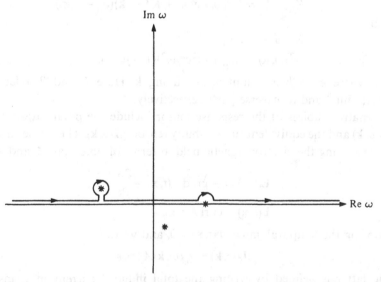

Fig. 1.3 The contour in Fig. 1.2(b) is redrawn in the complex $\omega(=ip)$ plane
showing that the contour must deviate into the upper half plane around any
poles there or on the real ω-axis.

infinitesimally above it. Equation (1.32) can also be interpreted in terms of generalized functions; it is $-i$ times the Fourier transform of the equation $H(t) = \frac{1}{2}(t/|t| + 1)$, where

$$H(t) = \begin{cases} 1 & \text{for } t > 0 \\ 0 & \text{for } t < 0 \end{cases} \tag{1.33}$$

is the step function.

The Landau prescription is that a singular integral be evaluated by giving ω a positive infinitesimal part $i0$ and using the Plemelj formula (1.32). The semi-residue (from the final term in (1.32)) leads to Landau damping.

Growing waves are described by a complex frequency with a positive imaginary part of ω. This corresponds to a singularity in the upper half complex ω-plane.

Exercise set 1

1.1 The response of a medium is described by the phenomenological relations

$$\mathbf{D} = \varepsilon \mathbf{E}, \qquad \mathbf{B} = \mu \mathbf{H},$$

with $\mathbf{D} = \varepsilon_0 \mathbf{E} + \mathbf{P}$ and $\mathbf{H} = \mathbf{B}/\mu_0 - \mathbf{M}$. The polarization $\mathbf{P}$ and magnetization $\mathbf{M}$ are related to the induced current by

$$\mathbf{J} = \frac{\partial \mathbf{P}}{\partial t} + \operatorname{curl} \mathbf{M}.$$

Show that this response may be described in terms of the equivalent dielectric tensor

$$K_{ij}(\omega, \mathbf{k}) = K^L(\omega, \mathbf{k})\kappa_i\kappa_j + K^T(\omega, \mathbf{k})(\delta_{ij} - \kappa_i\kappa_j),$$

with

$$K^L(\omega, \mathbf{k}) = \varepsilon/\varepsilon_0,$$
$$K^T(\omega, \mathbf{k}) = \varepsilon/\varepsilon_0 + (k^2 c^2/\omega^2)(1 - \mu_0/\mu),$$

and where $\boldsymbol{\kappa} = \mathbf{k}/k$ is a unit vector along $\mathbf{k}$. Here L and T refer to the longitudinal and transverse parts respectively.

1.2 Alternative choices of the response tensor include the polarization 3-tensor $\alpha_{ij}(\omega, \mathbf{k})$ and the equivalent susceptibility tensor $\chi_{ij}(\omega, \mathbf{k})$. The former is defined by describing the electromagnetic field in terms of potentials Φ and $\mathbf{A}$,

$$\mathbf{E}(t, \mathbf{x}) = -\operatorname{grad} \Phi(t, \mathbf{x}) - \frac{\partial \mathbf{A}(t, \mathbf{x})}{\partial t},$$

$$\mathbf{B}(t, \mathbf{x}) = \operatorname{curl} \mathbf{A}(t, \mathbf{x}),$$

choosing the temporal gauge $\Phi(t, \mathbf{x}) = 0$, and writing

$$J_i(\omega, \mathbf{k}) = \alpha_{ij}(\omega, \mathbf{k})A_j(\omega, \mathbf{k}).$$

The latter is defined by writing the total induced current in terms of an equivalent polarization $\mathbf{P}(t, \mathbf{x})$ by $\mathbf{J}(t, \mathbf{x}) = \partial \mathbf{P}(t, \mathbf{x})/\partial t$, and then writing

$$P_i(\omega, \mathbf{k}) = \chi_{ij}(\omega, \mathbf{k})E_j(\omega, \mathbf{k}).$$

Show that these response tensors are related to $\sigma_{ij}(\omega, \mathbf{k})$ and $K_{ij}(\omega, \mathbf{k})$ by

$$\alpha_{ij}(\omega, \mathbf{k}) = i\omega\sigma_{ij}(\omega, \mathbf{k}) = \varepsilon_0\omega^2\{K_{ij}(\omega, \mathbf{k}) - \delta_{ij}\}$$

$$\chi_{ij}(\omega, \mathbf{k}) = \frac{i}{\omega}\sigma_{ij}(\omega, \mathbf{k}) = \varepsilon_0\{K_{ij}(\omega, \mathbf{k}) - \delta_{ij}\}.$$

1.3   The response of a medium involves two independent processes, described by response functions $A/(\omega + i\alpha)$, i.e. a broad-band damping at a rate α, and $B/(\omega - \omega_0 + i\beta)$, i.e. damping at a rate β centred on a frequency ω_0. The total damping rate is found by convolving them:

$$K(\omega) = \int_{-\infty}^{\infty} \frac{d\omega'}{2\pi} \frac{A}{\omega - \omega' + i(\alpha + \Gamma)} \frac{B}{\omega' - \omega_0 + i(\beta + \Gamma)}$$

where Γ lies above the poles at $\omega = -i\alpha$ and $\omega = \omega_0 - i\beta$ for either α or β negative.

Show that one has

$$K(\omega) = \frac{-iAB}{\omega - \omega_0 + i(\alpha + \beta)}$$

irrespective of the signs of α and β.

Hints: Show

$$\int_{-\infty}^{\infty} \frac{d\omega'}{2\pi} \frac{1}{\omega' + i\alpha} = -i/2$$

for $\alpha > 0$, and deform the contour as indicated in Figure 1.3 for α or $\beta < 0$.

1.4   Establish the following properties of Fourier transforms.

(i) Reality Condition: For $f(t, \mathbf{x})$ real one has

$$\tilde{f}(\omega, \mathbf{k}) = \tilde{f}^*(-\omega, -\mathbf{k}).$$

(ii) Power Theorem: For any $f_1(t, \mathbf{x})$ and $f_2(t, \mathbf{x})$ one has

$$\int dt \, d^3x f_1(t, \mathbf{x}) f_2(t, \mathbf{x}) = \int \frac{d\omega d^3k}{(2\pi)^4} \tilde{f}_1(\omega, \mathbf{k}) \tilde{f}_2(-\omega, -\mathbf{k})$$

(iii) Convolution Theorem: The Fourier transform of $g(t, \mathbf{x}) = f_1(t, \mathbf{x}) f_2(t, \mathbf{x})$ is

$$\tilde{g}(\omega, \mathbf{k}) = \int \frac{d\omega' d^3k'}{(2\pi)^4} \tilde{f}_1(\omega', \mathbf{k}') \tilde{f}_2(\omega - \omega', \mathbf{k} - \mathbf{k}').$$

1.5   The Hilbert transform $\tilde{f}(\omega)$ of a function $f(\omega)$ is defined by

$$\tilde{f}(\omega) = -\frac{1}{\pi} P \int_{-\infty}^{\infty} d\omega' \frac{f(\omega')}{\omega - \omega'}.$$

Use the Plemelj formula to establish the identity

$$\int_{-\infty}^{\infty} d\omega' \left(\frac{P}{\omega'}\right)\left(\frac{P}{\omega - \omega'}\right) = -\pi^2\delta(\omega)$$

and hence the formula for the inverse Hilbert transform

$$f(\omega) = \frac{1}{\pi} P \int_{-\infty}^{\infty} d\omega' \frac{\tilde{f}(\omega')}{\omega - \omega'}.$$

1.6   An optically active isotropic medium (e.g. a solution of dextrose) has a
      dielectric tensor

$$K_{ij}(\omega, \mathbf{k}) = a(\omega)\delta_{ij} + \frac{icb(\omega)}{\omega}\varepsilon_{ijl}k_l$$

where $a(\omega)$ and $b(\omega)$ depend on the medium and where

$$\varepsilon_{ijl} = \begin{cases} 1 & ijl \text{ an even permutation of } 123 \\ -1 & ijl \text{ an odd permutation of } 123 \\ 0 & \text{otherwise} \end{cases}$$

is the permutation symbol. Show that the homogeneous wave equation, i.e.
(1.28) with the right hand side replaced by zero, implies the equation

$$\begin{pmatrix} a(\omega) - N^2 & ib(\omega)N \\ -ib(\omega)N & a(\omega) - N^2 \end{pmatrix}\begin{pmatrix} E_x \\ E_y \end{pmatrix} = 0$$

for the two transverse components E_x and E_y where $\mathbf{k}$ lies along the z-axis.
Hence show the allowed values of N^2 for transverse waves in the medium are

$$N^2 = a(\omega) + \tfrac{1}{2}b^2(\omega) \pm \{a(\omega)b^2(\omega) + \tfrac{1}{4}b^4(\omega)\}^{1/2}$$

and that the solutions for E_x and E_y have $E_x/E_y = \pm i$, i.e. the natural wave
modes are oppositely circularly polarized transverse waves.

2

The response of an unmagnetized plasma

2.1 The cold plasma approach

The simplest model for calculating the response tensor for a plasma is that of a cold plasma. Each species is described by its charge, mass, number density and velocity, and fluid equations are used to derive the induced current and hence the response tensor. Although this method would appear to be quite restrictive, due to the neglect of thermal motions, with a minor re-interpretation it can be made quite general for unmagnetized plasmas. This re-interpretation involves separating each species into elements in velocity or momentum space, calculating their contributions separately and then summing over them with a weighting function determined by the particle distribution function. This procedure is carried out in §2.2 below. For the present we consider the contribution to the response tensor from one species with given q, m, n and $\mathbf{v}$.

The relevant fluid equations are the equation of continuity

$$\frac{\partial}{\partial t} n(t, \mathbf{x}) + \text{div}\left[\mathbf{v}(t, \mathbf{x})n(t, \mathbf{x})\right] = 0 \tag{2.1}$$

and the equation of fluid motion

$$\left[\frac{\partial}{\partial t} + \mathbf{v}(t, \mathbf{x}) \cdot \text{grad}\right]\mathbf{p}(t, \mathbf{x}) = q[\mathbf{E}(t, \mathbf{x}) + \mathbf{v}(t, \mathbf{x}) \times \mathbf{B}(t, \mathbf{x})], \tag{2.2}$$

where $\mathbf{p}(t, \mathbf{x}) = \gamma(t, \mathbf{x})m\mathbf{v}(t, \mathbf{x})$ is the fluid momentum and $\gamma(t, \mathbf{x})$ is its Lorentz factor. The current density due to the fluid is

$$\mathbf{J}(t, \mathbf{x}) = qn(t, \mathbf{x})\mathbf{v}(t, \mathbf{x}). \tag{2.3}$$

Later we shall require the nonlinear as well as the linear response, and it is appropriate to retain the nonlinear terms for the present. Each of $n, \mathbf{v}, \gamma, \mathbf{p}$ and $\mathbf{J}$ has a zeroth order term and terms linear, quadratic etc. in the fields $\mathbf{E}$ and $\mathbf{B}$. For each of them we make a formal expansion, after Fourier transforming,

$$G(\omega, \mathbf{k}) = G^{(0)}(2\pi)^4 \delta(\omega)\delta^3(\mathbf{k}) + \sum_{n=1}^{\infty} G^{(n)}(\omega, \mathbf{k}) \tag{2.4}$$

where superscript (0) indicates the unperturbed term.

Each of (2.1) to (2.3) involves the product of functions, and the Fourier transform of a product is evaluated using the convolution theorem: the Fourier transform of $G(t, \mathbf{x}) = F_1(t, \mathbf{x})F_2(t, \mathbf{x}) \cdots F_n(t, \mathbf{x})$ is the convolution of the Fourier transforms:

$$G(\omega, \mathbf{k}) = \int d\lambda^{(n)} F_1(\omega_1, \mathbf{k}_1)F_2(\omega_2, \mathbf{k}_2) \cdots F_n(\omega_n, \mathbf{k}_n) \tag{2.5}$$

where

$$d\lambda^{(n)} = \frac{d\omega_1 d^3\mathbf{k}_1}{(2\pi)^4} \frac{d\omega_2 d^3\mathbf{k}_2}{(2\pi)^4} \cdots \frac{d\omega_n d^3\mathbf{k}_n}{(2\pi)^4}$$
$$(2\pi)^4\delta(\omega - \omega_1 - \omega_2 \cdots - \omega_n)\delta^3(\mathbf{k} - \mathbf{k}_1 - \mathbf{k}_2 \cdots - \mathbf{k}_n) \tag{2.6}$$

denotes the n-fold convolution integral. The Fourier transforms are

$$n(\omega, \mathbf{k}) = \int d\lambda^{(2)}\frac{\mathbf{k} \cdot \mathbf{v}(\omega_1, \mathbf{k}_1)}{\omega}n(\omega_2, \mathbf{k}_2), \tag{2.7}$$

$$- i\omega\mathbf{p}(\omega, \mathbf{k}) + \int d\lambda^{(2)}\{i\mathbf{k}_2 \cdot \mathbf{v}(\omega_1, \mathbf{k}_1)\mathbf{p}(\omega_2, \mathbf{k}_2)\}$$
$$= q\mathbf{E}(\omega, \mathbf{k}) + q\int d\lambda^{(2)}\mathbf{v}(\omega_1, \mathbf{k}_1) \times \frac{[\mathbf{k}_2 \times \mathbf{E}(\omega_2, \mathbf{k}_2)]}{\omega_2} \tag{2.8}$$

and, after summing over all species of particle,

$$\mathbf{J}(\omega, \mathbf{k}) = \sum q \int d\lambda^{(2)}n(\omega_1, \mathbf{k}_1)\mathbf{v}(\omega_2, \mathbf{k}_2), \tag{2.9}$$

where in (2.8), $\mathbf{B}(\omega_2, \mathbf{k}_2)$ has been written in terms of $\mathbf{E}(\omega_2, \mathbf{k}_2)$ using (1.15). We also have

$$\mathbf{p}(\omega, \mathbf{k}) = m\int d\lambda^{(2)}\gamma(\omega_1, \mathbf{k}_1)\mathbf{v}(\omega_2, \mathbf{k}_2). \tag{2.10}$$

Substituting in the zeroth order terms one readily calculates the first order terms and thence higher order terms by simply noting that $\mathbf{E}$ in (2.8) is of first order. One finds

$$\mathbf{p}^{(1)}(\omega, \mathbf{k}) = m[\gamma^{(0)}\mathbf{v}^{(1)}(\omega, \mathbf{k}) + \gamma^{(1)}(\omega, \mathbf{k})\mathbf{v}^{(0)}], \tag{2.11}$$

$$\gamma^{(1)}(\omega, \mathbf{k}) = \{\gamma^{(0)}\}^3\frac{\mathbf{v}^{(0)} \cdot \mathbf{v}^{(1)}(\omega, \mathbf{k})}{c^2}, \tag{2.12}$$

$$n^{(1)}(\omega, \mathbf{k}) = n^{(0)}\frac{\mathbf{k} \cdot \mathbf{v}^{(1)}(\omega, \mathbf{k})}{\omega - \mathbf{k} \cdot \mathbf{v}^{(0)}}, \tag{2.13}$$

$$- i(\omega - \mathbf{k} \cdot \mathbf{v}^{(0)})\mathbf{p}^{(1)}(\omega, \mathbf{k}) = q\mathbf{E}(\omega, \mathbf{k}) + q\frac{\mathbf{v}^{(0)}}{\omega} \times [\mathbf{k} \times \mathbf{E}(\omega, \mathbf{k})] \tag{2.14}$$

and

$$\mathbf{J}^{(1)}(\omega, \mathbf{k}) = \sum q[n^{(0)}\mathbf{v}^{(1)}(\omega, \mathbf{k}) + n^{(1)}(\omega, \mathbf{k})\mathbf{v}^{(0)}] \tag{2.15}$$

The remainder of the calculation involves using (2.11) to (2.14) to evaluate $\mathbf{J}^{(1)}$ in terms of $\mathbf{E}$, thus identifying σ_{ij} and hence K_{ij} using (1.26) and (1.27).

After summing over all species of particle one finds

$$K_{ij}(\omega, \mathbf{k}) = \delta_{ij} - \sum \frac{q^2 n}{\varepsilon_0 \gamma m \omega^2} \left[\delta_{ij} + \frac{k_i v_j + k_j v_i}{\omega - \mathbf{k} \cdot \mathbf{v}} + \frac{(k^2 - \omega^2/c^2) v_i v_j}{(\omega - \mathbf{k} \cdot \mathbf{v})^2} \right] \qquad (2.16)$$

where the superscripts (0) are now omitted.

2.2 The Vlasov approach

The most widely used method for calculating response tensors is kinetic theory. Each distribution of particles is described by a single particle distribution function $f(\mathbf{p}, t, \mathbf{x})$ normalized here according to

$$n(t, \mathbf{x}) = \int d^3\mathbf{p} \, f(\mathbf{p}, t, \mathbf{x}). \qquad (2.17)$$

The current density is then given by

$$\mathbf{J}(t, \mathbf{x}) = \sum q \int d^3\mathbf{p} \, \mathbf{v} f(\mathbf{p}, t, \mathbf{x}), \qquad (2.18)$$

where the sum is over all species of particle.

If $\mathbf{E}$ and $\mathbf{B}$ are external fields, and if collisions are neglected, then f satisfies the collisionless Boltzmann equation

$$\left\{ \frac{\partial}{\partial t} + \mathbf{v} \cdot \frac{\partial}{\partial \mathbf{x}} + q[\mathbf{E}(t, \mathbf{x}) + \mathbf{v} \times \mathbf{B}(t, \mathbf{x})] \cdot \frac{\partial}{\partial \mathbf{p}} \right\} f(\mathbf{p}, t, \mathbf{x}) = 0. \qquad (2.19)$$

In the Vlasov theory it is postulated that f may be determined by (2.19) with $\mathbf{E}$ and $\mathbf{B}$ the self-consistent fields determined from Maxwell's equations with $\mathbf{J}$ (and the associated ρ) given by (2.18) as the source term. Thus (2.19), which for external fields $\mathbf{E}$ and $\mathbf{B}$ is a linear first order partial differential equation for f, becomes a nonlinear integrodifferential equation for a coupled set of f's. Re-interpreted thus, (2.19) is called the Vlasov equation.

After Fourier transforming and eliminating $\mathbf{B}(\omega, \mathbf{k})$ using (1.15), the Vlasov eqution (2.19) becomes

$$- i(\omega - \mathbf{k} \cdot \mathbf{v}) f(\mathbf{p}, \omega, \mathbf{k}) +$$

$$q \int d\lambda^{(2)} \frac{1}{\omega_1} \{ (\omega_1 - \mathbf{k}_1 \cdot \mathbf{v}) \delta_{sj} + k_{1s} v_j \} \, E_j(\omega_1, \mathbf{k}_1) \frac{\partial f}{\partial p_s}(\mathbf{p}, \omega_2, \mathbf{k}_2) = 0. \qquad (2.20)$$

We solve (2.20) by formally expanding $f(\mathbf{p}, \omega, \mathbf{k})$ in powers of $\mathbf{E}$,

$$f(\mathbf{p}, \omega, \mathbf{k}) = f^{(0)}(\mathbf{p})(2\pi)^4 \delta(\omega) \delta^3(\mathbf{k}) + \sum_{n=1}^{\infty} f^{(n)}(\mathbf{p}, \omega, \mathbf{k}) \qquad (2.21)$$

and using a perturbation approach. The first order solution is

$$f^{(1)}(\mathbf{p}, \omega, \mathbf{k}) = \frac{-iq}{\omega(\omega - \mathbf{k} \cdot \mathbf{v})} \{ (\omega - \mathbf{k} \cdot \mathbf{v}) \delta_{sj} + k_s v_j \} E_j(\omega, \mathbf{k}) \frac{\partial f^{(0)}(\mathbf{p})}{\partial p_s}. \qquad (2.22)$$

The first order current from (2.18) is

$$\mathbf{J}^{(1)}(\omega, \mathbf{k}) = \sum q \int d^3\mathbf{p} \mathbf{v} f^{(1)}(\mathbf{p}, \omega, \mathbf{k}).$$ (2.23)

The conductivity tensor σ_{ij} may be identified by inserting (2.22) in (2.23) and K_{ij} then follows from the definition (1.27). One finds

$$K_{ij}(\omega, \mathbf{k}) = \delta_{ij} + \sum \frac{q^2}{\varepsilon_0 \omega^2} \int d^3\mathbf{p} \frac{v_i}{\omega - \mathbf{k} \cdot \mathbf{v}} \{(\omega - \mathbf{k} \cdot \mathbf{v})\delta_{sj} + k_s v_j\} \frac{\partial f(\mathbf{p})}{\partial p_s},$$ (2.24)

where the superscript (0) is now omitted. An alternative expression may be obtained by partially integrating with respect to $\mathbf{p}$. After some elementary but lengthy algebra one obtains

$$K_{ij}(\omega, \mathbf{k}) = \delta_{ij} - \sum \frac{q^2}{\varepsilon_0 m \omega^2} \int \frac{d^3\mathbf{p}}{\gamma} \left\{ \delta_{ij} + \frac{k_i v_j + k_j v_i}{\omega - \mathbf{k} \cdot \mathbf{v}} + \frac{(k^2 - \omega^2/c^2)v_i v_j}{(\omega - \mathbf{k} \cdot \mathbf{v})^2} \right\} f(\mathbf{p}).$$
(2.25)

As remarked at the beginning of §2.1, the cold-plasma and kinetic theory results are equivalent in the following sense. Suppose that in (2.16) n is re-interpreted as the number density $d^3\mathbf{p}f(\mathbf{p})$ of particles in the element $d^3\mathbf{p}$ of momentum space, and that the sum in (2.16) includes an integral over such elements. Then, by inspection, (2.16) reproduces (2.25).

In most applications one makes the nonrelativistic approximation. A paradox arises when one sets $\mathbf{p} = m\mathbf{v}$ and $\gamma = 1$ in (2.24): the two expressions (2.24) and (2.25) are then not equivalent! That is, setting $\mathbf{p} = m\mathbf{v}$ in (2.24) and partially integrating, one does not obtain (2.25) with $\mathbf{p} = m\mathbf{v}$ and $\gamma = 1$. The resolution of this paradox is that the nonrelativistic limit is technically $c \to \infty$, and it is the term ω^2/c^2 in (2.25) which is not reproduced by the foregoing procedure. The effect of the term ω^2/c^2 in (2.25) is not important in most applications for unmagnetized plasmas; however, an analogous term is of crucial importance in cyclotron maser emission (§11.3). It is important to be aware that inconsistencies can arise in making nonrelativistic approximations in electromagnetic theory; technically the nonrelativistic approximation can be justified a priori only when the phase speed (as well as the particle speeds) is nonrelativistic, i.e. for $\omega/k \ll c$.

2.3 Maxwellian distributions

Thermal nonrelativistic particles have the Maxwellian distribution

$$f(\mathbf{p}) = \frac{n}{(2\pi)^{3/2} m^3 V^3} \exp\left[-\frac{v^2}{2V^2} \right]$$ (2.26)

with $\mathbf{p} = m\mathbf{v}$ and $V = (T/m)^{1/2}$, where T is the temperature in energy units. On inserting (2.26) in (2.24) one obtains

$$K_{ij}(\omega, \mathbf{k}) = \delta_{ij} - \sum \frac{n q^2}{\varepsilon_0 \omega m V^2} \int d^3\mathbf{v} \frac{v_i v_j}{\omega - \mathbf{k} \cdot \mathbf{v}} \frac{\exp[-v^2/2V^2]}{(2\pi)^{3/2} V^3}.$$ (2.27)

The medium is isotropic and (2.27) may be rewritten in the form

$$K_{ij}(\omega, \mathbf{k}) = K^L(\omega, \mathbf{k})\kappa_i\kappa_j + K^T(\omega, \mathbf{k})(\delta_{ij} - \kappa_i\kappa_j) \tag{2.28}$$

where K^L and K^T denote the longitudinal and transverse parts respectively, and where $\mathbf{\kappa} = \mathbf{k}/k$ is a unit vector along $\mathbf{k}$. One finds

$$K^L(\omega, \mathbf{k}) = 1 - \sum \frac{\omega_p^2}{\omega^2 V^2} \int d^3\mathbf{v} \frac{\omega}{\omega - \mathbf{k}\cdot\mathbf{v}} \left(\frac{\mathbf{k}\cdot\mathbf{v}}{k}\right)^2 \frac{\exp[-v^2/2V^2]}{(2\pi)^{3/2} V^3} \tag{2.29a}$$

and

$$K^T(\omega, \mathbf{k}) = 1 - \sum \frac{\omega_p^2}{\omega^2 V^2} \int d^3\mathbf{v} \frac{\omega}{\omega - \mathbf{k}\cdot\mathbf{v}} \frac{1}{2}\left\{v^2 - \left(\frac{\mathbf{k}\cdot\mathbf{v}}{k}\right)^2\right\} \frac{\exp[-v^2/2V^2]}{(2\pi)^{3/2} V^3} \tag{2.29b}$$

with $\omega_{p\alpha}^2 = n_\alpha q_\alpha^2/\varepsilon_0 m_\alpha$ for each species α.

The integrals may be evaluated by choosing the z-axis along $\mathbf{k}$. The v_x- and v_y-integrals are then elementary. The v_z-integral may be evaluated in terms of a single transcendental function called the plasma dispersion function. There are several different notations for this function; here we define it by (cf. Appendix A)

$$\bar{\phi}(z) = -\frac{z}{\sqrt{\pi}} \int_{-\infty}^{\infty} \frac{dt\, e^{-t^2}}{t - z} \tag{2.30}$$

The imaginary part of $\bar{\phi}(z)$ for real ω is determined by the Plemelj formula (1.32). In (2.30) the denominator is $t - (\omega + i0)/\sqrt{2}kV = t - z - i0$ for $k > 0$ with $z = \omega/\sqrt{2}kV > 0$. The imaginary part of $\bar{\phi}(z)$ is then $-i\sqrt{\pi}z\exp(-z^2)$. Including this imaginary part, the resulting expressions for K^L and K^T are

$$K^L(\omega, \mathbf{k}) = 1 + \sum_\alpha \frac{\omega_{p\alpha}^2}{k^2 V_\alpha^2}\left\{1 - \phi(\omega/\sqrt{2}kV_\alpha) + i\left(\frac{\pi}{2}\right)^{1/2} \frac{\omega}{kV_\alpha}\exp\left(-\frac{\omega^2}{2k^2 V_\alpha^2}\right)\right\} \tag{2.31a}$$

and

$$K^T(\omega, \mathbf{k}) = 1 - \sum_\alpha \frac{\omega_{p\alpha}^2}{\omega^2}\left\{\phi(\omega/\sqrt{2}kV_\alpha) - i\left(\frac{\pi}{2}\right)^{1/2} \frac{\omega}{kV_\alpha}\exp\left(-\frac{\omega^2}{2k^2 V_\alpha^2}\right)\right\} \tag{2.31b}$$

where the label α for each species is now included explicitly.

2.4 The dispersion equation

The theory of weakly damped waves may be developed as follows. First, neglect the imaginary parts of K^L and K^T, as these relate to the damping of the waves. Then insert the resulting expression for K_{ij} from (2.28) in the homogeneous wave equation derived from (1.28) by neglecting the right hand term. The resulting equation

$$\Lambda_{ij}(\omega, \mathbf{k})E_j(\omega, \mathbf{k}) = 0 \tag{2.32}$$

may be interpreted as three real coupled linear equations for the three components of $\mathbf{E}(\omega, \mathbf{k})$. The condition for a solution to exist is that the determinant of the coefficients vanish. Let $\Lambda(\omega, \mathbf{k})$ be the determinant of the 3×3 matrix $\Lambda_{ij}(\omega, \mathbf{k})$.

The condition

$$\Lambda(\omega, \mathbf{k}) = 0 \qquad (2.33)$$

is called the *dispersion equation*.

Explicit evaluation of $\Lambda(\omega, \mathbf{k})$ is facilitated by choosing the z-axis to be along $\mathbf{k}$. Then (1.29) with (2.28) gives

$$\Lambda_{ij}(\omega, \mathbf{k}) = \begin{bmatrix} \mathrm{Re}[K^T(\omega, \mathbf{k})] - N^2 & 0 & 0 \\ 0 & \mathrm{Re}[K^T(\omega, \mathbf{k})] - N^2 & 0 \\ 0 & 0 & \mathrm{Re}[K^L(\omega, \mathbf{k})] \end{bmatrix}, \quad (2.34)$$

and we have

$$\Lambda(\omega, \mathbf{k}) = \{N^2 - \mathrm{Re}[K^T(\omega, \mathbf{k})]\}^2 \, \mathrm{Re}[K^L(\omega, \mathbf{k})]. \qquad (2.35)$$

There are two types of solution. Those with

$$\mathrm{Re}[K^L(\omega, \mathbf{k})] = 0 \qquad (2.36)$$

must have $\mathbf{E}$ along the z-axis, i.e. along $\mathbf{k}$, as may be seen from (2.32) and (2.34). Thus (2.36) is the dispersion equation for longitudinal waves. The solution

$$N^2 = \mathrm{Re}[K^T(\omega, \mathbf{k})] \qquad (2.37)$$

is a double solution and corresponds to two degenerate states of transverse polarization; (2.31b) in (2.37) gives the well known result $N^2 = 1 - \omega_p^2/\omega^2$.

The dispersion relation for Langmuir waves is obtained by neglecting the contribution of the ions in (2.31a) and inserting the resulting expression in (2.36), which then gives

$$1 + \frac{\omega_p^2}{k^2 V_e^2}\left\{1 - \phi\left(\frac{\omega}{\sqrt{2}kV_e}\right)\right\} = 0. \qquad (2.38)$$

The real part of $\bar{\phi}(z)$ may be written in the form (cf. Exercise 2.1)

$$\phi(z) = 2ze^{-z^2}\int_0^z dt\, e^{t^2} \qquad (2.39)$$

This function is plotted in Figure 2.1. By inspection (2.38) has solutions only for $\phi(\omega/\sqrt{2}kV_e) > 1$, which corresponds to $\omega/\sqrt{2}kV_e \gtrsim 1$. For $\omega \gg \sqrt{2}kV_e$ an asymptotic expansion of (2.39) gives

$$\phi(z) = 1 + \frac{1}{2z^2} + \frac{3}{4z^4} + \cdots \quad (z^2 \gg 1), \qquad (2.40)$$

and the solution of (2.38) reduces to the dispersion relation $\omega = \omega_L(\mathbf{k})$ given in (1.3), viz. $\omega_L(\mathbf{k}) = [\omega_p^2 + 3k^2 V_e^2]^{1/2}$.

The dispersion relation for ion sound waves, cf. (1.5), is obtained from (2.36) with (2.31a) by assuming $\omega/\sqrt{2}kV_e \ll 1$ for the electrons, with

$$\phi(z) = 2z^2 - \tfrac{4}{3}z^4 + \cdots \quad (z^2 \ll 1), \qquad (2.41)$$

and by evaluating the ionic contribution in (2.31a) for $\omega/\sqrt{2}kV_i \gg 1$ using

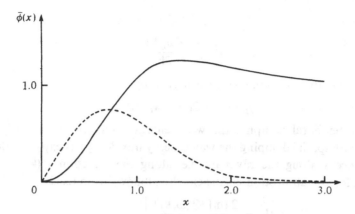

Fig. 2.1 The real (solid line) and imaginary (dashed line) parts of $\bar{\phi}(x)$ are plotted as a function of x.

(2.40). Then $\text{Re}[K^L]$ is approximated by $1 + 1/k^2\lambda_{De}^2 - \omega_{pi}^2/\omega^2$, and setting this to zero gives $\omega^2 = \omega_s^2(\mathbf{k})$ with $\omega_s(\mathbf{k})$ given by (1.5).

2.5 The absorption coefficient

Wave damping may be treated by allowing the frequency to have a small imaginary part $i\omega_i$ and the wave vector to have a small imaginary part $i\mathbf{k}_i$. The wave amplitude then varies secularly as $\exp[\omega_i t - \mathbf{k}_i \cdot \mathbf{x}]$. The wave energy is proportional to the square of the amplitude, and hence it varies as $\exp[2\omega_i t - 2\mathbf{k}_i \cdot \mathbf{x}]$.

Damping in time and space are not independent; the relation between them depends on the boundary conditions. Consider waves uniformly excited everywhere initially; these can only decay in time. Similarly, waves generated at a constant rate at a fixed point source can only decay in space away from the source. The damping process itself is included through an imaginary (Im) part (more generally, the antihermitian part) of the dielectric tensor. For longitudinal waves we may treat the damping by balancing the three small imaginary terms which arise when one makes the replacements $\omega, \mathbf{k} \to \omega + i\omega_i, \mathbf{k} + i\mathbf{k}_i$ and $\text{Re}\, K^L \to \text{Re}\, K^L + i\,\text{Im}\, K^L$ in (2.36). One finds

$$\left(\omega_i\frac{\partial}{\partial\omega} + \mathbf{k}_i\cdot\frac{\partial}{\partial\mathbf{k}}\right)\text{Re}[K^L(\omega, \mathbf{k})] + \text{Im}[K^L(\omega, \mathbf{k})] = 0, \tag{2.42}$$

which is to be evaluated at $\text{Re}[K^L(\omega, \mathbf{k})] = 0$.

The chain rule of partial differentiation implies

$$-\frac{\partial}{\partial\mathbf{k}}\text{Re}[K^L(\omega, \mathbf{k})]\bigg/\frac{\partial}{\partial\omega}\text{Re}[K^L(\omega, \mathbf{k})] = \frac{\partial\omega}{\partial\mathbf{k}},$$

where $\text{Re}[K^L(\omega, \mathbf{k})] = 0$ is implicit. The right hand member defines the group velocity

$$v_g^L(\mathbf{k}) = \frac{\partial \omega_L(\mathbf{k})}{\partial \mathbf{k}}. \tag{2.43}$$

Let us define the *absorption coefficient* $\gamma_L(\mathbf{k})$ by

$$\gamma_L(\mathbf{k}) = -2(\omega_i - \mathbf{k}_i \cdot \mathbf{v}_g^L(\mathbf{k})). \tag{2.44}$$

For purely temporal damping the wave energy then decays as $\exp[-\gamma_L(\mathbf{k})t]$, and for purely spatial damping the wave energy flux decays as $\exp[-\gamma_L(\mathbf{k})s/v_g^L(\mathbf{k})]$ with distance s along the ray path, i.e. along the direction $v_g^L(\mathbf{k})$. An explicit expression for the absorption coefficient then follows from (2.42):

$$\gamma_L(\mathbf{k}) = \frac{2\,\mathrm{Im}[K^L(\omega,\mathbf{k})]}{\dfrac{\partial}{\partial \omega}\,\mathrm{Re}[K^L(\omega,\mathbf{k})]}\Bigg|_{\mathrm{Re}[K^L(\omega,\mathbf{k})]=0} \tag{2.45}$$

Below we argue that the quantity

$$R_L(\mathbf{k}) = \left\{\omega\frac{\partial}{\partial \omega}\,\mathrm{Re}[K^L(\omega,\mathbf{k})]\right\}^{-1}\Bigg|_{\mathrm{Re}[K^L(\omega,\mathbf{k})]=0} \tag{2.46}$$

may be interpreted as the ratio of the electric to the total energy in the waves. With the notation (2.46), the result (2.45) becomes

$$\gamma_L(\mathbf{k}) = 2\omega_L(\mathbf{k})R_L(\mathbf{k})\,\mathrm{Im}[K^L(\omega_L(\mathbf{k}),\mathbf{k})]. \tag{2.47}$$

For Langmuir waves the imaginary part of K^L follows from the electron contribution to (2.31a). The ratio $R_L(\mathbf{k})$ may be evaluated by inserting (2.38) with (2.40) in (2.46). One finds

$$R_L(\mathbf{k}) = \frac{1}{2}\left\{\frac{\omega_L(\mathbf{k})}{\omega_p}\right\}^2 \tag{2.48}$$

and, for the Landau damping coefficient,

$$\gamma_L(\mathbf{k}) = \left(\frac{\pi}{2}\right)^{1/2}\frac{\{\omega_L(\mathbf{k})\}^4}{k^3 V_e^3}\exp\left[-\frac{\{\omega_L(\mathbf{k})\}^2}{2k^2 V_e^2}\right]. \tag{2.49}$$

The corresponding results for ion sound waves are

$$R_s(\mathbf{k}) = \frac{1}{2}\left(\frac{\omega_s(\mathbf{k})}{\omega_{pi}}\right)^2 \tag{2.50}$$

and

$$\gamma_s(\mathbf{k}) = \left(\frac{\pi}{2}\right)^{1/2}\omega_s(\mathbf{k})\left\{\frac{v_s}{V_e} + \left(\frac{\omega_s(\mathbf{k})}{kV_i}\right)^3\exp\left[-\frac{\{\omega_s(\mathbf{k})\}^2}{2k^2 V_i^2}\right]\right\} \tag{2.51}$$

The term v_s/V_e in (2.51) arises from Landau damping by electrons, and the other term arises from Landau damping by ions. For $\omega_s(\mathbf{k}) \approx kv_s$ the absorption coefficient is comparable with the wave frequency, which violates the weak damping assumption. ('Waves' which damp in about one period do not have a well-defined period and cannot be identified as waves.) The waves do not exist in a more rigorous sense in that the dispersion relation (1.5) is derived under the assumption

$\omega \gg \sqrt{2k}V_i$ when the ionic contribution

$$\frac{\omega_{pi}^2}{k^2 V_i^2}\left\{1 - \phi\left(\frac{\omega}{\sqrt{2k}V_i}\right)\right\}$$

is approximated by $-\omega_{pi}^2/\omega^2$; for $\phi(\omega/\sqrt{2k}V_i) < 1$ the ionic contribution to K^L is positive and there is no solution at all to $K^L = 0$. Thus there are no ion sound waves in any sense for $\omega \lesssim \sqrt{2k}V_i$. For ion sound waves to be weakly damped we require $\omega \approx kv_s \gg \sqrt{2k}V_i$; the implied condition $v_s \gg V_i$ corresponds to

$$Z_i T_e \gg T_i \tag{2.52}$$

in a charge neutral plasma ($Z_i n_i = n_e$) with ions of charge $Z_i e$. In a singly ionized plasma, ion sound waves exist only if the electron temperature is much greater than the ion temperature.

2.6 Formal theory of weakly damped waves

The quantity $R_L(\mathbf{k})$ defined by (2.40) is interpreted as the ratio of the electric to total energy in longitudinal waves. The formal justification for this interpretation is presented here. It is convenient to present the derivation for arbitrary waves in an arbitrary collisionless plasma.

Let $\omega = \omega_M(\mathbf{k})$ be an arbitrary solution (for the mode 'M') of the dispersion equation (2.33). The direction of the electric vector for waves in mode M may be described by a unimodular vector $\mathbf{e}_M(\mathbf{k})$,

$$\mathbf{e}_M(\mathbf{k}) \cdot \mathbf{e}_M^*(\mathbf{k}) = 1, \tag{2.53}$$

which is along $\mathbf{E}(\omega_M(\mathbf{k}), \mathbf{k})$, i.e. along the direction of the solution of the homogeneous wave equation (2.32) for $\omega = \omega_M(\mathbf{k})$. We call $\mathbf{e}_M(\mathbf{k})$ the *polarization vector* for waves in the mode M. It may be constructed as follows. Let $\lambda_{ij}(\omega, \mathbf{k})$ be the matrix of cofactors of $\lambda_{ij}(\omega, \mathbf{k})$:

$$\Lambda_{ij}(\omega, \mathbf{k})\lambda_{il}(\omega, \mathbf{k}) = \Lambda(\omega, \mathbf{k})\delta_{jl}. \tag{2.54}$$

Then one has

$$\lambda_{ij}(\omega_M(\mathbf{k}), \mathbf{k}) = \lambda_{ss}(\omega_M(\mathbf{k}), \mathbf{k})e_{Mi}(\mathbf{k})e_{Mj}^*(\mathbf{k}). \tag{2.55}$$

In treating the wave energetics the wave amplitude, $E_M(\mathbf{k})$ say, needs to be defined. The following definition is adopted:

$$\mathbf{E}(\omega, \mathbf{k}) = \mathbf{e}_M(\mathbf{k})E_M(\mathbf{k})2\pi[\delta(\omega - \omega_M(\mathbf{k})) + \delta(\omega - \omega_M(-\mathbf{k}))]. \tag{2.56}$$

Let $W_M^{(E)}(\mathbf{k})$ be the electric energy in waves in the mode M in the elemental range $d^3k/(2\pi)^3$. The time-averaged electric energy is

$$\lim_{T \to \infty} \frac{1}{T} \int_{-T/2}^{T/2} dt \int d^3\mathbf{x} \frac{\varepsilon_0}{2}|\mathbf{E}(t, \mathbf{x})|^2 = \lim_{T \to \infty} \frac{1}{T} \int \frac{d\omega d^3\mathbf{k}}{(2\pi)^4} \frac{\varepsilon_0}{2}|\mathbf{E}(\omega, \mathbf{k})|^2$$

$$= \int \frac{d^3\mathbf{k}}{(2\pi)^3} \varepsilon_0 |E_M(\mathbf{k})|^2, \tag{2.57}$$

where in the first step the power theorem for Fourier transforms is used, and in the second step (2.56) is inserted and the squares of the δ-functions are reduced using (cf. Exercise 2.6)

$$\lim_{T \to \infty} \frac{1}{T}[\delta(\omega - \omega_0)]^2 = \frac{1}{2\pi}\delta(\omega - \omega_0). \tag{2.58}$$

The identification

$$W_M^{(E)}(\mathbf{k}) = \varepsilon_0 |\mathbf{E}_M(\mathbf{k})|^2 \tag{2.59}$$

follows.

The total energy may be identified by calculating the rate mechanical work is done on the wave system and then identifying the quantity into which this work goes as the wave energy. The time-averaged rate work is done on the wave system is

$$-\lim_{T \to \infty} \frac{1}{T}\int_{-T/2}^{T/2} dt \int d^3\mathbf{x} \mathbf{J}(t, \mathbf{x}) \cdot \mathbf{E}(t, \mathbf{x}) = -\lim_{T \to \infty} \frac{1}{T}\int \frac{d\omega d^3\mathbf{k}}{(2\pi)^4}\mathbf{J}(\omega, \mathbf{k}) \cdot \mathbf{E}^*(\omega, \mathbf{k}), \tag{2.60}$$

where the power theorem is used again. The current in (2.60) is identified as that associated with dissipative processes in the plasma, i.e. as

$$J_i(\omega, \mathbf{k}) = -i\omega\varepsilon_0 K_{ij}^A(\omega, \mathbf{k})E_j(\omega, \mathbf{k}), \tag{2.61}$$

where the antihermitian (A) part is the relevant generalization of the imaginary part in general. The hermitian (H) and antihermitian parts satisfy

$$K_{ij}^H(\omega, \mathbf{k}) = K_{ji}^H(-\omega, -\mathbf{k}), \tag{2.62a}$$

$$K_{ij}^A(\omega, \mathbf{k}) = -K_{ji}^A(-\omega, -\mathbf{k}), \tag{2.62b}$$

with the reality condition (cf. Exercise 1.6) implying $K_{ij}(-\omega, -\mathbf{k}) = K_{ij}^*(\omega, \mathbf{k})$ quite generally. Then on inserting (2.56) in (2.60) with (2.61), the energy gained is set equal to the rate $-\gamma_M(\mathbf{k})W_M(\mathbf{k})$, integrated over $d^3\mathbf{k}/(2\pi)^3$, at which the total wave energy increases. One finds

$$-\int \frac{d^3\mathbf{k}}{(2\pi)^3}\gamma_M(\mathbf{k})W_M(\mathbf{k}) = -\int \frac{d^3\mathbf{k}}{(2\pi)^3}2\varepsilon_0 |\omega_M(\mathbf{k})\| \mathbf{E}_M(\mathbf{k})|^2 \operatorname{Im}[K^M(\omega, \mathbf{k})] \tag{2.63}$$

with

$$i\operatorname{Im}[K^M(\omega, \mathbf{k})] = e_{Mi}^*(\mathbf{k})e_{Mj}(\mathbf{k})K_{ij}^A(\omega, \mathbf{k}). \tag{2.64}$$

We also require the appropriate generalization of (2.49), which is

$$\gamma_M(\mathbf{k}) = \frac{2\lambda_{ss}(\omega, \mathbf{k})\operatorname{Im}[K^M(\omega, \mathbf{k})]}{\dfrac{\partial}{\partial\omega}\Lambda(\omega, \mathbf{k})}\Bigg|_{\omega = \omega_M(\mathbf{k})} \tag{2.65}$$

Collecting results, one finds

$$R_M(\mathbf{k}) = \frac{W_M^{(E)}(\mathbf{k})}{W_M(\mathbf{k})} = \frac{\lambda_{ss}(\omega, \mathbf{k})}{\omega\dfrac{\partial}{\partial\omega}\Lambda(\omega, \mathbf{k})}\Bigg|_{\omega = \omega_M(\mathbf{k})} \tag{2.66}$$

and
$$\gamma_M(\mathbf{k}) = 2\omega_M(\mathbf{k})R_M(\mathbf{k})\,\text{Im}\,[K^M(\omega_M(\mathbf{k}),\mathbf{k})]. \tag{2.67}$$
In the case of an isotropic plasma, cf. (2.54), one has
$$\lambda_{ss} = \{\text{Re}\,K^T - N^2\}\{\text{Re}\,K^T - N^2 + 2\,\text{Re}\,K^L\}, \tag{2.68}$$
with Λ given by (2.35). Then (2.66) reduces to (2.45). An alternative expression to (2.66) when the dispersion relation is expressed in terms of the refractive index N as a function of ω and of the direction $\boldsymbol{\kappa} = \mathbf{k}/k$, i.e. $kc/\omega = N_M(\omega, \boldsymbol{\kappa})$, is
$$R_M(\mathbf{k}) = \left\{[1 - |\boldsymbol{\kappa}\cdot\mathbf{e}_M(\mathbf{k})|^2]2N_M\frac{\partial}{\partial\omega}(\omega N_M)\right\}^{-1}. \tag{2.69}$$
This result may be used for transverse waves, i.e. for $\boldsymbol{\kappa}\cdot\mathbf{e}_M(\mathbf{k}) = 0$.

Exercise set 2

2.1 Starting from the definition (2.30) of the plasma dispersion function, viz
$$\bar{\phi}(z) = -\frac{z}{\sqrt{\pi}}\int_{-\infty}^{\infty}\frac{dt\,e^{-t^2}}{t - z},$$
derive (i) the differential equation
$$\frac{d\phi(z)}{dz} = \frac{\phi(z)}{z} + 2z\{1 - \phi(z)\},$$
and hence (ii) the alternative form
$$\phi(z) = 2ze^{-z^2}\int_0^z dt\,e^{t^2}$$

2.2 Show that integrals of the form
$$F(\omega, \mathbf{k}, V) = \int\frac{d^3\mathbf{v}}{(2\pi)^{3/2}V^3}e^{-v^2/2V^2}f(\omega, \mathbf{k}, \mathbf{v})$$
have the values listed in the table below ($z = \omega/\sqrt{2}kV$)

$f(\omega, \mathbf{k}, \mathbf{v})$	$F(\omega, \mathbf{k}, V)$
$\dfrac{1}{\omega - \mathbf{k}\cdot\mathbf{v}}$	$\dfrac{1}{\omega}\phi(z)$
$\dfrac{v_i}{\omega - \mathbf{k}\cdot\mathbf{v}}$	$\dfrac{k_i}{k^2}\{\phi(z) - 1\}$
$\dfrac{1}{(\omega - \mathbf{k}\cdot\mathbf{v})^2}$	$\dfrac{1}{k^2V^2}\{\phi(z) - 1\}$
$\dfrac{v_i}{(\omega - \mathbf{k}\cdot\mathbf{v})^2}$	$\dfrac{k_i}{k^2\omega}\left[\dfrac{\omega^2}{k^2V^2}\{\phi(z) - 1\} - \phi(z)\right]$
$\dfrac{v_iv_j}{(\omega - \mathbf{k}\cdot\mathbf{v})^2}$	$\dfrac{\delta_{ij}}{k^2}\{\phi(z) - 1\} + \dfrac{k_ik_j}{k^4}\left[\dfrac{\omega^2}{k^2V^2}\{\phi(z) - 1\} - 3\phi(z) + 2\right]$

2.3   The two expressions (2.24) and (2.25) are equivalent in the fully relativistic case, and in the case where the nonrelativistic approximation is made to the particles, (2.25) includes additional terms of order ω^2/c^2. Evaluate these terms for nonrelativistic particles, and show that the additional terms are, to K^L in (2.31a)

$$\sum \frac{\omega_p^2}{k^2 c^2} \left\{ 1 - 2\phi(z) + \frac{\omega^2}{k^2 V^2} [\phi(z) - 1] \right\},$$

and to K^T in (2.31b)

$$\sum \frac{\omega_p^2}{k^2 c^2} [\phi(z) - 1],$$

with $z = \omega/\sqrt{2kV}$, and where subscripts α are omitted.

2.4   (a) Show that when relativistic effects are retained the longitudinal and transverse parts of K_{ij}, e.g. as given by (2.25) for an isotropic plasma, reduce to

$$K^L(\omega, \mathbf{k}) = 1 - \sum \frac{q^2}{\varepsilon_0 m} \int \frac{d^3 \mathbf{p}}{\gamma} \frac{(1 - (\mathbf{k} \cdot \mathbf{v})^2/k^2 c^2) f(p)}{(\omega - \mathbf{k} \cdot \mathbf{v})^2},$$

and

$$K^T(\omega, \mathbf{k}) = 1 - \sum \frac{q^2}{\varepsilon_0 m \omega^2} \int \frac{d^3 \mathbf{p}}{\gamma} \left[1 + \frac{k^2 - \omega^2/c^2}{(\omega - \mathbf{k} \cdot \mathbf{v})^2} \frac{|\mathbf{k} \times \mathbf{v}|^2}{2k^2} \right] f(p)$$

    (b) Show, using the form (2.24), that the imaginary parts for real ω are given by

$$\operatorname{Im} K^L(\omega, \mathbf{k}) = \sum \frac{2\pi^2 q^2}{\varepsilon_0 \omega k} \left\{ p_\phi^2 f(p_\phi) + 2 \frac{v_\phi^2}{c^2} \int_{p_\phi}^{\infty} dp \, p f(p) \right\}$$

and

$$\operatorname{Im} K^T(\omega, \mathbf{k}) = \sum \frac{2\pi^2 q^2}{\varepsilon_0 \omega k} \frac{1}{\gamma_\phi^2} \int_{p_\phi}^{\infty} dp \, p f(p)$$

with $v_\phi = \omega/k$, $\gamma_\phi = (1 - v_\phi^2/c^2)^{-1/2}$, $p_\phi = \gamma_\phi m v_\phi$.
Remark: The imaginary parts are strictly zero for $v_\phi > c$, when γ_ϕ and p_ϕ are imaginary; the non-vanishing of the non-relativistic forms (2.31a, b) for $\omega/k > c$ is due entirely to the effects of unphysical particles with $v > c$ in the Maxwellian distribution (2.26).

2.5   (a) Include the effect of friction between the electrons and ions in a cold plasma treatment of the response of a nonstreaming electron gas, i.e. consider the equation of motion

$$m_e \frac{d\mathbf{v}}{dt} = -e\mathbf{E} - m_e \nu \mathbf{v},$$

where ν is the collision frequency. Show that the dielectric tensor is

$$K_{ij}(\omega, \mathbf{k}) = \delta_{ij} \left\{ 1 - \frac{\omega_p^2}{\omega(\omega + i\nu)} \right\}.$$

    (b) Evaluate the imaginary part of the dielectric tensor to first order in $i\nu/\omega$ and use (2.67) to estimate the collisional damping rate for Langmuir waves and transverse waves in an isotropic plasma.

(c) The collision frequency is a weak function of ω and may be approximated by

$$\nu(\omega) = \frac{1}{12\pi} \left(\frac{2}{\pi}\right)^{1/2} \frac{e^4 Z_i^2 n_i}{\varepsilon_0^2 m_e^2 V_e^3} \frac{\pi}{\sqrt{3}} G(T_e, \omega)$$

with the Gaunt factor approximated by (Scheuer 1960)

$$\frac{\pi}{\sqrt{3}} G(T_e, \omega) = \ln\left(\frac{4\sqrt{2} V_e}{\omega b_0}\right) - \tfrac{5}{2}\gamma_E$$

with $\gamma_E = 0.577\ldots$ and with $b_0 = Z_i e^2/4\pi\varepsilon_0 m_e V_e^2$ a characteristic impact parameter for thermal electrons. Show that the damping rate per unit length $(\gamma^c/v_g \approx \gamma^c/c)$ for transverse waves with $\omega \gg \omega_p$ is given by

$$\mu^c = 9.8 \times 10^{-13} \frac{Z_i^2 n_i n_e}{f^2 T_e^{3/2}} \frac{\pi}{\sqrt{3}} G(T_e, \omega),$$

$$\frac{\pi}{\sqrt{3}} G(T_e, \omega) \approx 17.7 + \tfrac{3}{2}\ln T_e - \ln f,$$

where μ^c is in m^{-1}, n_e and n_i in m^{-3}, $f = \omega/2\pi$ in Hz and T_e in kelvins. In astrophysical applications μ^c is called the free-free absorption coefficient.

2.6 The identity (2.58) arises from truncating Fourier transforms. The truncated δ-function is defined by

$$\lim_{T\to\infty} \int_{-T/2}^{T/2} dt\, e^{i\omega t} = 2\pi\delta(\omega).$$

Derive (2.58) by evaluating

$$\lim_{T\to\infty} \int_{-T/2}^{T/2} dt \int_{-T/2}^{T/2} dt'\, e^{i\omega(t+t')}$$

by changing variables of integration from t and t' to $t+t'$ and $t-t'$ and by equating the integral over $-T/2 < (t, t') < T/2$ to half the integral over $-T < (t+t', t-t') < T$.

PART II
Instabilities in unmagnetized plasmas

3

Reactive instabilities

3.1 Effect of a beam on wave dispersion

Consider a beam of electrons with mean velocity v_b and number density n_1 moving through a background electron distribution which is at rest with number density n_0. In the limit $n_1 \ll n_0$ the effect of the beam on the waves in the background electron gas may be treated as a perturbation. Let $K_1^L(\omega, \mathbf{k})$ be the contribution of the beam to the longitudinal part of the dielectric tensor, with $K_0^L(\omega, \mathbf{k})$ the contribution from the background electrons. (The unit term, from the vacuum, is included in K_0^L and not in K_1^L.) In the absence of the beam let the Langmuir waves have frequency $\omega_L(\mathbf{k})$. The beam causes a frequency shift $\Delta\omega_L$ determined by

$$\Delta\omega_L \frac{\partial}{\partial\omega} \text{Re}[K_0^L(\omega, \mathbf{k})] + K_1^L(\omega, \mathbf{k}) = 0 \qquad (3.1)$$

evaluated at $\text{Re} \, K_0^L(\omega, \mathbf{k}) = 0$. The real part of K_1^L gives a real frequency shift, and the imaginary part of K_1^L contributes to the damping of the waves. Growth can occur when $\text{Im} \, K_1^L$ is negative; the instability is then due to negative absorption and is called a kinetic instability. Kinetic instabilities are discussed in Chapter 4.

For a cold beam K_1^L is real, and (3.1) is a real equation for $\Delta\omega_L$. Complex solutions can appear only as complex conjugate pairs. One of this pair has $\text{Im} \, \omega > 0$ and is growing and the other has $\text{Im} \, \omega < 0$ and is damping. The instability in this case has been given a variety of names, such as 'hydrodynamic', 'cold-beam', 'bunching' and 'reactive-medium'. Bohm & Gross (1949a) identified the instability as being due to a feedback mechanism involving bunching of the electrons leading to a growth in the space charge density. Briggs (1964) emphasized that this is a feature of a reactive-medium response. An appropriate name for this kind of instability is a *reactive instability*.

Suppose we start from a value of $\mathbf{k}$ for which all solutions of $\text{Re}[K^L(\omega, \mathbf{k})] = 0$, with $K^L = K_0^L + K_1^L$, are real. Then as $\mathbf{k}$ is varied, complex solutions can appear only when $\mathbf{k}$ passes through a value at which two solutions merge to become a double solution. The condition for a double solution of $\text{Re}[K^L(\omega, \mathbf{k})] = 0$ is that the condition

$$\frac{\partial}{\partial \omega} \text{Re}[K^L(\omega, \mathbf{k})] = 0 \tag{3.2}$$

also be satisfied. This condition implies $R_L(\mathbf{k}) = \infty$, cf. (2.46); that is, when (3.2) is satisfied the total energy $W_L(\mathbf{k})$ in the waves is zero. The electric energy is positive by definition, and zero energy implies that the contribution from particle motions is negative. A negative contribution of this kind may be interpreted in terms of a negative potential energy for the particles. For beam particles, with a beam energy $\frac{1}{2}mv_b^2$, the wave perturbation can reduce the energy of the particles on average, implying a negative contribution to the wave energy.

This argument implies that in a reactive instability the growing waves have *negative energy*. The concept of negative energy is useful for some qualitative purposes. For example, positive damping (Im $K^L > 0$) extracts energy from negative energy waves and hence causes their amplitude to increase. This follows from (2.45) in that Im $K^L > 0$ and $(\partial/\partial \omega) \text{Re}\, K^L < 0$ imply $\gamma_L < 0$, i.e. negative damping. However, no entirely satisfactory quantitative treatment of the energetics of intrinsically growing waves is available, and thus the concept of negative energy waves remains a largely qualitative one.

3.2 The counter-streaming instability

The dispersion equation $\text{Re}[K^L(\omega, \mathbf{k})] = 0$ is a quartic equation for ω even in the simplest case of a cold electron beam passing through a cold plasma. Although one may solve a quartic equation the solution is so cumbersome that it is rarely useful. To investigate the appearance of complex solutions it is convenient to study a special system in which the quartic equation reduces to a bi-quadratic equation, i.e. a quadratic equation for ω^2. The full solution is then relatively simple.

An appropriate system is for two interpenetrating beams of equal density viewed from the centre of mass frame. Let each beam have number density $n_0/2$ and velocity $\pm \mathbf{v}_b$. Let $\omega_{p0} = (e^2 n_0/\varepsilon_0 m)^{1/2}$ be the plasma frequency for the total system. We refer to this as the counter-streaming system.

The dielectric tensor for the stream follows directly from (2.16):

$$K_{ij}(\omega, \mathbf{k}) = \delta_{ij} - \frac{\omega_{p0}^2}{\gamma_b \omega^2}\left[\delta_{ij} + \frac{\mathbf{k} \cdot \mathbf{v}_b(k_i v_{bj} + k_j v_{bi})}{\omega^2 - (\mathbf{k} \cdot \mathbf{v}_b)^2} \right.$$
$$\left. + \frac{(k^2 - \omega^2/c^2)\{\omega^2 + (\mathbf{k} \cdot \mathbf{v}_b)^2\} v_{bi} v_{bj}}{\{\omega^2 - (\mathbf{k} \cdot \mathbf{v}_b)^2\}^2} \right], \tag{3.3}$$

with $\gamma_b = (1 - v_b^2/c^2)^{-1/2}$. The longitudinal part of (3.3) is

$$K^L(\omega, \mathbf{k}) = 1 - \frac{\omega_{p0}^2\{\omega^2 + (\mathbf{k} \cdot \mathbf{v}_b)^2\}}{\gamma_b\{\omega^2 - (\mathbf{k} \cdot \mathbf{v}_b)^2\}^2}\left(1 - \frac{\omega^2}{k^2 c^2}\right). \tag{3.4}$$

Neglecting relativistic effects corresponds to setting $\gamma_b = 1$ and omitting the

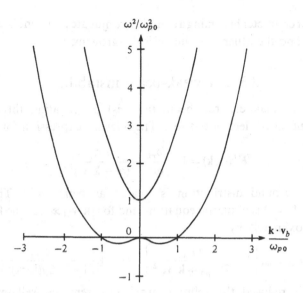

Fig. 3.1 The solutions of $\omega^4 - 2\omega^2((k \cdot v_b)^2 + \frac{1}{2}\omega_{p0}^2) + (k \cdot v_b)^2 ((k \cdot v_b)^2 - \omega_{p0}^2) = 0$ for ω^2/ω_{p0}^2 as a function of $k \cdot v_b/\omega_{p0}$. Negative values of ω^2 correspond to wave growth.

term ω^2/k^2c^2 in (3.4). The dispersion equation $K^L(\omega, k) = 0$ may then be solved for the four solutions

$$\omega = \pm [(k \cdot v_b)^2 + \tfrac{1}{2}\omega_{p0}^2 \pm \tfrac{1}{2}\omega_{p0}^2 \{1 + 8(k \cdot v_b)^2/\omega_{p0}^2\}^{1/2}]^{1/2}. \qquad (3.5)$$

One readily determines that all four solutions (3.5) are real for sufficiently large $|k \cdot v_b|$ and that two of the solutions are imaginary for sufficiently small $|k \cdot v_b|$. The double solution is at $\omega = 0$ and occurs for

$$(k \cdot v_b)^2 = \omega_{p0}^2. \qquad (3.6)$$

The solutions for ω^2 as a function of $k \cdot v_b$ are illustrated in Figure 3.1. The maximum negative value of ω^2 is $-\omega_{p0}^2/8$ at $(k \cdot v_b)^2 = 3\omega_{p0}^2/8$. Hence the maximum growth rate $|\gamma_{\max}|$, defined for the wave energy (i.e. twice the growth rate for the amplitude), is

$$|\gamma_{\max}| = \frac{\omega_{p0}}{\sqrt{2}} \quad \text{at} \quad |k \cdot v_b| = (\tfrac{3}{8})^{1/2}\omega_{p0}. \qquad (3.7)$$

In the limit of large $|k \cdot v_b|/\omega_{p0}$ the four solutions (3.5) reduce to $k \cdot v_b \pm \omega_{p0}/\sqrt{2}$ and $-(k \cdot v_b \pm \omega_{p0}/\sqrt{2})$. These may be interpreted as Doppler shifted either forward or backward waves propagating in either one beam or the other. In this limit one beam has little effect on the Langmuir waves in the other beam. Let us consider the effect of reducing $|k \cdot v_b|/\omega_{p0}$; then the forward propagating waves in the backward propagating beam and the backward propagating waves in the forward propagating beam approach each other in frequency. In the centre of mass frame they both approach zero as $|k \cdot v_b|/\omega_{p0}$ approaches unity. The two modes then merge together and change their character becoming a complex-conjugate pair. One is an intrinsically damped mode and the other is an intrinsically growing mode.

3.3 The weak-beam instability

Now consider a weak electron beam ($n_1 \ll n_0$) propagating through a background distribution of electrons at rest. The relevant expression for K^L is

$$K^L(\omega, \mathbf{k}) = 1 - \frac{\omega_{p0}^2}{\omega^2} - \frac{\omega_{p1}^2}{(\omega - \mathbf{k} \cdot \mathbf{v}_b)^2}, \tag{3.8}$$

where the background distribution is treated as being cold. The dispersion equation $K^L(\omega, \mathbf{k}) = 0$ is a quartic equation, and for $|\mathbf{k} \cdot \mathbf{v}_b| \gg \omega_{p0}$ the four solutions are given approximately by

$$\omega = \pm \omega_{p0} \left\{ 1 + \frac{\omega_{p1}^2}{2(\omega_{p0} \mp \mathbf{k} \cdot \mathbf{v}_b)^2} \right\}, \quad \mathbf{k} \cdot \mathbf{v}_b \pm \frac{\omega_{p1}}{[1 - \omega_{p0}^2/(\mathbf{k} \cdot \mathbf{v}_b)^2]^{1/2}} \tag{3.9}$$

As $|\mathbf{k} \cdot \mathbf{v}_b|/\omega_{p0}$ is reduced the solution $\omega \approx -\omega_{p0}$ remains well separated from the other three. The other three become comparable with each other as $\mathbf{k} \cdot \mathbf{v}_b/\omega_{p0}$ approaches unity. This suggests that we should seek a cubic approximation to the dispersion equation.

Let us write

$$\Delta\omega = \omega - \omega_{p0}, \quad \Delta\omega_0 = \omega_{p0} - \mathbf{k} \cdot \mathbf{v}_b, \tag{3.10a, b}$$

and approximate $1 - \omega_{p0}^2/\omega^2$ by $2\Delta\omega/\omega_{p0}$. An appropriate approximation to the dispersion equation is then

$$\Delta\omega(\Delta\omega + \Delta\omega_0)^2 - \tfrac{1}{2}\omega_{p0}\omega_{p1}^2 = 0. \tag{3.11}$$

The three solutions are

$$\Delta\omega = \alpha \left(\frac{n_1}{n_0} \right)^{1/3} \omega_{p0} \tag{3.12}$$

where α is a solution of

$$\alpha(\alpha + \alpha_0)^2 = 1/2, \quad \alpha_0 = \frac{\Delta\omega_0}{\omega_{p0}(n_1/n_0)^{1/3}}. \tag{3.13}$$

For $\Delta\omega_0 = 0$ the three solutions of (3.13) are $2^{-1/3}$ times the three cube roots of unity. The growth rate of the growing mode is

$$|\gamma_{max}| = \sqrt{3} \left(\frac{n_1}{2n_0} \right)^{1/3} \omega_{p0}, \tag{3.14}$$

which applies for $\mathbf{k} \cdot \mathbf{v}_b \approx \omega_{p0}$. Further details are indicated in Exercise 3.1.

3.4 Suppression due to increasing velocity spread

A thermal spread in the velocities of the beam particles tends to suppress the reactive instability. This has an important implication even when the velocity spread is initially small because the development of the instability itself causes the velocity spread to increase. This must ultimately lead to suppression of the reactive instability (O'Neil & Malmberg 1968).

Consider a nonrelativistic streaming Maxwellian distribution for the beam with a thermal spread V_b:

$$f(\mathbf{p}) = \frac{n_1}{(2\pi)^{3/2} m^3 V_b^3} \exp\left[-\frac{(\mathbf{v} - \mathbf{v}_b)^2}{2V_b^2} \right], \tag{3.15}$$

with $\mathbf{p} = m\mathbf{v}$ here. The longitudinal part of the dielectric tensor becomes, cf. Exercise 3.2,

$$\mathrm{Re}[K^L(\omega, \mathbf{k})] = 1 - \frac{\omega_{p0}^2}{\omega^2} - \frac{\omega_{p1}^2}{k^2 V_b^2}\left\{ \phi\left(\frac{\omega - \mathbf{k}\cdot\mathbf{v}_b}{\sqrt{2}\,kV_b}\right) - 1 \right\}, \tag{3.16}$$

where the background electrons are again assumed to be cold. For $(\omega - \mathbf{k}\cdot\mathbf{v}_b)^2 \gg 2k^2 V_b^2$ the asymptotic expansion (2.40) of the function $\phi(z)$ shows that (3.16) reproduces (3.8). In the reactive instability we have $|\omega - \mathbf{k}\cdot\mathbf{v}_b|$ of order $(n_1/n_0)^{1/3}\omega_{p0}$, cf. (3.14), and hence the condition for the neglect of the thermal spread in the beam is $(n_1/n_0)^{1/3}\omega_{p0} \gg \sqrt{2}kV_b$, which with $k \approx \omega_{p0}/v_b$, implies

$$\left(\frac{n_1}{n_0}\right)^{1/3} \gg \frac{V_b}{v_b}. \tag{3.17}$$

For $(\omega - \mathbf{k}\cdot\mathbf{v}_b)^2 \lesssim 2k^2 V_b^2$, $K^L(\omega, \mathbf{k})$ cannot be approximated as in (3.8) and no reactive instability is possible.

Suppose that (3.17) is satisfied and that the instability develops. As the wave amplitude grows the spread in the velocities of the electrons increases. Approximately half the wave energy is in forced motions of the electrons and hence the mean spread V_b in the velocities of the electrons may be estimated by setting $\frac{1}{2}n_1 m_e V_b^2$ equal to $\frac{1}{2}W_L$, where W_L is the energy density in the waves. Suppression then occurs for (ignoring factors of order unity)

$$W_L \approx n_1 m_e V_b^2 \approx n_1 m_e v_b^2 \left(\frac{n_1}{n_0}\right)^{2/3}. \tag{3.18}$$

That is, the reactive instability tends to suppress itself when approximately $(n_1/n_0)^{2/3}$ of the initial energy density in the beam has been transferred to the waves. This suppression effect was pointed out by Shapiro (1963). When this suppression occurs the kinetic version of the instability may take over, as discussed in §4.1 below, cf. however O'Neil, Winfrey & Malmberg (1971).

The reactive version of the weak-beam instability is the relevant one in laboratory applications. However in space plasmas, e.g. in the interplanetary medium, the beams tend to be very weak and long times are available for them to relax. Consequently the kinetic version and not the reactive version of the instability is likely to be the more relevant in applications outside the laboratory.

3.5 The Buneman instability

An instability which is closely related to the weak-beam instability was pointed out by Buneman (1958, 1959). A current with current density J implies a relative motion of the electrons and the ions at a drift speed

$$v_d = \frac{J}{n_e e}.$$ (3.19)

The longitudinal part of the dielectric tensor in the rest frame of the electrons is

$$K^L(\omega, \mathbf{k}) = 1 - \frac{\omega_p^2}{\omega^2} - \frac{\omega_{pi}^2}{(\omega - \mathbf{k} \cdot \mathbf{v}_d)^2},$$ (3.20)

where the electrons are assumed to be cold. The analogy between (3.20) and (3.8) is obvious. By analogy with (3.14) an instability occurs with growth rate

$$|\gamma_{\max}| = \sqrt{3} \left(\frac{Z_i m_e}{m_i} \right)^{1/3} \omega_p,$$ (3.21)

where the ions are assumed to have charge $Z_i e$ and where charge neutrality implies $Z_i n_i = n_e$. This growth rate (3.21) is large, e.g. $|\gamma_{\max}| \approx 0.1 \omega_p$.

The maximum growth (3.21) occurs at $\mathbf{k} \cdot \mathbf{v}_d \approx \omega_p$. The neglect of the thermal motions of the electrons is justified only for $\omega/kV_e \gg 1$ which requires $v_d \gg V_e$ here. Hence the electron drift speed must exceed about the thermal speed of electrons before the Buneman instability can develop.

As the Buneman instability develops the velocity spread of the electrons increases, just as in the case of the weak-beam instability. Here this corresponds to heating the electrons, and thus increasing V_e. Thus the Buneman instability tends to suppress itself by reducing the ratio v_d/V_e (by increasing V_e) until it falls below the threshold for the instability.

In an unmagnetized plasma there is a competing instability which involves the growth of ion sound waves (§4.3). The threshold drift velocity for this instability is $v_d > v_s$, which is a much weaker condition than for the development of the Buneman instability. Thus one would expect the ion sound instability to develop first and to restrict v_d to $\approx v_s$ and hence prevent the Buneman instability from ever occurring. However the ion sound instability can develop only for $T_e \gg T_i$, which is the condition (2.52) for ion sound waves to exist. Suppose this condition is not satisfied initially, and that the Buneman instability develops, i.e. that the threshold condition $v_d > V_e$ is satisfied. As a result the electrons should be heated and the conditions for the ion sound instability to develop should become satisfied. As the ion sound instability develops the drift velocity should reduce towards v_s.

These ideas are related to the concept of anomalous conductivity in current-carrying plasmas. A general difficulty in applying this concept to specific applications is in relating the local development of an instability, and hence of anomalous conductivity, to the global conditions which cause the current to flow.

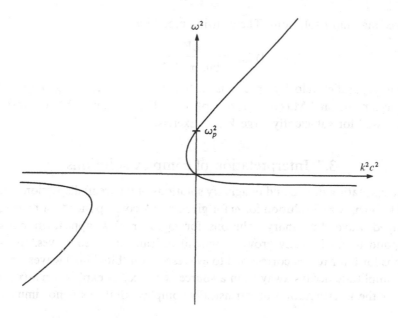

Fig. 3.2 The solutions of the dispersion equation (3.24) for ω^2 as a function of $k^2 c^2$ are plotted schematically. The Weibel instability corresponds to the region of negative ω^2 for real positive $k^2 c^2$.

3.6 The Weibel instability

So far only instabilities involving longitudinal waves have been considered. Weibel (1959) pointed out an example where transverse waves grow in an unmagnetized plasma. The example considered is rather specialized. The distribution chosen is nonrelativistic with

$$f(\mathbf{p}) = \frac{n}{2\pi V_\perp^2 m^3} \delta(v_\parallel) \exp\left[-\frac{v_\perp^2}{2V_\perp^2} \right] \tag{3.22}$$

where $\perp$ and $\parallel$ refer to an axis in the plasma. In addition $\mathbf{k}$ is assumed to lie along this axis. The transverse part K^T in this case may be evaluated using (2.25), e.g. using a result quoted in Exercise (2.4). Setting $\gamma = 1$ and $\mathbf{p} = m\mathbf{v}$ one finds

$$K^T(\omega, \mathbf{k}) = 1 - \frac{\omega_p^2}{\omega^2}\left\{ 1 + \frac{V_\perp^2}{\omega^2}\left(k^2 - \frac{\omega^2}{c^2} \right) \right\}. \tag{3.23}$$

To be consistent with the nonrelativistic approximation the term ω^2/c^2 in (3.23) should also be omitted. The dispersion equation $N^2 = K^T(\omega, \mathbf{k})$ then reduces to

$$\omega^4 - \omega^2(\omega_p^2 + k^2 c^2) - \omega_p^2 k^2 V_\perp^2 = 0. \tag{3.24}$$

The dispersion curve for (3.24) is illustrated in Figure 3.2. There are two real and two imaginary solutions. The growing mode has

$$|\gamma_{\max}| = \frac{kV_\perp \omega_p}{(\omega_p^2 + k^2 c^2)^{1/2}} \tag{3.25}$$

When a parallel velocity spread is included, e.g. by replacing $\delta(v_\parallel)$ in (3.22) by a one-dimensional Maxwellian $\propto \exp[-v_\parallel^2/2V_\parallel^2]$, one finds that the instability is suppressed for sufficiently large $V_\parallel$, cf. Exercise 3.3.

3.7 Interpretation of complex solutions

The interpretation of real and imaginary solutions of the (real) dispersion equation is well known. A real solution for ω for given real $\mathbf{k}$ corresponds to a propagating undamped wave. Imaginary solutions for ω for real $\mathbf{k}$ appear in pairs and correspond to intrinsically growing and intrinsically damped waves. Imaginary solutions for $\mathbf{k}$ for real ω correspond to evanescent (or 'blocking') waves, i.e. waves whose amplitude decays away from a source at $\mathbf{x} = \mathbf{x}_0$ as $\exp[-|\operatorname{Im}\mathbf{k}\cdot(\mathbf{x} - \mathbf{x}_0)|]$. However the interpretation of intrinsically complex solutions is not immediately obvious.

To discuss this interpretation consider a one-dimensional system in which there is a localized source of electromagnetic disturbances with a spatial distribution $g(z)$ about some point $z = z_0$ and with a temporal variation $f(t)$. Let $R(\omega, k)$ be a response function for the one-dimensional system such that the response is of the form $\psi(\omega, k) = R(\omega, k) f(\omega) g(k)$. We then have

$$\psi(t, z) = \int_{-\infty + i\Gamma}^{\infty + i\Gamma} \frac{d\omega}{2\pi} \int \frac{dk}{2\pi} e^{-i(\omega t - kz)} R(\omega, k) f(\omega) g(k). \tag{3.26}$$

Here Γ is larger than the maximum imaginary part of ω at which the integrand has a pole, in accord with the criterion discussed in §1.5.

The response function $R(\omega, k)$ has the form $H(\omega, k)/\Lambda(\omega, k)$, where $\Lambda(\omega, k) = 0$ is the dispersion equation. This is due to the fact that $\psi(\omega, k) = R(\omega, k) f(\omega) g(k)$ is obtained by solving a wave equation. We may assume that $H(\omega, k)$ is a non-singular function. Let $k = k_M$ be a solution of the dispersion equation for a mode M. Summing over all modes, the response (3.26) becomes

$$\psi(t, z) = \sum_M \int_{-\infty + i\Gamma}^{\infty + i\Gamma} \frac{d\omega}{2\pi} \int \frac{dk}{2\pi} e^{-i(\omega t - kz)} \frac{R(\omega, k) f(\omega) g(k)}{\left[\dfrac{\partial}{\partial k}\Lambda(\omega, k)\right](k - k_M)}. \tag{3.27}$$

Let us denote by $M+$ and $M-$ poles in the upper half k-plane and in the lower half k-plane respectively. For large z, the assumption that the source is localized in z implies that

$$g(k) = \int dz\, e^{ikz} g(z)$$

is well behaved for $\operatorname{Im}(kz) > 0$. Thus for positive z outside the source region

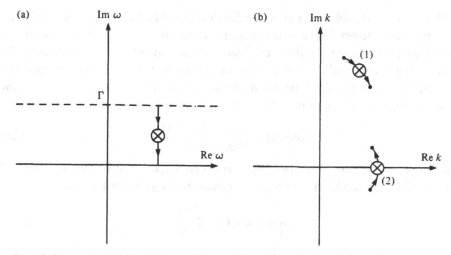

Fig. 3.3 As Im ω is reduced from Γ to 0 for fixed Re ω (Fig. 3.3(a)) the solutions $k_M(\omega)$ move along paths in the complex-k plane (Fig. 3.3(b)). Solutions such as (1) correspond to evanescent modes, and solutions such as (2) which cross the Re k axis corresponding to amplifying modes.

we may close the contour of k-integration in the upper half plane. Similarly for z negative and outside the source region we may close the k-contour in the lower half k-plane. In either case the value of the integral is determined by the residues at the poles inside the contour. Hence one finds

$$\psi(t,z) = \mp i \sum_{M\pm} \int_{-\infty+i\Gamma}^{\infty+i\Gamma} \frac{d\omega}{2\pi} \exp[-i\omega t + ik_{M\pm}z] \frac{R(\omega, k_{M\pm})f(\omega)g(k_{M\pm})}{\left[\frac{\partial}{\partial k}\Lambda(\omega, k)\right]_{k=k_{M\pm}}}$$

(3.28)

where the upper sign applies for $z > 0$ and the lower sign for $z < 0$. It follows that in both cases the disturbance decays in space away from the source, i.e. Re$[ik_{M+}z]$ is negative for $z > 0$ and Re$[ik_{M-}z]$ is negative for $z < 0$.

Suppose there are no solutions to the dispersion equation with complex ω for real k; then we may choose Γ infinitesimally small and the foregoing argument implies that all modes with complex k decay in space. Hence complex k for real ω implies evanescent waves.

Now consider a solution with complex k and complex ω. For Im $\omega < 0$ the argument is unaltered, and complex k implies evanescence. For Im $\omega > 0$ we must choose $\Gamma > \text{Im}\,\omega$ in (3.28). As Im ω varies, for Re $\omega = $ constant say, the solution for k moves around the complex k plane. Suppose we reduce Im ω from its initial value to zero. Then the foregoing argument continues to apply provided none of the solutions for complex k cross the real k axis, i.e. provided none of them moves from one half plane to the other. If, on the contrary, one solution does cross the real k axis (Figure 3.3) then (3.28) implies that it is a spatially growing disturbance for real ω.

This leads to the following criterion for deciding whether a wave is amplifying or evanescent (Briggs 1964): If the imaginary part of $k_M(\omega)$ changes sign as the imaginary part of ω varies from arbitrarily large and positive to zero, then the mode M is growing, and if this sign does not change then the mode M is evanescent.

A related analytic technique for determining stability is based on use of the Nyquist diagram. The idea is as follows. Consider the function

$$G(\omega, \mathbf{k}) = \frac{1}{\Lambda(\omega, \mathbf{k})} \frac{\partial \Lambda(\omega, \mathbf{k})}{\partial \omega}. \tag{3.29}$$

The wave modes are described by zeros of $\Lambda(\omega, \mathbf{k})$ and hence by poles of $G(\omega, \mathbf{k})$. Growing modes have $\mathrm{Im}\,\omega > 0$. Consider the contour integral

$$\oint d\omega\, G(\omega, \mathbf{k}) = \oint \frac{d\Lambda}{\Lambda} \tag{3.30}$$

where the contour encloses the entire upper half complex ω-plane. The residue theorem implies that the integral is equal to $2\pi i$ times the sum of the residues at the poles. In this case each residue contributes unity, and hence the integral to $2\pi i$ times the number of growing modes. This condition for instability is called the *Penrose criterion* (Penrose 1960).

Another distinction which is often made is between *absolute* and *convective* instabilities (Sturrock 1958). Briefly an instability is absolute if it leads to unlimited growth at a fixed point, and it is convective if the amplitude at a fixed point remains bounded. This distinction can be of practical significance, but it is not an important physical distinction because an instability which is absolute in one inertial frame may be convective in another frame moving relative to the first. Sturrock (1958) gave rules for distinguishing between absolute and convective instabilities.

Exercise set 3

3.1 Find the three solutions of the cubic equation, cf. (3.13),

$$\alpha(\alpha + \alpha_0)^2 = \tfrac{1}{2}$$

and show that the complex solutions have

$$|\mathrm{Im}\,\alpha| = \frac{\sqrt{3}}{2}(\alpha_+ - \alpha_-)$$

$$\alpha_\pm = \left[\frac{1}{4} + \left(\frac{\alpha_0}{3} \right)^3 \pm \left\{ \frac{1}{2} \left(\frac{\alpha_0}{3} \right)^3 + \frac{1}{16} \right\}^{1/2} \right]^{1/3}.$$

Hence show that the maximum growth rate for the weak beam instability occurs for $\Delta\omega_0 = \alpha_0(\omega_{p0}\omega_{p1}^2)^{1/3} = 0$.

3.2 (a) Using the results of Exercise 2.2 show that the dielectric tensor (2.25) for the distributions of the form

$$f(\mathbf{p}) = \frac{n}{(2\pi)^{3/2} m^3 V_b^3} \exp\left[-\frac{(\mathbf{v} - \mathbf{v}_b)^2}{2V_b^2}\right].$$

with $\mathbf{p} = m\mathbf{v}$ reduces to

$$K_{ij}(\omega, \mathbf{k}) = \delta_{ij} - \sum \frac{\omega_p^2}{\omega^2}\left[\delta_{ij} + (k_i v_{bj} + k_j v_{bi})\frac{\phi(\bar{z})}{\bar{\omega}} + 2\kappa_i\kappa_j[\phi(\bar{z}) - 1]\right.$$

$$+ \left(1 - \frac{\omega^2}{k^2 c^2}\right)\left\{\frac{v_{bi}v_{bj}}{V_b^2}[\phi(\bar{z}) - 1] + \frac{(k_i v_{bj} + k_j v_{bi})}{\bar{\omega}}\left[\phi(\bar{z})\left(\frac{\bar{\omega}^2}{k^2 V_b^2} - 1\right) - \frac{\bar{\omega}^2}{k^2 V_b^2}\right]\right.$$

$$\left.\left. + \delta_{ij}[\phi(\bar{z}) - 1] + \kappa_i\kappa_j\frac{\bar{\omega}^2}{k^2 V_b^2}[\phi(\bar{z}) - 1] - \kappa_i\kappa_j[3\phi(\bar{z}) - 2]\right\}\right]$$

with $\bar{\omega} = \omega - \mathbf{k}\cdot\mathbf{v}_b$, $\bar{z} = \bar{\omega}/\sqrt{2}kV_b$, and where the sum is over all species.

(b) Evaluate the longitudinal part of $K_{ij}(\omega, \mathbf{k})$ and show that it reduces to (3.16) when the terms proportional to ω^2/c^2 are neglected.

(c) Show that when the terms proportional to ω^2/c^2 are retained (3.8) is replaced by

$$K^L(\omega, \mathbf{k}) = 1 - \frac{\omega_{p0}^2}{\omega^2} - \frac{\omega_{p1}^2}{(\omega - \mathbf{k}\cdot\mathbf{v}_b)^2} + \frac{\omega_{p1}^2}{k^2 c^2}\left(\frac{\mathbf{k}\cdot\mathbf{v}_b}{\omega - \mathbf{k}\cdot\mathbf{v}_b}\right)^2.$$

3.3 (a) Show that when the dielectric tensor is evaluated as in Exercise 3.2a, but for the distribution

$$f(\mathbf{p}) = \frac{n}{(2\pi)^{3/2} m^3 V_\perp^2 V_\parallel} \exp\left[-\frac{v_\perp^2}{2V_\perp^2} - \frac{v_\parallel^2}{2V_\parallel^2}\right],$$

the result is

$$K_{ij}(\omega, \mathbf{k}) = \delta_{ij} - \sum \frac{\omega_p^2}{\omega^2}\left[\delta_{ij} + 2\kappa_i\kappa_j[\phi(z) - 1]\right.$$

$$+ \left(1 - \frac{\omega^2}{k^2 c^2}\right)\left\{\frac{V_\perp^2}{V_\parallel^2}(\delta_{ij} - \kappa_i\kappa_j)[\phi(z) - 1]\right.$$

$$\left.\left. + \kappa_i\kappa_j\left(\frac{\omega^2}{k^2 V_\parallel^2} - 2\right)\phi(z) - \kappa_i\kappa_j\left(\frac{\omega^2}{k^2 V_\parallel^2} - 1\right)\right\}\right]$$

with $z = \omega/\sqrt{2}kV_\parallel$.

(b) Hence show that the Weibel instability exists only for $\omega/\sqrt{2}kV_\parallel \ll 1$.

3.4 (a) Show that the equation of fluid motion of nonrelativistic electrons in the field of a transverse wave with electric and magnetic amplitudes $\mathbf{E}$ and $\mathbf{B}$ gives

$$m\frac{d\mathbf{v}^{(1)}}{dt} = -e\mathbf{E}, \qquad \mathbf{v}^{(2)} = \frac{\mathbf{k}}{2\omega}|\mathbf{v}^{(1)}|^2$$

to second order in an expansion in $\mathbf{E}$.

(b) Hence show that the mean kinetic energy density $\frac{1}{2}mn|\mathbf{v}^{(1)}|^2$ in the forced motion of the electrons is a fraction ω_p^2/ω^2 of the electric energy density in the waves and that the mean momentum density $mn\mathbf{v}^{(2)}$ in this forced motion is $\omega_p^2/2\omega^2$ times the electromagnetic momentum density.

(c) Hence agrue that as $|\mathbf{E}|^2$ increases due to the growth of transverse waves, the spread in velocities along $\mathbf{k}$ increases proportional to $|\mathbf{E}|^2$. (According to Exercise 3.3 this effect can lead to self-suppression of the Weibel instability.)

3.5 A plasma consists of thermal electrons and two cold beams of oppositely directed ions. The beams have equal densities $\frac{1}{2}n_i$, and velocities $\pm\mathbf{v}_b$.

(a) Show that for $\omega \ll \sqrt{2kV_e}$ the longitudinal part of the dielectric tensor reduces to

$$K^L(\omega, \mathbf{k}) = 1 + \frac{1}{k^2\lambda_{De}^2} - \frac{\omega_{pi}^2\{\omega^2 + (\mathbf{k}\cdot\mathbf{v}_b)^2\}}{\{\omega^2 - (\mathbf{k}\cdot\mathbf{v}_b)^2\}^2}.$$

(b) Show that ion sound waves are subject to a reactive instability, and determine the maximum growth rate.

3.6 (a) Find the solutions for $k(\omega)$ of
$$\omega^2 = \omega_0^2 + k^2c^2$$

and show that they do not cross the real $-k$ axis as Im ω varies from 0 to ∞.

(b) Show that the solutions of
$$(\omega - kv)^2 = -\omega_0^2 + k^2c^2$$

for $k(\omega)$ describe real waves for $v^2 < c^2$ and that for $v^2 > c^2$ and $\omega^2 < \omega_0^2(v^2/c^2 - 1)$ one of the solutions corresponds to an amplifying wave. *Hint*: It suffices to consider the case Re $\omega = 0$.

4

Kinetic instabilities

4.1 The bump-in-tail instability

Many microinstabilities have both reactive and kinetic forms. From a mathematical viewpoint one treats the reactive form by ignoring the imaginary part of the dielectrie tensor and solving a real dispersion equation to find complex solutions, and one treats the kinetic form by assuming real frequencies to a first approximation and then including weak (negative) damping. To be more specific, in the weak-beam instability (§3.3) thermal motions are neglected and the correction to the real part of the dielectric tensor due to the presence of the beam leads to a cubic equation (3.11) for the frequency shift of the Langmuir waves; this cubic equation has a real solution and a pair of complex conjugate solutions in the regime of interest. The kinetic version of this instability is known as the bump-in-tail instability. It is treated by first finding the imaginary contribution of the beam to the dielectric tensor and using this to evaluate the imaginary part of the frequency shift.

It is apparent from the foregoing discussion that the reactive and kinetic versions should be limiting cases of a single instability. This may be shown by finding both the real and imaginary parts of the frequency shift simultaneously. The real and imaginary parts of the frequency can be found as a complex solution for ω as a function of real $\mathbf{k}$ to the complex dispersion equation $K^L(\omega, \mathbf{k}) = 0$, where both real and imaginary parts of K^L are retained. In practice a relatively sophisticated root-finding procedure is required to find such solutions. Such solutions are informative from a formal viewpoint in that they show how the reactive and kinetic versions pass over into each other. However they are rarely of much practical interest due to the smallness of the range of parameter space where the reactive and kinetic versions pass over into each other.

The growth rate for the bump-in-tail instability may be found using the expression (2.47) for the absorption coefficient for Langmuir waves. A general expression for $\mathrm{Im}\,[K^L(\omega, \mathbf{k})]$ may be obtained from (2.24), giving

$$\gamma_L(\mathbf{k}) = -\frac{\pi q^2}{\varepsilon_0 \omega_p^2} \frac{\{\omega_L(\mathbf{k})\}^3}{k^2} \int d^3\mathbf{p}\,\delta\{\omega_L(\mathbf{k}) - \mathbf{k}\cdot\mathbf{v}\}\mathbf{k}\cdot\frac{\partial f(\mathbf{p})}{\partial \mathbf{p}} \tag{4.1}$$

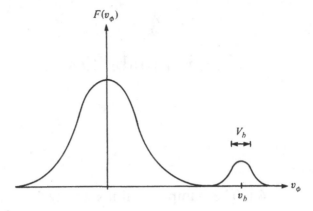

Fig. 4.1 The projected distribution $F(v_\phi)$, cf. Equation (4.2), is plotted for a distribution composed of a central Maxwellian with thermal speed V and a Maxwellian bump-in-tail distribution with beam velocity v_b and spread V_b, cf. (4.5). It is assumed that **k** is along $\mathbf{v}_b$. Growth occurs at $v_\phi \lesssim v_b$ where $dF(v_\phi)/dv_\phi$ is positive.

for an arbitrary distribution $f(\mathbf{p})$ of particles with charge q. The sign of $\gamma_L(\mathbf{k})$ can be negative if $\mathbf{k} \cdot \partial f(\mathbf{p})/\partial \mathbf{p}$ is positive. However, although this is a necessary condition for negative damping, it is not a sufficient condition. In particular, $\gamma_L(\mathbf{k})$ cannot be negative for an isotropic distribution of particles, as may be shown using the result of Exercise 2.4. Even if the isotropic distribution has $\partial f(p)/\partial p > 0$ over some range of p, it must have $\partial f(p)/\partial p < 0$ at sufficiently large p in order that $f(p)$ be normalizable, and it turns out that for an isotropic distribution the contribution from regions where $\partial f(p)/\partial p$ is negative always dominates the contribution from regions where $\partial f(p)/\partial p$ is positive.

In the nonrelativistic limit we may reduce (4.1) by defining a one-dimensional distribution function

$$F(v_x) = \frac{1}{n_1} \int d^3\mathbf{p}\, \delta(v_x - \mathbf{k} \cdot \mathbf{v}/k) f(\mathbf{p}) \qquad (4.2)$$

where the normalization is now

$$\int dv_x F(v_x) = 1, \qquad (4.3)$$

where n_1 is the number density of particles in the distribution $f(\mathbf{p})$ or $F(v_x)$, and where we now interpret **p** as $m\mathbf{v}$ in (4.2). Then (4.1) becomes

$$\gamma_L(\mathbf{k}) = -\pi \frac{\omega_{p1}^2}{\omega_p^2} \omega_L(\mathbf{k}) v_\phi^2 \frac{dF(v_\phi)}{dv_\phi} \qquad (4.4)$$

with $v_\phi = \omega_L(\mathbf{k})/k$ and $\omega_{p1}^2 = n_1 q^2/m\varepsilon_0$.

A distribution of the form illustrated in Figure 4.1 has a region with $dF(v_\phi)/dv_\phi > 0$ is called a bump-in-tail distribution. According to (4.4) Langmuir waves grow due to negative absorption at phase speeds which correspond to the range where $dF(v_x)/dv_x$ is positive.

For the streaming Maxwellian distribution, (3.15), viz.

$$f(\mathbf{p}) = \frac{n_1}{(2\pi)^{3/2} m^3 V_b^3} \exp\left[-\frac{(\mathbf{v} - \mathbf{v}_b)^2}{2V_b^2} \right] \qquad (4.5)$$

the one-dimensional distribution (4.2) is

$$F(v_x) = \frac{1}{(2\pi)^{1/2} V_b} \exp\left[-\frac{(v_x - \mathbf{k}\cdot\mathbf{v}_b/k)^2}{2V_b^2} \right]. \qquad (4.6)$$

The absorption coefficient (4.4) then becomes

$$\gamma_L(\mathbf{k}) = \left(\frac{\pi}{2}\right)^{1/2} \frac{\omega_{p1}^2}{\omega_p^2} \omega_L(\mathbf{k}) \frac{v_\phi^2}{V_b^2} (v_\phi - \mathbf{k}\cdot\mathbf{v}_b/k) \exp\left[-\frac{(v_\phi - \mathbf{k}\cdot\mathbf{v}_b/k)^2}{2V_b^2} \right]. \qquad (4.7)$$

For $v_b \gg V_b$ maximum growth occurs at $v_\phi \approx v_b - V_b$ with $\mathbf{k}$ along $\mathbf{v}_b$, and with the growth rate

$$|\gamma_{\max}| \approx \left(\frac{\pi}{2e}\right)^{1/2} \omega_p \frac{n_1}{n_0} \left(\frac{v_b}{V_b}\right)^2. \qquad (4.8)$$

(In (4.8) e denotes $\exp[1]$.)

The exponent in (4.7) may be written as $-(\omega_L(\mathbf{k}) - \mathbf{k}\cdot\mathbf{v}_b)^2/2k^2 V_b^2$. The range of frequencies $\Delta\omega$ over which the growth rate is close to its maximum value is given by $(\Delta\omega)^2 \approx 2k^2 V_b^2$, or

$$\frac{\Delta\omega}{\omega_p} \approx \frac{V_b}{v_b}. \qquad (4.9)$$

We refer to $\Delta\omega$ as the *bandwidth of the growing waves*. Comparison of (4.8) and (4.9) shows that the growth rate is less than the bandwidth of the growing waves provided the condition $n_1/n_0 \lesssim (V_b/v_b)^3$ is satisfied, where factors of order unity are ignored. This inequality is the reverse of the inequality (3.17), confirming that the kinetic version takes over from the reactive version for $n_1/n_0 \approx (V_b/v_b)^3$.

The dependence of the growth rate on the angle, θ say, between $\mathbf{k}$ and $\mathbf{v}_b$ also follows from (4.7). The growth rate remains near its maximum value at fixed $k \approx \omega_p/v_b$ for

$$\theta \lesssim \frac{1}{2} \frac{V_b}{v_b}. \qquad (4.10)$$

However because the growth rate depends on $\mathbf{k}$ (apart from the weak dependence through $\omega_L(\mathbf{k})$) only in the form $\mathbf{k}\cdot\mathbf{v}_b$, it is almost independent of the component of $\mathbf{k}$ orthogonal to $\mathbf{v}_b$. Including the dependence on θ, (4.8) may be replaced by

$$|\gamma_{\max}| \approx \left(\frac{\pi}{2e}\right)^{1/2} \omega_p \left(\frac{v_b \cos\theta}{V_b}\right)^2 \qquad (4.11)$$

at $k \approx \omega_p/(v_b \cos\theta) - V_b$.

4.2 Quasilinear relaxation: the time-asymptotic state

Saturation of the bump-in-tail instability is attributed either to some nonlinear process or to quasilinear relaxation. These processes are discussed from a more formal viewpoint in §6.4 and §6.3 respectively. Here a one-dimensional version of the quasilinear equations is inferred using a heuristic argument, and these equations are then solved to find the time-asymptotic (i.e. $t \to \infty$) state of the instability.

The results derived in §4.1 may be rewritten as follows. Let $W(v_\phi)$ be the energy density in Langmuir waves in the range dv_ϕ of phase speeds at v_ϕ. Then the growth is described by

$$\frac{\partial W(v_\phi)}{\partial t} = - \gamma(v_\phi) W(v_\phi) \tag{4.12}$$

with

$$\gamma(v_\phi) = - \pi \frac{n_1}{n_e} \omega_p v_\phi^2 \frac{\partial F(v_\phi)}{\partial v_\phi}. \tag{4.13}$$

We now assume that only Langmuir waves along the streaming direction need be considered (this is the "one-dimensional" assumption). Our immediate objective is to determine how $F(v)$ evolves due to the emission or absorption of the Langmuir waves. First we note three conservation laws:–

(a) the number of particles is conserved

$$n_1 \int dv \frac{\partial F(v)}{\partial t} = 0; \tag{4.14a}$$

(b) the total energy is conserved

$$\int dv_\phi \frac{\partial W(v_\phi)}{\partial t} + n_1 \int dv \tfrac{1}{2} m v^2 \frac{\partial F(v)}{\partial t} = 0, \tag{4.14b}$$

and (c) the total momentum is conserved

$$\int dv_\phi \frac{1}{v_\phi} \frac{\partial W(v_\phi)}{\partial t} + n_1 \int dv \, mv \frac{\partial F(v)}{\partial t} = 0. \tag{4.14c}$$

In (4.14c) use has been made of the fact that the ratio of the momentum of waves to the energy of waves is $\mathbf{k} : \omega$. If one tries the form

$$\frac{\partial F(v)}{\partial t} = \frac{\partial G(v)}{\partial v},$$

then (4.14a) is satisfied automatically. Now inserting (4.12) and (4.13) in (4.14b) and (4.14c) one identifies

$$G(v) = \frac{\pi \omega_p}{n_1 m} v \frac{\partial F(v)}{\partial v} W(v).$$

That is, one finds

$$\frac{\partial F(v)}{\partial t} = \frac{\partial}{\partial v} D(v) \frac{\partial F(v)}{\partial v}, \tag{4.15}$$

which is a diffusion equation in velocity space, with diffusion coefficient

$$D(v) = \frac{\pi \omega_p}{n_e m} v W(v). \tag{4.16}$$

Equations (4.12), (4.13), (4.15) and (4.16) are the quasilinear equations in the one-dimensional case.

Suppose initially ($t = 0$) we have a distribution

$$F_0(v) = \delta(v - v_b). \tag{4.17}$$

At very large times we require that the system be relaxed, and hence that $\partial F(v)/\partial t$ approach zero. The time-asymptotic solution for the particles is the plateau distribution $\partial F(v)/\partial v = 0$ according to (4.15). (The second solution of $(\partial/\partial v)[vW(v)\partial F(v)/\partial v] = 0$ with $\partial F(v)/\partial v \neq 0$ is unstable.) Now (4.15) with (4.16) does not allow $F(v)$ to change except when $W(v)$ is non-zero, and (4.12) and (4.13) do not allow $W(v)$ to grow at $v > v_b$ if $\partial F(v)/\partial v$ is always negative at $v > v_b$. The time-asymptotic solution ($t \to \infty$) must have $\partial F_\infty(v)/\partial v = 0$ for $W_\infty(v) \neq 0$, and hence is of the form $F_\infty(v) = \text{constant}$ over a range $v_1 < v < v_b$; for the present we assume $F_\infty(v) = 0$ for $v < v_1$ and later we set $v_1 = 0$. Thus our time-asymptotic solution is

$$F_\infty(v) = \begin{cases} \dfrac{1}{v_b - v_1} & \text{for } v_1 < v < v_b \\ 0 & \text{otherwise} \end{cases} \tag{4.18}$$

Note that (4.18) is consistent with the fact that there are waves only at $v_\phi < v_b$ and hence particles can diffuse to $v < v_b$ but not to $v > v_b$.

By combining (4.12), (4.13), (4.15) and (4.16) one can derive the equation

$$\frac{\partial F(v)}{\partial t} = \frac{\partial}{\partial t}\left[\frac{\partial}{\partial v}\left\{\frac{W(v)}{n_1 m v}\right\}\right], \tag{4.19}$$

which may be integrated to give

$$F_\infty(v) - F_0(v) = \frac{\partial}{\partial v}\left\{\frac{W_\infty(v) - W_0(v)}{n_1 m v}\right\}. \tag{4.20}$$

The waves grow from an initial thermal level which is negligible in comparison with the final level. Hence we set $W_0(v) \approx 0$. Also assuming $W_\infty(v_1) = 0$, (4.20) is integrated to find

$$\int_{v_1}^{v} dv'\{F_\infty(v') - F_0(v')\} = \frac{v - v_1}{v_b - v_1} = \frac{W_\infty(v)}{n_1 m v} \tag{4.21}$$

for $v_1 < v < v_b$. Thus we find

$$W_\infty(v) = n_1 m v \frac{v - v_1}{v_b - v_1}, \tag{4.22}$$

which may be integrated to find the time-asymptotic energy density in the waves:

$$\int_{v_1}^{v_b} dv\, W_\infty(v) = \frac{m n_1}{v_b - v_1}\left\{\tfrac{1}{3}(v_b^3 - v_1^3) - \frac{v_1}{2}(v_b^2 - v_1^2)\right\}. \tag{4.23}$$

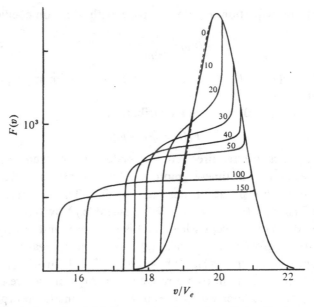

Fig. 4.2 Evolution of an initially gaussian bump-in-tail distribution centred at 20 thermal speeds and with a width of one thermal speed. Numbers on the curves indicate time in units of the inverse of the initial maximum growth rate. (Grognard 1975).

The time-asymptotic energy density in the particles is

$$n_1 \int_{v_1}^{v_b} dv \, \tfrac{1}{2} mv^2 F_\infty(v) = \frac{mn_1}{2(v_b - v_1)} \frac{1}{3}(v_b^3 - v_1^3). \tag{4.24}$$

As required the sum of these two energy densities is equal to the initial energy density $\tfrac{1}{2} n_1 mv_b^2$.

In the limit $v_1 \to 0$ one third of the initial energy density remains in the particles and the remaining two thirds is in the waves.

The evolution of a one-dimensional gaussian stream of particles towards the time-asymptotic state is illustrated in Figure 4.2 following Grognard (1975).

4.3 Current and ion-beam driven growth of ion sound waves

The growth rate for ion sound waves may be written in a form analogous to (4.4), i.e. in terms of a one-dimensional distribution function, as defined by (4.2). In this case both electron and ion contributions may be relevant.

The absorption coefficient for ion sound waves follows from (2.47) with (2.50):

$$\gamma_s(\mathbf{k}) = \frac{\{\omega_s(\mathbf{k})\}^3}{\omega_{pi}^2} \mathrm{Im} \, [K^L(\omega_s(\mathbf{k}), \mathbf{k})]. \tag{4.25}$$

Retaining both electron and ion contributions in the dielectric tensor (2.24),

(4.25) reduces to

$$\gamma_s(\mathbf{k}) = -\frac{\pi e^2}{\varepsilon_0 \omega_{pi}^2} \frac{\{\omega_s(\mathbf{k})\}^3}{k^2} \int d^3\mathbf{p}\,\delta\{\omega_s(\mathbf{k}) - \mathbf{k}\cdot\mathbf{v}\}\,\mathbf{k}\cdot\frac{\partial}{\partial\mathbf{p}}[f_e(\mathbf{p}) + Z_i^2 f_i(\mathbf{p})],$$

(4.26)

where $Z_i e$ is the charge on the ion. Now we make the nonrelativistic approximation and define the one-dimensional electron distribution $F_e(v_x)$ and ion distribution $F_i(v_x)$ as in (4.2). Then (4.26) reduces to

$$\gamma_s(\mathbf{k}) = -\pi\omega_s(\mathbf{k})v_\phi^2 \frac{d}{dv_\phi}\left[\frac{\omega_{p1e}^2}{\omega_{pi}^2}F_e(v_\phi) + \frac{\omega_{p1e}^2}{\omega_{pi}^2}F_i(v_\phi)\right],$$

(4.27)

where $\omega_{p1e}^2 = e^2 n_{1e}/m_e\varepsilon_0$ and $\omega_{p1i}^2 = Z_i^2 e^2 n_{1i}/\varepsilon_0 m_i$ denote the electron and ion plasma frequencies in the distributions $F_e(v_x)$ and $F_i(v_x)$ respectively. The dispersion relation $\omega = \omega_s(\mathbf{k}) = kv_s/\{1 + k^2\lambda_{De}^2\}^{1/2}$ is approximated by $\omega \approx kv_s$ for ion sound waves, i.e. for $k^2\lambda_{De}^2 \ll 1$, and by $\omega \approx \omega_{pi}$ for ion plasma waves, i.e. for $k^2\lambda_{De}^2 \gtrsim 1$. In these two cases the phase speed $v_\phi = \omega_s(\mathbf{k})/k$ in (4.27) reduces to $v_\phi \approx v_s$ and $v_\phi \approx \omega_{pi}/k$ respectively.

An important example is current-driven ion sound turbulence. Let the electrons be drifting relative to the ions at a rate $\mathbf{v}_d$, so that the current density is $J = n_e e v_d$. For $n_{1e} = n_e$ and with

$$f_e(\mathbf{p}) = \frac{n_e}{(2\pi)^{3/2}m_e^3 V_e^3}\exp\left[-\frac{(\mathbf{v} - \mathbf{v}_d)^2}{2V_e^2}\right],$$

(4.28)

the electronic contribution to (4.27) becomes

$$\gamma_s(\mathbf{k}) = \left(\frac{\pi}{2}\right)^{1/2}\omega_s(\mathbf{k})\frac{\omega_p^2}{\omega_{pi}^2}\frac{\{\omega_s(\mathbf{k}) - \mathbf{k}\cdot\mathbf{v}_d\}}{k}\frac{v_\phi^2}{V_e^3}\exp\left[-\frac{\{\omega_s(\mathbf{k}) - \mathbf{k}\cdot\mathbf{v}_d\}^2}{2k^2V_e^2}\right]$$

$$\approx \left(\frac{\pi}{2}\right)^{1/2}\frac{v_s}{V_e}\{kv_s - \mathbf{k}\cdot\mathbf{v}_d\},$$

(4.29)

where the approximate expression applies for $k^2\lambda_{De}^2 \ll 1$ and $v_d \ll V_e$. It follows from (4.29) that growth is possible for $v_d > v_s$.

In this instability ion sound waves grow on a cone at an angle $\theta = \arccos(v_s/v_d)$ relative to the drift velocity $\mathbf{v}_d$. As with the bump-in-tail instability, saturation may result from quasilinear relaxation. However there are several nonlinear processes which may saturate the instability before quasilinear relaxation occurs. These processes cause the ion sound waves to be removed from the regions of $\mathbf{k}$-space where they are growing to regions where they are damping, e.g. to angles away from $\theta = \arccos(v_s/v_d)$.

Current-driven instabilities can produce ion sound waves only if the condition $Z_i T_e \gg T_i$ is satisfied, i.e. only if ion sound waves exist as weakly damped waves. In a magnetized plasma an alternative current-driven instability is available: this involves electrostatic ion cyclotron waves (§12.1), and this alternative instability does not require $Z_i T_e \gg T_i$. In the absence of a magnetic field the only alternative current-driven instability is the Buneman instability.

Ion sound waves can be driven unstable by an ion beam with a flow speed $v_{bi} > v_s$. For a streaming Maxwellian distribution

$$f_i(\mathbf{p}) = \frac{n_{1i}}{(2\pi)^{3/2} m_i^3 V_{bi}^3} \exp\left[-\frac{(\mathbf{v} - \mathbf{v}_{bi})^2}{2V_{bi}^2} \right] \tag{4.30}$$

one finds

$$\gamma_s(\mathbf{k}) = \left(\frac{\pi}{2}\right)^{1/2} \frac{n_{1i}}{n_i} \left(\frac{\omega_s(\mathbf{k})}{kV_{bi}}\right)^3 \{\omega_s(\mathbf{k}) - \mathbf{k} \cdot \mathbf{v}_{bi}\} \exp\left[-\frac{\{\omega_s(\mathbf{k}) - \mathbf{k} \cdot \mathbf{v}_{bi}\}^2}{2k^2 V_{bi}^2} \right]. \tag{4.31}$$

The maximum growth rate is

$$|\gamma_s|_{\max} \approx \left(\frac{\pi}{2e}\right)^2 kv_s \frac{n_{1i}}{n_i} \left(\frac{v_s}{V_{bi}}\right)^2 \tag{4.32}$$

for $k^2\lambda_{De}^2 \ll 1$. There is a close analogy between this instability and the bump-in-tail instability for Langmuir waves, cf. (4.8) and (4.32).

4.4 Heat conduction and ion sound turbulence

Ion sound turbulence seems to be virtually a universal feature of the solar wind (e.g. Gurnett et al. 1979). The mechanism by which this turbulence is generated has not been identified unambiguously. Possibilities include generation by ion beams, and by a mechanism suggested by Forslund (1970). Forslund's mechanism relates to heat conduction by the electrons. There is a gradient in the electron temperature T_e away from the Sun and this drives a heat flux $Q_\parallel$ along the magnetic field lines. Thermo-electric effects lead to a coupling between the electric and heat currents ($J_\parallel$ and $Q_\parallel$ here) and their driving forces, namely the effective parallel electric field $E_\parallel'$ and the parallel component of the radial temperature gradient $(dT_e/dr)_\parallel$. Qualitatively the important point is that if there is no external e.m.f. then the vanishing of $J_\parallel$ requires $E_\parallel' \neq 0$ and this thermo-electric field causes a relative motion between the electrons and ions which can cause ion sound waves to grow.

The transport coefficients for a plasma are determined by Coulomb interaction between the various species of particles. A detailed discussion of transport coefficients has been given by Braginskii (1965); some of the relevant results are summarized in §9.3 and in Appendix D. Here the relevant equations are

$$J_\parallel = \sigma_\parallel E_\parallel' + \zeta_\parallel (dT_e/dr)_\parallel, \tag{4.33a}$$

$$Q_\parallel = \xi_\parallel E_\parallel' - \kappa_\parallel (dT_e/dr)_\parallel, \tag{4.33b}$$

where $\sigma_\parallel$ and $\kappa_\parallel$ are parallel electric and thermal, respectively, conductivities and $\zeta_\parallel$ and $\xi_\parallel$ are parallel thermo-electric conductivities. (The Onsager relations imply $\xi_\parallel = -\zeta_\parallel T_e$.) Here we require $J_\parallel = 0$ and then $e\zeta_\parallel/\sigma_\parallel \approx 0.71$ implies $eE_\parallel' \approx -0.71(dT_e/dr)_\parallel$.

The parallel electric field accelerates the electrons and ions in opposite directions, and electron-ion collisions tend to impede this. When these two effects balance the relative drift speed of the electrons and ions is $v_d \approx |eE'_\parallel|/2m_e\nu_{ei}$, i.e.

$$v_d \approx \frac{0.35}{m_e\nu_{ei}}\left(\frac{dT_e}{dr}\right)_\parallel. \tag{4.34}$$

For semiquantitative purposes one may treat this instability by inserting the value (4.34) for v_d in (4.28) et seq.

Gurnett et al. (1979) used observational data on the solar wind to estimate whether the conditions for growth are satisfied or not. Their results were inconclusive in that although growth should occur some of the time it did not seem possible to account for the almost universal presence of ion sound turbulence throughout the solar wind in terms of this mechanism.

4.5 Loss cone instability for Langmuir waves

A loss-cone distribution $f(p_\perp, p_\parallel)$ is one which has a deficiency of particles at small $p_\perp$ or small pitch angle α, where α is defined by $p_\perp = p\sin\alpha, p_\parallel = p\cos\alpha$. Loss-cone distributions can be formed when particles are confined by a magnetic bottle, cf. Figure 4.3. Within the magnetic bottle or magnetic trap the first adiabatic invariant is conserved, i.e.

$$\frac{p_\perp^2}{B} = \text{constant}. \tag{4.35}$$

In an idealized case, let B_{max} be the value of B at the ends of the magnetic trap; then at an arbitrary point in the trap there are no particles in the loss cones at $\alpha < \alpha_0, \alpha > \pi - \alpha_0$ with

$$\alpha_0 = \arcsin(B/B_{max})^{1/2}. \tag{4.36}$$

In practice the loss cone is not entirely empty, due to scattering of particles into the loss cone. A loss-cone distribution is shown schematically in Figure 4.4.

It might seem inconsistent to ignore the magnetic field in treating the growth of the Langmuir waves when the presence of a magnetic field is essential in forming the loss-cone distribution. The conditionds under which it is valid to neglect the magnetic field in treating the particle-wave interaction are that (i) the wave frequency be much greater than the cyclotron frequency Ω, and (ii) the product of the perpendicular wavenumber $k_\perp$ and the gyroradius $v_\perp/\Omega$ of the particle be much greater than unity. These conditions are usually satisfied, and for quantitative purposes one can justify neglecting the magnetic field. However, there are qualitative effects, notably relating to polarization and fine structure in resulting radiation, for which it is essential to retain the magnetic field. Here we treat a loss-cone distribution as a particular example of an unmagnetized axisymmetric distribution.

Consider the absorption coefficient (4.1) for Langmuir waves for an axially symmetric distribution $f(p_\perp, p_\parallel)$. We later identify $f(p_\perp, p_\parallel)$ as a loss-cone distribution. Let the axis of symmetry be the z-axis. Without loss of generality we may

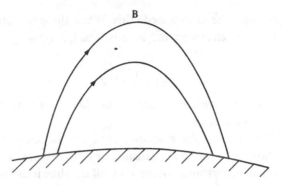

Fig. 4.3 A magnetic flux tube (in a planetary magnetosphere or in the solar corona) can act as a magnetic bottle. Particles reaching the shaded region are lost (they are said to 'precipitate') due to collisions in the denser regions of the atmosphere. As a consequence there is a deficiency of particles with small pitch angles in the trap.

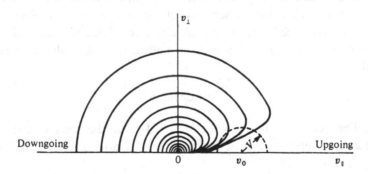

Fig. 4.4 Contours of constant distribution are drawn in $v_\perp - v_\parallel$ space for a distribution function with a loss cone at $\alpha = \alpha_0$. The 'sharpness' of the loss cone depends on how closely the contours are spaced just inside the loss cone.

assume that $\mathbf{k}$ is in the x–z plane. The argument of the δ-function in (4.1) becomes $\omega_L(\mathbf{k}) - k_x v_\perp \cos \phi - k_\parallel v_\parallel$, where ϕ is an aximuthal angle. We use cylindrical polar coordinates $p_\perp$, ϕ, $p_\parallel$ to evaluate the integral:

$$\gamma_L(\mathbf{k}) = -\frac{\pi q^2}{\varepsilon_0 \omega_p^2} \frac{\{\omega_L(\mathbf{k})\}^3}{k^2} \int_0^{2\pi} d\phi \int_0^\infty dp_\perp p_\perp \int_{-\infty}^\infty dp_\parallel \delta\{\omega_L(\mathbf{k}) - k_x v_\perp \cos \phi - k_\parallel v_\parallel\}$$

$$\times \left[\frac{\omega_L(\mathbf{k}) - k_\parallel v_\parallel}{v_\perp} \frac{\partial}{\partial p_\perp} + k_\parallel \frac{\partial}{\partial p_\parallel} \right] f(p_\perp, p_\parallel) \qquad (4.37)$$

where we write

$$\frac{\partial}{\partial p_x} f(p_\perp, p_\parallel) = \frac{v_x}{v_\perp} \frac{\partial}{\partial p_\perp} f(p_\perp, p_\parallel). \qquad (4.38)$$

The ϕ-integral is performed over the δ-function and the remaining integrals are then subject to the condition $\cos^2 \phi \leqslant 1$, i.e. to

$$\{\omega_L(\mathbf{k}) - k_{\parallel} v_{\parallel}\}^2 \leqslant k_{\perp}^2 v_{\perp}^2. \tag{4.39}$$

The result is

$$\gamma_L(\mathbf{k}) = -\frac{2\pi q^2}{\varepsilon_0 \omega_p^2} \frac{\{\omega_L(\mathbf{k})\}^3}{k^2} \int dp_{\perp} p_{\perp} \int dp_{\parallel} \frac{1}{[k_{\perp}^2 v_{\perp}^2 - \{\omega_L(\mathbf{k}) - k_{\parallel} v_{\parallel}\}^2]^{1/2}}$$
$$\times \left[\frac{\omega_L(\mathbf{k}) - k_{\parallel} v_{\parallel}}{v_{\perp}} \frac{\partial}{\partial p_{\perp}} + k_{\parallel} \frac{\partial}{\partial p_{\parallel}} \right] f(p_{\perp}, p_{\parallel}), \tag{4.40}$$

where the integrals are subject to the condition (4.39).

For a loss-cone distribution the favourable angle of emission in most cases is perpendicular to the magnetic field, i.e. $k_{\parallel} = 0$. In this case (4.40) reduces to

$$\gamma_L(\mathbf{k}) = -\frac{2\pi q^2}{\varepsilon_0 \omega_p^2} \omega_L(\mathbf{k}) v_{\phi}^3 \int_{p_{\phi}}^{\infty} dp_{\perp} p_{\perp} \int_{-\infty}^{\infty} dp_{\parallel} \frac{1}{(v_{\perp}^2 - v_{\phi}^2)^{1/2}} \frac{1}{v_{\perp}} \frac{\partial}{\partial p_{\perp}} f(p_{\perp}, p_{\parallel}) \tag{4.41}$$

with

$$v_{\phi} = \frac{\omega_L(\mathbf{k})}{k_{\perp}}, \qquad p_{\phi} = \frac{m v_{\phi}}{(1 - v_{\phi}^2/c^2)^{1/2}} \tag{4.42}$$

The nonrelativistic version of (4.41) may be rewritten in the form (4.4). The relation between the two expressions involves the identity

$$\frac{dF(v_x)}{dv_x}\bigg|_{v_x = v_{\phi}} = \frac{m^3}{n_1} \int_{v_{\phi}}^{\infty} dv_{\perp} \int_{-\infty}^{\infty} dv_{\parallel} \frac{v_{\phi}}{(v_{\perp}^2 - v_{\phi}^2)^{1/2}} \frac{\partial}{\partial v_{\perp}} f(p_{\perp}, p_{\parallel}) \tag{4.43}$$

with $p_{\perp} = mv_{\perp}$ and $p_{\parallel} = mv_{\parallel}$ here. The function $F(v_x)$ is defined by (4.2).

Let us suppose the loss-cone distribution falls off sharply at the loss cone. The profile may then be represented by a step function, e.g. by $H(v_{\perp}^2 - v_{\parallel}^2/(\sigma - 1))$ which implies no particles at $v_{\perp}^2 < v_{\parallel}^2/(\sigma - 1)$ with the loss-cone angle determined by $\alpha_0 = \arctan 1/(\sigma - 1)^{1/2}$. For a nonrelativistic distribution of electrons the full distribution function is

$$f(\mathbf{p}) = \frac{n_1}{(2\pi)^{3/2} m^3 V^3} \left(\frac{\sigma}{\sigma - 1} \right)^{1/2} \exp\left[-\frac{v^2}{2V^2} \right] H\left(v_{\perp}^2 - \frac{v_{\parallel}^2}{\sigma - 1} \right). \tag{4.44}$$

In differentiating $f(\mathbf{p})$ one uses

$$\frac{\partial}{\partial v_{\perp}} H\left(v_{\perp}^2 - \frac{v_{\parallel}^2}{\sigma - 1} \right) = 2 v_{\perp} \delta\left(v_{\perp}^2 - \frac{v_{\parallel}^2}{\sigma - 1} \right). \tag{4.45}$$

A lengthy calculation then gives

$$\gamma_L(\mathbf{k}) = -\frac{1}{(2\pi)^{1/2}} \frac{n_1}{n_e} \frac{\{\omega_L(\mathbf{k})\}^4}{k^3 V^3} \sigma^{1/2} \left\{ e^{-q} K_0(q) - \frac{2\sqrt{\pi}}{(\sigma - 1)^{1/2}} \exp\left(-\frac{\omega^2}{2k^2 V^2} \right) \right.$$
$$\left. \times \int_0^{\infty} dx \, e^{-x^2} \mathrm{erf}[(\sigma - 1)^{1/2} (x^2 + \omega^2/2k^2 V^2)^{1/2}] \right\}, \tag{4.46}$$

where $K_0(q)$ is a modified Bessel function or Macdonald function of argument $q = (\sigma/4)(\omega/kV)^2$. The result (4.46) (due to R.G. Hewitt) corrects a form used by

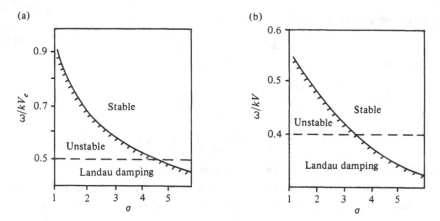

Fig. 4.5 The regions of growth estimated by Zaitsev & Stepanov (1983) for (a) a form close to (4.46) and (b) the form (4.48).

Zaitsev & Stepanov (1983); for semi-quantitative purposes the two forms do not differ greatly and the conditions for growth to occur are illustrated in Figure 4.5a using the uncorrected form of Zaitsev & Stepanov. These authors also considered the 'plateau' distribution at $U < v < V$ with a loss cone:

$$f_1(\mathbf{p}) = \frac{n_1}{(2\pi)^{3/2} m_e^3 V^3} \exp\left(-\frac{v^2}{2V^2}\right) H(U - v), \tag{4.47a}$$

$$f(\mathbf{p}) = \frac{3n}{4\pi(V^3 - U^3)} \left(\frac{\sigma}{\sigma - 1}\right)^{1/2} [H(v - U) - H(v - V)] H\left(v_\perp^2 - \frac{v_\parallel^2}{\sigma - 1}\right) \tag{4.47b}$$

with $f_1(p) = f(p)$ at $v = U$. They found

$$\gamma_L(\mathbf{k}) = -\frac{3n}{n_1} \frac{\{\omega_L(\mathbf{k})\}^4}{k^3(V^3 - U^3)} \left(\frac{\sigma}{\sigma - 1}\right)^{1/2} \Big[(\sigma - 1)^{1/2} \ln|z + (z^2 - 1)^{1/2}|$$

$$- \arctan\left((\sigma - 1)^{1/2} \frac{z}{(z^2 - 1)^{1/2}}\right)\Big]. \tag{4.48}$$

The region of growth is illustrated in Figure 4.5b.

More generally, growth of Langmuir waves due to a loss-cone distribution is favourable only when special conditions are satisfied. The foregoing examples show that growth is possible for sharp (step-function) loss-cone distributions. A standard analytic form for a loss-cone distribution is the DGH form; it is shown in Exercise 4.1 that Langmuir waves are stable for DGH distributions. Growth is favourable under two types of condition: (i) a sharp loss cone when the instability is driven by the very large gradient $\partial f / \partial v_\perp$ in the loss cone, and (ii) a less sharp loss-cone distribution with a large loss-cone angle ($\alpha_0 \gtrsim 45°$) with $\partial f / \partial p$ either positive or nearly zero.

4.6 Axisymmetric distributions: general case

In §4.5 we have treated axisymmetric distributions assuming perpendicular propagation ($k_\parallel = 0$). In the general case, where we make neither the perpendicular nor the nonrelativistic approximation, the resonance condition $\omega - \mathbf{k} \cdot \mathbf{v} = 0$ reduces to the equation for a hyperboloid of two sheets in momentum space. Let $p_L = \mathbf{p} \cdot \mathbf{k}/k$ be the longitudinal component of $\mathbf{p}$, and $p_T = |\mathbf{k} \times \mathbf{p}|/k$ be the transverse component. Then the resonance condition is

$$\left(\frac{k^2 c^2}{\omega^2} - 1\right)\left(\frac{p_L}{mc}\right)^2 - \left(\frac{p_T}{mc}\right)^2 = 1. \tag{4.49}$$

Only the sheet with $p_L > 0$ is physical. The physical sheet comes closest to the origin of momentum space at

$$p_L = |p_L|_{\min} = \left(\frac{\omega^2}{k^2 c^2 - \omega^2}\right)^{1/2} mc, \qquad p_T = 0, \tag{4.50}$$

and its asymptotic cone is

$$p_T = \left(\frac{k^2 c^2 - \omega^2}{\omega^2}\right)^{1/2} p_L. \tag{4.51}$$

The resonance hyperboloid is illustrated in Figure 4.6.

In the expression (4.1) for the absorption coefficient for Langmuir waves the resonance condition is represented by a δ-function; one may re-interpret the integral over $\mathbf{p}$-space with a δ-function as a surface integral over the resonance hyperboloid. It is convenient to define a projected distribution $F(P)$ by (Hewitt & Melrose 1985)

$$F(P) = \frac{1}{n_1} \int d^3\mathbf{p} \left[1 + \left(\frac{p_T}{mc}\right)^2 \right]^{1/2}$$

$$\times \delta\left(p_L - mc\left\{ \frac{\omega^2}{k^2 c^2 - \omega^2}\left[1 + \left(\frac{p_T}{mc}\right)^2 \right]^{1/2} \right\} - P \right) f(\mathbf{p}), \tag{4.52}$$

where we use

$$\delta(\omega - \mathbf{k} \cdot \mathbf{v}) = \frac{m}{k}\left(\frac{k^2 c^2}{k^2 c^2 - \omega^2} \right)^{3/2} \left[1 + \left(\frac{p_T}{mc}\right)^2 \right]^{1/2}$$

$$\times \delta\left(p_L - mc\left\{ \frac{\omega^2}{k^2 c^2 - \omega^2}\left[1 + \left(\frac{p_T}{mc}\right)^2 \right]^{1/2} \right\} \right). \tag{4.53}$$

The projected distribution (4.52) is a relativistic counterpart of the nonrelativistic form (4.2). It is then straightforward to evaluate the absorption coefficient (4.1):

$$\gamma_L(\mathbf{k}) = -\frac{\omega_{p1}^2}{\omega_p^2}\omega_L(\mathbf{k})\gamma_\phi p_\phi^2 \frac{dF(P)}{dP}\bigg|_{P=0}, \tag{4.54}$$

with $v_\phi = \omega_L(\mathbf{k})/k$, $\gamma_\phi = (1 - v_\phi^2/c^2)^{-1/2}$, $p_\phi = \gamma_\phi m v_\phi$. Equation (4.54) is an alternative form of (4.41).

The projected distribution may be evaluated by introducing variables P_0, ψ and ϕ in momentum space:

$$p_L = P + \left(\frac{\omega^2}{k^2 c^2 - \omega^2}\right)^{1/2} P_0 \cosh\psi, \qquad p_T = P_0 \sinh\psi. \tag{4.55a, b}$$

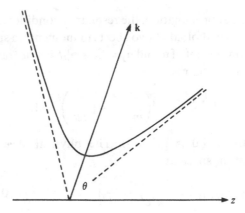

Fig. 4.6 The resonance hyperboloid is shown schematically in momentum space. This hyperboloid is the locus of all particles which satisfy $\omega - \mathbf{k} \cdot \mathbf{v} = 0$ for given ω and $\mathbf{k}$. The horizontal axis is the z-axis. (Hewitt & Melrose 1985).

Writing $f(\mathbf{p})$ as $f(P_0, \psi, \phi)$ one finds

$$F(P) = m^2 c^2 \int_0^{2\pi} d\phi \int_0^{\infty} d\psi \, \sinh \psi \cosh^2 \psi \, f(mc, \psi, \phi). \qquad (4.56)$$

The dependence on P is now implicit in the dependence on ψ. The derivative $dF(P)/dP$ for $P = 0$ which appears in (4.54) may be evaluated by expanding (4.55a) with $P_0 = mc$ to lowest order in P.

The advantage of using (4.56) to evaluate $F(P)$ is that the ϕ-integral can usually be evaluated relatively simply. However, the algebra involved remains quite cumbersome in general.

A simplified class of distributions is that of separate functions

$$f(\mathbf{p}) = f(p)\Phi(\alpha), \qquad (4.57)$$

with

$$\frac{1}{2} \int_{-1}^{+1} d\cos\alpha \, \Phi(\alpha) = 1. \qquad (4.58)$$

For such distributions it is possible to calculate the absorption coefficient in the form

$$\gamma_L(\mathbf{k}) = \frac{4\pi p_0^2}{n_1} \int_0^{\infty} dp_0 f(p_0)\gamma_{L0}(\mathbf{k}) \qquad (4.59)$$

where $\gamma_{L0}(\mathbf{k})$ is evaluated for $f(p) = f_0(p)$,

$$f_0(p) = \frac{n_1}{4\pi p_0^2} \delta(p - p_0). \qquad (4.60)$$

One finds

$$\gamma_{L0}(\mathbf{k}) = \gamma_L^I(k, \theta) + \gamma_L^R(k, \theta) + \gamma_L^A(k, \theta) \qquad (4.61)$$

with

$$\gamma_L^I(k, \theta) = \frac{\pi}{2} \frac{n_1}{n_e} \omega_p m v_0 \delta(p_0 - p_\phi)\Phi(\theta), \qquad (4.62a)$$

$$\gamma_L^R(k, \theta) = \frac{\pi}{2}\frac{n_1}{n_e}\omega_p\frac{2mv_\phi^3}{p_0 c^2}g(\theta, \chi),\tag{4.62b}$$

$$\gamma_L^A(k, \theta) = -\frac{\pi}{2}\frac{n_1}{n_e}\omega_p\frac{m^3 v_\phi^2 v_0}{p_0^3}\{1 + (\gamma_0^2 - 1)\sin^2\chi\}\frac{\partial g(\theta, \chi)}{\partial\cos\chi}.\tag{4.62c}$$

The function $g(\theta, \chi)$ is defined by

$$g(\theta, \chi) = \frac{1}{\pi}\int_{\cos(\theta+\chi)}^{\cos(\theta-\chi)} d\cos\alpha\frac{\Phi(\alpha)}{F_+(\alpha, \theta, \chi)},\tag{4.63}$$

with χ defined as $\arccos(\omega_L(k)/kc)$ and with

$$F_\pm(\alpha, \theta, \chi) = [1 \pm 2\cos\alpha\cos\theta\cos\chi - \cos^2\alpha - \cos^2\theta - \cos^2\chi]^{1/2}\tag{4.64}$$

The contribution $\gamma_L^I(k, \theta)$ is the only one which appears for a nonrelativistic isotropic distribution; it may be attributed to waves at a given v_ϕ being absorbed by electrons with $v = v_\phi$. The contribution $\gamma_L^R(k, \theta)$ is intrinsically relativistic and involves waves at v_ϕ being absorbed by particles with $v > v_\phi$. Both these contributions are strictly positive. The final contribution arises from the anisotropy of the distribution.

Some examples of axisymmetric distributions are the following:
(i) Forward-cone distribution:

$$\Phi(\alpha) = \begin{cases} \dfrac{2}{1 - \cos\alpha_0} & \alpha < \alpha_0 \\ 0 & \text{otherwise} \end{cases}\tag{4.65}$$

In this case one finds

$$g(\theta, \chi) = \frac{2}{1 - \cos\alpha_0}\begin{cases} 1 & \text{for } v_+ \leqslant v_\phi \leqslant v_0, \\ \dfrac{1}{2} - \dfrac{1}{\pi}\arcsin\left(\dfrac{\cos\alpha_0 - \cos\theta\cos\chi}{\sin\theta\sin\chi}\right) & \text{for } v_- \leqslant v_\phi \leqslant v_+, \\ 0 & \text{otherwise} \end{cases}\tag{4.66a}$$

and

$$\frac{\partial g(\theta, \chi)}{\partial\cos\chi} = \frac{2}{1 - \cos\alpha_0}\begin{cases} \dfrac{1}{\pi}\dfrac{(\cos\theta - \cos\alpha_0\cos\chi)}{\sin^2\chi F(\alpha, \theta, \chi)} & \text{for } v_- \leqslant v_\phi \leqslant v_+, \\ 0 & \text{otherwise} \end{cases}\tag{4.66b}$$

with $v_\pm = v_0\cos(\alpha_0 \pm \theta)$.

(ii) P_1-anisotropy:

$$\Phi(\alpha) = 1 + \frac{3v_d}{v_0}\cos\alpha.\tag{4.67}$$

In this case complete evaluation of $\gamma_{L0}(k)$ is straightforward:

$$\gamma_{L0}(k) = -\frac{3\pi}{2}\frac{n_1}{n_e}\omega_L(k)\frac{v_\phi^2}{v_0^3}\left(v_d - \frac{2v_\phi^3}{3c^2}\right)\cos\theta.\tag{4.68}$$

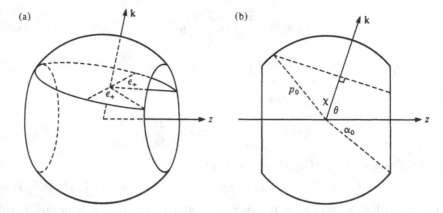

Fig. 4.7(a) The angle ε_+, cf. (4.70), is shown in the case where the resonance hyperboloid intersects the shell at $p = p_0$ such that it passes through one loss cone and touches the other. The angle ε_- is defined in an analogous manner when the other loss cone is also intersected.
(b) A section of the plane containing $\mathbf{k}$ and the axis of symmetry, with the angles θ, α_0 and χ shown.
(After Hewitt & Melrose 1985).

(iii) Loss-cone anisotropy:

$$\Phi(\alpha) = \begin{cases} \dfrac{1}{\cos \alpha_0} & \alpha_0 \leqslant \alpha \leqslant \pi - \alpha_0 \\ 0 & \text{otherwise} \end{cases} \tag{4.69}$$

One finds $\gamma_L^A(k, \theta) = 0$ if the resonance hyperboloid does not pass through either loss cone, with the range of the ϕ-integral being 2π in this case, $\pi + 2\varepsilon_+$ when one loss cone is intersected and $2\varepsilon_- + 2\varepsilon_+$ when both loss cones are intersected, cf. Figure 4.7, with

$$\varepsilon_{\pm} = \arcsin \frac{\cos \alpha_0 \pm \cos \theta \cos \psi}{\sin \theta \sin \chi}. \tag{4.70}$$

The value of $\gamma_{L0}(\mathbf{k})$ is

$$\gamma_{L0}(\mathbf{k}) = \frac{n_1}{n_e} \frac{\omega_L(\mathbf{k}) m v_\phi^3}{p_0 c^2 \cos \alpha_0} \left\{ \frac{\pi}{2} + \varepsilon_+ - \frac{1}{2} \frac{c^2}{\gamma_\phi^2 v_0 v_\phi} \left[\frac{-\cos \theta + \cos \alpha \cos \psi}{\sin^2 \chi F_+(\alpha_0, \theta, \chi)} \right] \right\} \tag{4.71a}$$

when one loss cone is intersected and

$$\gamma_{L0}(\mathbf{k}) = \frac{n_1}{n_e} \frac{\omega_L(\mathbf{k}) m v_\phi^3}{p_0 c^2 \cos \alpha_0} \left\{ \varepsilon_- + \varepsilon_+ - \frac{1}{2} \frac{c^2}{\gamma_0^2 v_0 v_\phi} \right.$$
$$\left. \times \left[\frac{\cos \theta + \cos \alpha_0 \cos \chi}{\sin^2 \chi F_-(\alpha_0, \theta, \chi)} + \frac{-\cos \theta + \cos \alpha_0 \cos \chi}{\sin^2 \chi F_+(\alpha_0, \theta, \chi)} \right] \right\} \tag{4.71b}$$

when both loss cones are intersected.

Exercise set 4

4.1 A class of idealized loss-cone distributions introduced by Dory, Guest & Harris (1965) (DGH distributions) are nonrelativistic and of the form

$$f(p_\perp, p_\parallel) = \frac{n_1}{j!(2\pi)^{3/2} m^3 V_\perp^2 V_\parallel} \left(\frac{v_\perp^2}{2V_\perp^2}\right)^j \exp\left[-\frac{v_\perp^2}{2V_\perp^2} - \frac{v_\parallel^2}{2V_\parallel^2}\right]$$

with j a non-negative integer. The parameter j relates to the sharpness with which the distribution falls off in the loss cone: large j corresponds to a sharp gradient inside the loss cone.

Evaluate $dF(v_x)/dv_x$ for the DGH distribution for the case $k_\parallel = 0$ using (4.2) and show that one has

$$\frac{dF(v_x)}{dv_x} = -\frac{1}{j!} \frac{v_x}{(2\pi)^{1/2} V_\perp^2} \left(\frac{v_x^2}{2V_\perp^2}\right)^j \exp\left[-\frac{v_x^2}{2V_\perp^2}\right]$$

Hence show that such a distribution cannot lead to growth of Langmuir waves for $k_\parallel = 0$.

4.2 (a) Calculate the absorption coefficient for Langmuir waves for a P_2-distribution

$$\Phi(\alpha) = 1 + f_2(p_0) P_2(\cos\alpha)$$

where $P_2(x) = (3x^2 - 1)/2$ is a Legendre polynomial.

  (b) An initially isotropic distribution of particles in a magnetic field is distorted due to the magnitude of B increasing slowly (so that $\sin^2\alpha/B = $ constant). Approximate the resulting distortion by a P_2-distribution and hence show that it cannot lead to growth of Langmuir waves.

4.3 The final state of evolution of the one-dimensional bump-in-tail instability is the plateau distribution

$$f(v) = \begin{cases} \text{constant} & v < v_0 \\ 0 & v > v_0 \end{cases}$$

Show that this distribution is unstable to growth of Langmuir waves at oblique angles θ to the one-dimensional axis with absorption coefficient

$$\gamma_L(k, \theta) = -6\pi \frac{n_1}{n_e} \omega_L(\mathbf{k}) \left(\frac{v_\phi}{v_0 \cos\theta}\right)^3$$

for $v_\phi < v_0 \cos\theta$.

4.4 Confirm that the time-asymptotic solution (4.18) and (4.22) for the bump-in-tail instability conserves momentum; i.e. show that the final momentum of particles and waves equals the initial momentum $n_1 m_e v_b$.

4.5 Calculate the absorption coefficient for Langmuir waves due to a distribution of the form

$$f(p, \alpha) = f_0(p) \left\{1 + \frac{3U}{v} \cos\alpha\right\}$$

with

$$f_0(p) = \frac{n_1}{4\pi p_0^2} \delta(p - p_0).$$

In particular show that for $\omega_L \approx \omega_p$, $v_\phi = \omega_p/k$, one has

$$\gamma_L(k, \theta) = -\frac{3\pi}{2} \frac{n_1}{n_e} \omega_p \frac{v_\phi^2}{v_0^3} \left(U - \frac{2v_0^2}{3c^2} \right) \cos \theta.$$

5

Particle motions in waves

5.1 Perturbations in the orbit of a single particle

In plasma physics it is traditional to use a mixture of the collective-medium approach and the single-particle approach in treating various processes. These two approaches can complement each other in providing physical insight. A relevant example is in the treatment of Landau damping. In the collective-medium approach this is treated by allowing the frequency to have an imaginary part which is determined by the anti-hermitian part of the response tensor (§2.5). In the single-particle approach, one calculates Cerenkov emission by a single particle, relates absorption to emission using the Einstein coefficients (or Kirchhoff's law) and hence finds that Landau damping is the absorption process corresponding to Cerenkov emission by thermal electrons (§6.3). Although it is possible in principle to use the collective-medium approach to treat spontaneous, e.g. Cerenkov, emission it is cumbersome to do so and it is difficult to build up a physical understanding using this approach. Thus spontaneous emission is treated using a single-particle approach. This approach may be extended to calculate the response tensors (§5.5) which are the basis of the collective-medium approach.

In this Chapter we discuss several aspects of the single-particle approach applied to particle-wave interactions. The object is to identify the physical processes involved in reactive and kinetic instabilities, and in their saturation. In this section we treat perturbations in the orbit of a single particle (mass m, charge q) due to an arbitrary electromagnetic disturbance. The detailed analysis is carried out for the simpler case of a nonrelativistic particle interacting with a single plane wave in the electrostatic and one-dimensional (v along k) limits.

Quite generally the *orbit* of a particle may be described by an equation of the form

$$x = X(t), \tag{5.1}$$

where $X(t)$ also depends on the initial conditions, say $x = x_0$, $v = v_0$ at $t = t_0$. The instantaneous velocity is

$$v = \dot{X}(t), \tag{5.2}$$

where the dot denotes differentiation with respect to time. The acceleration is $\ddot{X}(t)$.

The orbit is found by solving the equation of motion

$$\frac{d\mathbf{p}}{dt} = \mathbf{F}(t, \mathbf{x}, \mathbf{v}), \tag{5.3}$$

where $\mathbf{F}(t, \mathbf{x}, \mathbf{v})$ is the applied force. Here we identify the force as the Lorentz force

$$\mathbf{F}(t, \mathbf{x}, \mathbf{v}) = q[\mathbf{E}(t, \mathbf{x}) + \mathbf{v} \times \mathbf{B}(t, \mathbf{x})] \tag{5.4}$$

due to the electromagnetic field in the wave. With $\mathbf{p} = \gamma m \mathbf{v}$ and $\gamma = (1 - v^2/c^2)^{-1/2}$, (5.3) may be rewritten in the form

$$\frac{d\mathbf{v}}{dt} = \frac{1}{m\gamma}\left\{\mathbf{F}(t, \mathbf{x}, \mathbf{v}) - \frac{\mathbf{v}}{c^2}\mathbf{v}\cdot\mathbf{F}(t, \mathbf{x}, \mathbf{v})\right\}. \tag{5.5}$$

Then on writing $d\mathbf{v}/dt = \ddot{\mathbf{X}}$, $\mathbf{v} = \dot{\mathbf{X}}$ and $\mathbf{x} = \mathbf{X}$, (5.5) becomes a differential equation for $\mathbf{X}$. In practice one needs to make approximations to solve this equation.

One procedure for solving (5.5) is to expand the orbit in powers of the amplitude of the wave

$$\mathbf{X}(t) = \mathbf{X}^{(0)}(t) + \mathbf{X}^{(1)}(t) + \mathbf{X}^{(2)}(t) + \cdots, \tag{5.6}$$

and to use perturbation theory. The zeroth order orbit describes the unperturbed motion

$$\mathbf{X}^{(0)}(t) = \mathbf{x}_0 + \mathbf{v}_0 t - \mathbf{v}_0 t_0. \tag{5.7}$$

All perturbations vanish at $t = t_0$. The first order equation is found by setting $\ddot{\mathbf{X}}(t) = \ddot{\mathbf{X}}^{(1)}(t)$ on the left hand side of (5.5) and $\mathbf{X}(t) = \mathbf{X}^{(0)}(t)$ on the right hand side. Once this equation has been solved one can write down the second order equation; it has $\ddot{\mathbf{X}}^{(2)}(t)$ on the left hand side and with the right hand side being the second order terms proportional to $\mathbf{X}^{(1)}$ and to $\mathbf{E}$ or $\mathbf{B}$.

First let us solve for the orbit in the case where the motion is nonrelativistic and one-dimensional in a monochromatic electrostatic wave. We assume that the electric field in the wave is of the form

$$E(t, z) = E(t)e^{-i\psi(t, z)} + \text{c.c.}, \tag{5.8}$$

where 'c.c.' denotes complex conjugate, and where the phase function is

$$\psi(t, z) = \omega_0 t - k_0 z. \tag{5.9}$$

We allow the amplitude to vary secularly with time according to

$$E(t) = E_0 G(t), \quad G(t) = \exp\int_{t_0}^{t} dt'\omega_i(t'), \tag{5.10}$$

where ω_i is slowly varying in the sense $|\dot{\omega}_i| \ll |\omega_i|^2$. The equation of motion can then be written as an equation for the phase

$$\ddot{\psi} = \frac{2qk_0 E_0}{m} G(t) \cos\psi. \tag{5.11}$$

A perturbation expansion $\psi = \psi^{(0)} + \psi^{(1)} + \ldots$ has

$$\psi^{(0)} = \Omega(t - t_0) + \psi_0, \tag{5.12a}$$

$$\Omega = \omega_0 - k_0 v_0, \qquad \psi_0 = \Omega t_0 - k_0 z_0. \tag{5.12b,c}$$

The first and second order equations of motion are

$$\ddot{\psi}^{(1)} = \frac{2qk_0 E_0}{m} G(t) \cos \psi^{(0)} \tag{5.13}$$

and

$$\ddot{\psi}^{(2)} = -\frac{2qk_0 E_0}{m} G(t) \psi^{(1)} \sin \psi^{(0)} \tag{5.14}$$

respectively.

In integrating (5.13) we use [Exercise 5.1]

$$\int_{t_0}^{t} dt' G(t') \exp\left[-i\psi^{(0)}(t')\right] = \frac{i}{\Omega + i\omega_i} \left\{ G(t) \exp\left[-i\psi^{(0)}(t)\right] - \exp\left[-i\psi_0\right] \right\}. \tag{5.15}$$

Then integrating (5.13) gives

$$\dot{\psi}^{(1)} = -\frac{qk_0 E_0}{m} \left[\frac{i}{\Omega + i\omega_i} \left\{ G(t) \exp\left[-i\psi^{(0)}(t)\right] - \exp\left[-i\psi_0\right] \right\} + \text{c.c.} \right], \tag{5.16}$$

and integrating again gives

$$\psi^{(1)} = \frac{qk_0 E_0}{m} \left[\left(\frac{i}{\Omega + i\omega_i} \right)^2 \left\{ G(t) \exp\left[-i\psi^{(0)}(t)\right] - \exp\left[-i\psi_0\right] \right\} \right.$$
$$\left. - \frac{i(t - t_0)}{\Omega + i\omega_i} \exp\left[-i\psi_0\right] + \text{c.c.} \right]. \tag{5.17}$$

We shall not require the solutions of the second order equation (5.14).

Before discussing the implications of this solution, let us return to the three-dimensional, electromagnetic case without the nonrelativistic approximation. We ignore the gain factor $G(t)$, setting it equal to unity. The electric field is expressed in terms of its Fourier transform,

$$\mathbf{E}(t, \mathbf{x}) = \int \frac{d\omega' d^3 \mathbf{k}'}{(2\pi)^4} e^{-i(\omega' t - \mathbf{k}' \cdot \mathbf{x})} \mathbf{E}(\omega', \mathbf{k}'), \tag{5.18}$$

and $\mathbf{B}(\omega, \mathbf{k})$ is expressed in terms of $\mathbf{E}(\omega, \mathbf{k})$ using (1.15). Then (5.5) with (5.4) reduces to

$$X_i(t) = \frac{q}{m\gamma} \left(\delta_{ir} - \frac{v_i v_r}{c^2} \right) \int \frac{d\omega' d^3 \mathbf{k}'}{(2\pi)^4} \exp\left[-i\{\omega' t - \mathbf{k}' \cdot \mathbf{X}(t)\}\right] g_{rj}(\omega', \mathbf{k}'; \mathbf{v}) \frac{E_j(\omega', \mathbf{k}')}{\omega'}, \tag{5.19}$$

with

$$g_{ij}(\omega, \mathbf{k}; \mathbf{v}) = (\omega - \mathbf{k} \cdot \mathbf{v}) \delta_{ij} + k_i v_j. \tag{5.20}$$

The first order equation of motion and its integrals are of the form

$$
\begin{bmatrix} \dot{X}_i(t) \\ \dot{X}_i^{(1)}(t) \\ X_i^{(1)}(t) \end{bmatrix} = \frac{q}{m\gamma} \left(\delta_{ir} - \frac{v_i v_r}{c^2} \right) \int \frac{d\omega' d^3 k'}{(2\pi)^4} \begin{bmatrix} G(\omega', \mathbf{k}', t) \\ H(\omega', \mathbf{k}', t) \\ L(\omega', \mathbf{k}', t) \end{bmatrix} g_{rj}(\omega', \mathbf{k}'; \mathbf{v}) \frac{E_j(\omega', \mathbf{k}')}{\omega'},
$$

$$(5.21)$$

with

$$
G(\omega, \mathbf{k}, t) = \exp[-i\psi(t)], \tag{5.22a}
$$

$$
H(\omega, \mathbf{k}, t) = \frac{i}{\omega - \mathbf{k} \cdot \mathbf{v}_0} \{\exp[-i\psi(t)] - \exp[-i\psi_0]\}, \tag{5.22b}
$$

$$
L(\omega, \mathbf{k}, t) = \left(\frac{i}{\omega - \mathbf{k} \cdot \mathbf{v}_0} \right)^2 \{\exp[-i\psi(t)] - \exp[-i\psi_0]\}
$$
$$
- \frac{i(t - t_0)}{\omega - \mathbf{k} \cdot \mathbf{v}_0} \exp[-i\psi_0], \tag{5.22c}
$$

with

$$
\psi(t) = (\omega - \mathbf{k} \cdot \mathbf{v}_0)t - \mathbf{k} \cdot (\mathbf{x}_0 - \mathbf{v}_0 t_0), \tag{5.23a}
$$

$$
\psi_0 = \omega t_0 - \mathbf{k} \cdot \mathbf{x}_0 \tag{5.23b}
$$

Apart from the obvious increase in complexity, this general case exhibits essentially the same temporal behavior as the simpler special case. On making the electrostatic, one-dimensional and nonrelativistic approximation (5.21) with (5.22) may be used to reproduce (5.13), (5.16) and (5.17). In reproducing these results $\omega - \mathbf{k} \cdot \mathbf{v}_0$ in (5.22) is re-interpreted as $\Omega + i\omega_i$, i.e., the imaginary part $i\omega_i$ is implicit in (5.22).

5.2 Phase bunching and reactive growth

We may use the one-dimensional model developed in §5.1 as the basis for a model of the reactive version of the weak-beam instability (§3.2). This model allows us to identify the instability mechanism as due to phase bunching. In this one-dimensional case phase bunching involves a group of particles initially uniformly distributed in z, and hence with respect to the phase of the wave, tending to bunch around a particular point moving with the phase velocity of the wave. The bunching which causes reactive growth is oscillatory. There is also a systematic phase bunching implied by the secular term in (5.17). We discuss this systematic effect first.

The systematic effect in (5.17) arises from the term proportional to $t - t_0$. It gives rise to a change in the relative phase between an electron and the wave:

$$
\psi^{(1)} = -\frac{ek_0 E_0(t - t_0)}{m_e(\Omega^2 + \omega_i^2)} [\Omega \sin \psi_0 + \omega_i \cos \psi_0]. \tag{5.24}
$$

Suppose we have a number of electrons distributed uniformly in z_0, and hence in $\psi_0 = \omega_0 t_0 - k_0 z_0$ at $t = t_0$. Further let us suppose that all these electrons have the same v_0 which is such that the frequency difference Ω from resonance

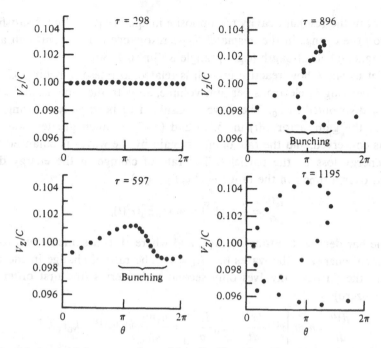

Fig. 5.1 An initial group of particles distributed uniformly in phase (here uniformly along the one-dimensional axis) tends to phase bunch due to the effects of a wave. The curves plotted are actually for axial bunching of electrons in a whistler wave (calculated by R.M. Winglee). τ is a parametrized time and the phase θ is that between the particle and the wave. At $\tau = 0$ The particles (represented by dots) are distributed uniformly in θ.

$(\Omega = \omega_0 - k_0 v_0 = 0)$ is much greater than ω_i in magnitude. We distinguish between two cases.

(i) $\Omega > 0$.

In this case $\psi^{(1)}$ and $\sin \psi_0$ have opposite signs. Hence those electrons with $\sin \psi_0 > 0$ slip backwards in phase ($\psi^{(1)} < 0$) and those with $\sin \psi_0 < 0$ advance in phase ($\psi^{(1)} > 0$). Hence the electrons tend to bunch around zero relative phase, i.e. at $\psi = (\omega_0 - k_0 v_0)(t - t_0)$.

(ii) $\Omega < 0$

In this case $\psi^{(1)}$ and $\sin \psi_0$ have the same signs. This leads to 'anti-bunching' in which the electrons drift away from zero relative phase.

This phase bunching effect, which is illustrated in Figure 5.1, has been used in the laboratory to enhance the efficiency of amplifiers, notably gyrotrons. In such devices the efficiency with which electrons give up energy to radiation in a resonant cavity depends on the relative phase between the electron and the wave when the electron enters the cavity. The efficiency can therefore be enhanced by a preliminary bunching of the electrons to the appropriate relative phase. This can be achieved in a region where the mismatch Ω from resonance is much larger than the imginary part of the frequency, and with the external parameters changing (here changing

k_0) so that in the resonant cavity the opposite inequality $|\Omega| < |\omega_i|$ is satisfied. In a gyrotron the change in the mismatch from resonance is made through a taper in the magnetic field strength (e.g. Sprangle & Smith 1980).

Now let us show how reactive growth is related to phase bunching. Consider a model involving n_1 electrons per unit volume, all with the same v_0 and initially uniformly distributed in z_0. The wave is assumed to be rapidly growing in the sense that the gain factor $G(t)$ in (5.16) and (5.17) is much greater than unity. Energy is conserved, and the rate of energy gain by the waves is balanced by the rate of energy loss by the particles. The rate of change in the energy density (averaged over a time t in the range $\omega^{-1} \ll t \ll \omega_i^{-1}$) is

$$\frac{dW}{dt} = \frac{d}{dt} \varepsilon_0 \overline{|E^2|} = 4\omega_i \varepsilon_0 E_0^2 G^2(t), \tag{5.25}$$

where the bar denotes the time-average, and where W is assumed equal to twice the electrical energy in the waves (i.e. $R_L = \frac{1}{2}$). The rate of change in the energy density in the particles involves only second order terms (the first order terms average to zero)

$$-\frac{dW}{dt} = n_1 m_e \left| v^{(1)} \frac{dv^{(1)}}{dt} + v_0 \frac{dv^{(2)}}{dt} \right| = \frac{n_1 m_e}{k_0^2} |\dot{\psi}^{(1)} \dot{\psi}^{(1)} - k_0 v_0 \ddot{\psi}^{(2)}|. \tag{5.26}$$

Now using (5.14) and (5.17), with only the terms proportional to $G(t)$ retained in (5.16) and (5.17), (5.26) gives

$$-\frac{dW}{dt} = 2\omega_{p1}^2 \varepsilon_0 E_0^2 G^2(t) \left\{ \frac{\omega_i}{\Omega^2 + \omega_i^2} + \frac{2k_0 v_0 \Omega \omega_i}{(\Omega^2 + \omega_i^2)^2} \right\}$$

$$= \frac{4\omega_{p1}^2 \omega_0 \Omega \omega_i}{(\Omega^2 + \omega_i^2)^2} \varepsilon_0 E_0^2 G^2(t), \tag{5.27}$$

where we assume $\omega_0 \gg \Omega, \omega_i$ and write $\omega_{p1}^2 = e^2 n_1 / \varepsilon_0 m_e$. The dominant contribution in (5.27) arises from the term

$$\psi^{(1)} = \frac{2ek_0 t_0}{m_e} \frac{2\Omega \omega_i}{(\Omega^2 + \omega_i^2)^2} \sin\left[\psi^0(t)\right] + \cdots \tag{5.28}$$

contained in (5.17).

Comparison of (5.25) and (5.27) implies

$$(\Omega^2 + \omega_i^2)^2 = \omega_{p1}^2 \omega_0 \Omega. \tag{5.29}$$

One readily confirms that (5.29) is consistent with Ω and ω_i being the real and imaginary parts of the solution of a cubic equation, i.e. that with $\Omega = r \cos\theta$, $\omega_i = r \sin\theta$, $r^2 = \Omega^2 + \omega_i^2$ one has $\cos\theta = \frac{1}{2}$, $\sin\theta = \sqrt{3}/2$ and $r^3 = \omega_{p1}^2 \omega_p / 2$ for $\omega_0 = \omega_p$. The growth rate $2\omega_i$ then corresponds to the result (3.14), confirming that (5.29) does indeed describe the (reactive) weak-beam instability. Note that the solution requires $|\Omega| = \omega_i / \sqrt{3}$ and $\Omega < 0$.

As already noted the dominant term in (5.27) arises from (5.17), in which we have $\Omega < 0$. Then (5.28) implies that the signs of $\psi^{(1)}$ and $\sin\left[\psi^{(0)}(t)\right]$ are opposite, which is the condition found above for phase bunching. Consider an instant when $\Omega(t - t_0)$ is an integral multiple of 2π so that $\sin[\psi^{(0)}(t)]$ is equal to $\sin\psi_0$. At such an instant the electrons are bunched around an electron with phase ψ_0, i.e. one which started at $z_0 = \omega_0 t_0/k_0$ at $t = t_0$. This bunch oscillates periodically, producing an electric field which adds in phase to the electric field in the wave, thereby amplifying it.

More generally, the mechanism driving all reactive instabilities may be identified as phase bunching. In the Weibel instability (§3.6) for example, the bunching is due to the Lorentz force rather than the electric force, and in some cyclotron instabilities the dominant bunching can be in azimuthal angle rather than with distance along the axis, e.g. in z analogous to the case discussed here. The details aside, reactive instabilities are due to bunching of the particles by the growing wave, with the bunches producing a field which enhances that of the wave.

Another feature of reactive instabilities is that the growing wave is phase coherent. The initial phase chosen here is zero for $\Omega = 0$ and $z_0 = 0$, cf. (5.12c), and we find phase bunching around relative phase zero. Suppose there is a second wave with initial phase separated from the first by $\Delta\psi_0$. One wave tends to cause bunching around zero phase and the other tends to cause bunching around $\Delta\psi_0$. These two tendencies are competing, and the presence of the second wave impedes the growth of the first. However if the bunching due to one wave is slightly the stronger, then it grows slightly the faster, and as its amplitude increases relative to the other wave, it will become the dominant wave in causing the bunching. After several growth times the slightly faster growing wave will dominate completely; the growth of the second wave will slow due to it being ineffective in competing with the dominant wave in bunching the electrons. Generalizing this argument, reactive growth starting from random-phase noise will pick out one preferential phase and the growing disturbance will rapidly become phase coherent with this phase. In other words reactive growth produces a phase coherent growing wave.

5.3 Wave trapping

We have already noted that when a reactive instability develops, there is an associated increase in the velocity spread; this increasing spread tends to suppress the growth. Alternatively saturation may be attributed to trapping of particles in the field of the growing wave.

Consider (5.11) in the case where the gain G is approximated by unity. Omitting the subscript zero on ω and k, we have

$$\frac{d^2\psi}{dt^2} = \frac{eE_0 k}{m_e}\cos\psi, \tag{5.30}$$

where we now assume that the particles are electrons ($q = -e$, $m = m_e$). In §5.1 we solve (5.30) using a perturbation approach. However we may solve (5.30) exactly as follows. First we need the additional equation

$$\frac{d\psi}{dt} = \omega - kv \tag{5.31}$$

for the rate of change of the phase seen by a particle. By differentiating (5.31) twice and using (5.30) one finds

$$\frac{d^2}{dt^2}(v - v_\phi) = -(\omega_T^2 \sin \psi)(v - v_\phi) \tag{5.32}$$

with $v_\phi = \omega/k$ and where

$$\omega_T = \left(\frac{eE_0 k}{m_e}\right)^{1/2} \tag{5.33}$$

is called the *trapping frequency* or the *bounce frequency*. For $\sin \psi > 0$, $v - v_\phi$ oscillates with a frequency close to ω_T.

Let us consider the motion of an electron in the wave frame, i.e. in a frame moving with velocity v_ϕ along the wave direction. The wave field may be described by its electrostatic potential $\Phi(z)$ which is a function of z but not of t in this frame. The phase ψ is also a function of z but not of t in this frame. We require $E = - \text{grad} \, \Phi = E_0 \cos \psi$, and an acceptable solution for Φ is

$$\Phi = \frac{E_0}{k} \sin \psi. \tag{5.34}$$

In this frame the kinetic energy of an electron is $\frac{1}{2}m(v - v_\phi)^2$, and conservation of energy implies

$$\varepsilon = \tfrac{1}{2}m(v - v_\phi)^2 - \frac{eE_0}{k} \sin \psi = \text{constant}. \tag{5.35}$$

If ε is negative then the range of ψ is restricted. Such electrons are said to be *trapped* by the wave in the sense that the relative phase ψ between the particle and the wave remains bounded. If a particle has $\varepsilon > 0$ then ψ is unbounded and the particle is untrapped, cf. Figure 5.2.

The orbit of a trapped particle in $v - \psi$ or in $v - z$ space may be found by solving (5.32) explicitly. Equation (5.35) is a first integral of (5.32). Let us write

$$\zeta = \tfrac{1}{2}\left(\psi - \frac{\pi}{2}\right), \tag{5.36}$$

and note that (5.31) then implies

$$\tfrac{1}{2}m_e(v - v_\phi)^2 = \frac{2m_e}{k^2}\left(\frac{d\zeta}{dt}\right)^2 \tag{5.37}$$

Thus (5.32) may be rewritten in the form

$$\left(\frac{d\zeta}{dt}\right)^2 = \omega_T^2\{\alpha^2 - \sin^2 \zeta\} \tag{5.38}$$

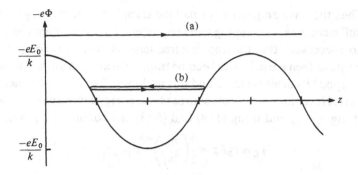

Fig. 5.2 The potential energy for an electron in the field of a monochromatic wave is plotted in the wave frame. The paths (a) of an untrapped particle, and (b) of a trapped particle are indicated.

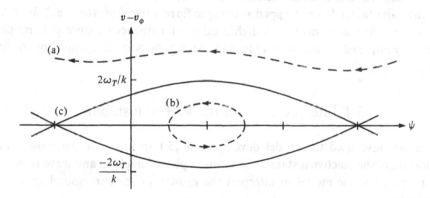

Fig. 5.3 The orbit of (a) an untrapped, and (b) a trapped particle in $(v - v_\phi) - \psi$ space are illustrated. The solid curve (c) is the separatrix.

with

$$\alpha^2 \omega_T^2 = \frac{\varepsilon k^2}{2m_e} + \frac{\omega_T^2}{2}. \tag{5.39}$$

The solution of (5.38) involves an elliptic integral. Some solutions are plotted schematically in Figure 5.3.

The case $\alpha^2 < 1$ in (5.38) corresponds to closed orbits in $(v - v_\phi) - \psi$ space, and these correspond to trapped particles. The velocity of a trapped particle oscillates with amplitude

$$(v - v_\phi)_{\text{max}} = 2\alpha \omega_T/k. \tag{5.40}$$

The orbits of trapped and untrapped particles are separated by a curve in $(v - v_\phi) - \psi$ space called the *separatrix*.

The argument that trapping can lead to suppression of reactive growth is as follows. When the wave amplitude is small the trapping frequency ω_T is small and there is insufficient time for trapping to be important. However when the amplitude increases to a level such that the trapping frequency ω_T becomes comparable with the growth rate ω_i then particles can become trapped in about a growth time. Once a particle is trapped its orbit is closed and it is no longer free to supply energy to the growing wave. Hence for $\omega_T \gtrsim \omega_i$ the source of free energy for the growing wave gets cut off. Setting $v_\phi = v_b$ and using (3.14) and (5.33), the condition $\omega_T = \omega_i$ implies

$$W_L = \varepsilon_0 E_0^2 = \frac{9}{2}\left(\frac{n_1}{2n_e}\right)^{2/3} n_1 m_e v_b^2. \tag{5.41}$$

Apart from a factor of order unity (which is actually ill-determined) (5.41) is equivalent to the condition (3.18) obtained in §3.4 based on suppression due to an increase in the velocity spread. It follows that the suppression mechanism discussed in §3.4 for the reactive instability may be attributed to the effects of wave trapping.

Trapping requires a single monochromatic wave. This is obvious from the fact that the particles tend to be trapped about specific relative phases ($\psi = \pi/2, 5\pi/2, \ldots$). If two monochromatic waves with slightly different frequencies are present, trapping is severely impeded. A second wave tends to detrap the particles trapped by the first wave.

5.4 Interpretation of the kinetic instability

So far we have used the model developed in §5.1 to interpret the growth and saturation of the reactive instability in terms of phase bunching and wave trapping. Let us now use the model to interpret the growth in the corresponding kinetic instability.

The growth in the kinetic instability is assumed to be weak. Recall that the absorption coefficient (which becomes negative in the instability) is derived in §2.5, and more generally in §2.6, by first neglecting the dissipative processes in determining the wave properties and then including the dissipative processes as a perturbation. Applying the same procedure here, we retain the imaginary part of ω only in describing the wave dissipation, that is in (5.25), viz.

$$\frac{dW}{dt} = 4\omega_i \varepsilon_0 E_0^2. \tag{5.42}$$

We have also set $G = 1$ in (5.42) in view of the fact that in our perturbation approach G would be replaced according to, cf. (5.10),

$$G(t) = 1 + \int_{t_0}^{t} dt' \omega_i(t') + \cdots \tag{5.43}$$

and the first order (in ω_i) term in (5.34) would give a second order (in ω_i) term in (5.42). In evaluating the rate of energy gain by the electrons using (5.26) we set $\omega_i = 0$ and $G = 1$ in (5.16) and (5.17). Then we have

$$m_e \left[v^{(1)} \frac{dv^{(1)}}{dt} + v_0 \frac{dv^{(2)}}{dt} \right] = \frac{m_e}{k^2} [\dot{\psi}^{(1)} \ddot{\psi}^{(1)} - k v_0 \ddot{\psi}^{(2)}]$$

$$= \frac{4e^2 E_0^2}{m_e} \left[\frac{\cos \psi^{(0)}}{\Omega} \{\sin \psi^{(0)} - \sin \psi_0\} - \frac{k v_0 \sin \psi^{(0)}}{\Omega^2} \right.$$

$$\left. \times \{\cos \psi^{(0)} - \cos \psi_0 + \Omega(t - t_0) \sin \psi_0\} \right] \tag{5.44}$$

with $\psi^{(0)} = \Omega(t - t_0) + \psi_0$.

We now average (5.44) over an initial distribution of electrons. The electrons are assumed to be distributed uniformly in z_0, and the average over z_0 is equivalent to an average over ψ_0 in (5.44). The one-dimensional velocity distribution is described by $F(v_0)$ with

$$1 = \int dv_0 F(v_0), \tag{5.45}$$

as in (4.2). Then (5.44) in (5.26) gives

$$\frac{-dW}{dt} = 2\omega_{p1}^2 \varepsilon_0 E_0^2 \int dv_0 F(v_0) \left[\frac{\omega}{\Omega^2} \sin \{\Omega(t - t_0)\} - \frac{k v_0 (t - t_0)}{\Omega} \cos \{\Omega(t - t_0)\} \right] \tag{5.46}$$

where we use $\Omega = \omega - k v_0$ in combining terms proportional to $\sin \{\Omega(t - t_0)\}$.

The kinetic instability applies in the time-asymptotic limit, which we discuss explicitly in §5.5 below. This limit corresponds to $t - t_0 \to \infty$ in (5.46). The integrand then oscillates rapidly and gives zero except for $\Omega = 0$. Hence only resonant electrons with $v_0 = \omega/k$ contribute in the time-asymptotic limit. In the time-asymptotic limit we have [Exercise 5.2]

$$\lim_{t \to \infty} \int_{-\infty}^{\infty} dv_0 \frac{\sin \Omega t}{\Omega} G(v_0) = \int_{-\infty}^{\infty} dv_0 \pi \delta(\Omega) G(v_0) \tag{5.47a}$$

and

$$\lim_{t \to \infty} \int_{-\infty}^{\infty} dv_0 \frac{1 - \cos \Omega t}{\Omega} G(v_0) = P \int_{-\infty}^{\infty} dv_0 \frac{G(v_0)}{\Omega}, \tag{5.47b}$$

where P denotes the Cauchy principal value.

The final term in (5.46) describes a transient response and is neglected in the time-asymptotic limit. The remaining term is evaluated in this limit using (5.47a) after making the expansion

$$F(v_0) = F(v_\phi - \Omega/k) = F(v_\phi) - \frac{\Omega}{\omega} v_\phi \frac{dF(v_\phi)}{dv_\phi} + \cdots \tag{5.48}$$

with $v_\phi = \omega/k$. The term $F(v_\phi)$ in (5.48) does not contribute. Then evaluating (5.46) and equating dW/dt to the value (5.42), one finds

$$\omega_i = \frac{\pi}{2} \frac{n_1}{n_e} \omega v_\phi^2 \frac{dF(v_\phi)}{dv_\phi} \tag{5.49}$$

This reproduces the result (4.4) when one identifies $\gamma_L(\mathbf{k})$ as $-2\omega_i$ and $\omega_L(\mathbf{k})$ as ω.

One feature of the kinetic instability is readily deduced from the sign of Ω and dW/dt in (5.46) (with the final term omitted). Electrons with $v_0 > v_\phi$ have $\Omega < 0$ and lose energy to the wave; electrons with $v_0 < v_\phi$ have $\Omega > 0$ and gain energy from the wave. Growth occurs if there are more electrons with energy $\frac{1}{2}m_e v_0^2 > \frac{1}{2}m_e v_\phi^2$ than there are with energy $\frac{1}{2}m_e v_0^2 < \frac{1}{2}m_e v_\phi^2$. This condition implies that the mechanism for the kinetic instability is essentially the same as in a maser or laser. Growth is due to negative absorption and occurs if there are more particles in the higher energy state than in the lower energy state. This point can be made explicit in terms of quantum mechanical ideas, as discussed in §6.1 and §6.3.

Another feature of the kinetic instability is that it is independent of the phase of the wave. This follows from the fact that phase bunching plays no role; the electrons initially distributed uniformly in z_0 remain uniformly distributed in z for all time. Hence if two different waves are present initially they both grow at the same rate, and a distribution of waves with random phases grows at the same rate as an initially phase-coherent wave. One says that the kinetic instability applies in the *random phase approximation*. Random phases are not required; the phase of the wave is simply irrelevant.

5.5 The forward-scattering method and the time-asymptotic assumption

In the previous section the derivation of the growth rate for the kinetic instability required that we make the time-asymptotic assumption. This corresponds to setting $t_0 = -\infty$ and hence ignoring all effects related to the initial conditions. In this Section we discuss a formal point relating to the time-asymptotic assumption and then discuss implications of relaxing the assumption.

The formal point concerns the inclusion of $t_0 \neq -\infty$ in the response of the plasma and the way in which one then takes the limit $t_0 = -\infty$. For this purpose it is appropriate to use the fully relativistic, electromagnetic theory rather than the nonrelativistic, one-dimensional, electrostatic theory. First we use (5.21) to derive an expression for the dielectric tensor using the *forward-scattering method*. The idea is to calculate the first order current $\mathbf{J}^{(1)}(\omega, \mathbf{k})$ due to an electromagnetic field $\mathbf{E}(\omega', \mathbf{k}')$ and then to take the limit $\omega' = \omega$, $\mathbf{k}' = \mathbf{k}$. This method is an alternative to the methods used in §2.1 and in §2.2 to derive the linear response tensor.

The current due to a single particle moving along its orbit (5.1) is

$$\mathbf{J}(t, \mathbf{x}) = q\dot{\mathbf{X}}(t)\delta^3\{\mathbf{x} - \mathbf{X}(t)\}, \tag{5.50}$$

and its Fourier transform is

$$\mathbf{J}(\omega, \mathbf{k}) = q \int dt\, \dot{\mathbf{X}}(t) \exp\left[i\{\omega t - \mathbf{k}\cdot\mathbf{X}(t)\}\right] \tag{5.51}$$

On making the perturbation expansion (5.6), the first-order terms in the current are

$$\mathbf{J}^{(1)}(\omega, \mathbf{k}) = q \int dt \{ \dot{\mathbf{X}}^{(1)}(t) - iv_0 \mathbf{k} \cdot \mathbf{X}^{(1)}(t) \} \exp \left[i(\omega - \mathbf{k} \cdot \mathbf{v}_0)t - i\mathbf{k} \cdot (\mathbf{x}_0 - \mathbf{v}_0 t_0) \right]$$

(5.52)

where we use (5.7). The next step is to insert the explicit expressions (5.21) with (5.22) for $\dot{\mathbf{X}}^{(1)}(t)$ and $\mathbf{X}^{(1)}(t)$, and then to set $\omega' = \omega$ and $\mathbf{k}' = \mathbf{k}$. The integral over $d\omega' d^3\mathbf{k}'/(2\pi)^4$ is then replaced by $1/VT$ where V is the normalization volume. The time integral in (5.52) and the factor $1/T$ are equivalent to a time average, and as we are interested in secular changes we omit this average. This factor $1/V$ is combined with a sum over particles to replace the sum by an integral over a distribution function. In this way we find a linear response [cf. Exercise 5.3]

$$K_{ij}(\omega, \mathbf{k}) = \delta_{ij} - \Sigma \frac{q^2}{\varepsilon_0 m \omega^2} \int \frac{d^3 \mathbf{p}}{\gamma} f(\mathbf{p}) \left[\left\{ \delta_{ij} + \frac{k_i v_j + k_j v_i}{\omega - \mathbf{k} \cdot \mathbf{v}} + \frac{(|\mathbf{k}|^2 - \omega^2/c^2) v_i v_j}{(\omega - \mathbf{k} \cdot \mathbf{v})^2} \right\} \right.$$

$$\times \{ 1 - \exp \left[i(\omega - \mathbf{k} \cdot \mathbf{v})(t - t_0) \right] \} + i \left\{ v_i k_j + \frac{(|\mathbf{k}|^2 - \omega \mathbf{k} \cdot \mathbf{v}/c^2) v_i v_j}{\omega - \mathbf{k} \cdot \mathbf{v}} \right\}$$

$$\left. \times (t - t_0) \exp \left[i(\omega - \mathbf{k} \cdot \mathbf{v})(t - t_0) \right] \right]$$

(5.53)

where we include a sum over all species.

The time asymptotic limit corresponds to $t - t_0 = \infty$ in (5.53). The causal condition that ω has an infinitesimal positive imaginary part $i0$ implies that the argument of the exponential functions has a negative real contribution proportional to $t - t_0$, and hence they do not contribute for arbitrarily large $t - t_0$. Then (5.53) reproduces the general results for $K_{ij}(\omega, \mathbf{k})$ found in §§2.1 and 2.2. Thus the forward-scattering method in the time-asymptotic limit is equivalent to the Vlasov theory for the purpose of calculating response tensors.

Let us refer to the terms proportional to the exponential functions in (5.53) as the transient (TR) terms. The *transient response* $K_{ij}^{(TR)}(\omega, \mathbf{k})$, which is a function of $(t - t_0)$ is included in (5.53). The relevant terms may be rewritten in the form

$$K_{ij}^{(TR)}(\omega, \mathbf{k}) = - \Sigma \frac{q^2}{\varepsilon_0 \omega^2} \int d^3 \mathbf{p} \frac{v_i}{\omega - \mathbf{k} \cdot \mathbf{v}} \{ (\omega - \mathbf{k} \cdot \mathbf{v}) \delta_{rj} + k_r v_j \}$$

$$\times \frac{\partial f(\mathbf{p})}{\partial p_r} \exp \left[i(\omega - \mathbf{k} \cdot \mathbf{v})(t - t_0) \right]$$

(5.54)

A partial integration in (5.54) reproduces the form (5.53) for the transient terms.

The transient response can lead to growth of waves. This is known in the context of the gyrotron (e.g. Sprangle & Smith, 1980). Such growth is qualitatively different from that in either a reactive instability or in a kinetic instability. One could refer to it as *transitory growth*. There are technical difficulties in treating this growth because ω_i is necessarily a function of $t - t_0$, and care is required in identifying and separating relevant timescales. However, these technical difficulties should not be allowed to obscure the fact that transitory growth can occur, and that this effect is artificially excluded by making the time-asymptotic assumption.

Transitory growth is limited by 'phase mixing'. For a sufficiently narrow momentum distribution the integral in (5.54) is slowly varying over a time $t - t_0 \leqslant 1/\Delta\omega$ with $\Delta\omega$ of order the spread in $\omega - \mathbf{k} \cdot \mathbf{v}$ due to the spread in velocities. For $\Delta\omega(t - t_0) \gg 1$ the exponential function oscillates rapidly leading to 'phase mixing'. For a Maxwellian distribution this timescale for 'phase mixing' is $1/kV_\alpha$, cf. Exercise 5.4. In many applications $\Delta\omega$ is relatively large and the time $1/\Delta\omega$ is too short to be of any physical interest. However, in such cases as the gyrotron (§11.2) and triggered VFL emissions in the magnetosphere (§13.3) timescales of interest can be shorter than $1/\Delta\omega$, and then transitory growth can be important. This topic has received little attention in the literature.

5.6 The ponderomotive force

Another application of the perturbation theory developed in §5.1 is in deriving the ponderomotive force. This force is due to high-frequency waves whose energy density is inhomogeneous. The energy density acts like a pressure and the pressure gradient acts on the plasma providing a force per unit area called the ponderomotive force.

Let us consider the one-dimensional case in which the waves are electrostatic and the particles are nonrelativistic. We assume that the electric field is of a form different from that assumed in (5.8), specifically we assume

$$E(t, z) = \mathscr{E}(t, z)e^{-i\omega_0 t} + \text{c.c.}. \tag{5.55}$$

That is we no longer assume that the spatial dependence is as $\exp[ikz]$. The reason is that we wish to model a situation in which the time average of $|E(t, z)|^2$ varies with z. The quantity $\mathscr{E}(t, z)$ introduced in (5.55) is called the *envelope* of the wave. For simplicity we assume that the envelope does not change shape as a function of time. That is, we assume that $\mathscr{E}(z)$ is not an explicit function of t.

The orbit $z = Z(t)$ for an electron is then determined by solving

$$\ddot{Z}(t) = -\frac{e}{m_e}[\mathscr{E}(Z(t))e^{-i\omega_0 t} + \text{c.c.}]. \tag{5.56}$$

Making the perturbation expansion (in powers of $\mathscr{E}$)

$$Z(t) = Z^{(0)}(t) + Z^{(1)}(t) + Z^{(2)}(t) + \cdots \tag{5.57}$$

the zeroth order motion is

$$Z^{(0)}(t) = z_0 + v_0(t - t_0) \tag{5.58}$$

and the first and second order motions are determined by

$$\ddot{Z}^{(1)}(t) = -\frac{e}{m_e}[\mathscr{E}(Z^{(0)}(t))e^{-i\omega_0 t} + \text{c.c.}] \tag{5.59}$$

and

$$\ddot{Z}^{(2)}(t) = -\frac{e}{m_e} Z^{(1)}(t) \left[\left\{ \frac{d}{dz} \mathcal{E}(z) \right\}^{(0)} e^{-i\omega_0 t} + \text{c.c.} \right] \tag{5.60}$$

where $\{d\mathcal{E}(z)/dz\}^{(0)}$ implies evaluation at $z = Z^{(0)}(t)$.

For an electron initially at rest ($v_0 = 0$) it is elementary to integrate (5.59) to find

$$Z^{(1)}(t) = \frac{e}{m_e \omega_0^2} [\mathcal{E}(z_0)e^{-i\omega_0 t} + \text{c.c.}]. \tag{5.61}$$

On inserting this in (5.60) and averaging over a period $2\pi/\omega_0$ of the waves, one finds

$$\overline{\ddot{Z}^{(2)}} = -\frac{e^2}{m_e^2 \omega_0^2} \frac{d}{dz_0} |\mathcal{E}(z_0)|^2 \tag{5.62}$$

This acceleration is attributed to the *ponderomotive force*.

The neglect of the velocity of the particle in the derivation of (5.62) is unimportant provided that the characteristic distance over which $|\mathcal{E}(z)|^2$ changes is much greater than the characteristic distance v_0/ω_0 that an electron propagates in a wave period.

The generalization of (5.62) to the three-dimensional case is

$$\overline{\ddot{\mathbf{X}}^{(2)}} = -\frac{q^2}{m^2 \omega_0^2} \text{grad}_0 |\mathcal{E}(\mathbf{x}_0)|^2 \tag{5.63}$$

where q and m refer to a particle of arbitrary species, and where grad_0 implies the gradient at $\mathbf{x} = \mathbf{x}_0$. The magnetic field does not contribute, essentially because the Lorentz force $\mathbf{v}_0 \times \mathbf{B}$ is zero for $\mathbf{v}_0 = 0$.

Exercise set 5

5.1 (a) Show that the equality (5.15) is satisfied when ω_i is independent of t.
   (b) By differentiating (5.15) with respect to t confirm that (5.15) is satisfied when terms proportional to $d\omega_i(t)/dt$ are neglected.
5.2 Derive the identity (5.47a), viz.

$$\lim_{t \to \infty} \int_{-\infty}^{\infty} dv_0 \frac{\sin \Omega t}{\Omega} G(v_0) = \int_{-\infty}^{\infty} dv_0 \pi \delta(\Omega) G(v_0),$$

with $\Omega = \omega - kv_0$, for the particular function

$$G(v_0) = \frac{1}{(2\pi)^{1/2} V_\alpha} \exp[-v_0^2/2V_\alpha^2]$$

by (i) evaluating

$$\int_{-\infty}^{\infty} dv_0 \cos(\Omega t) G(v_0)$$

and (ii) integrating over t from $-\infty$ to ∞.

5.3 Carry out the intermediate steps in the derivation of the dielectric tensor (5.53) using the forward-scattering method.

5.4 The one-dimensional nonrelativistic counterpart of the transient response (5.54) is

$$K^{(TR)}(\omega, k) = -\sum \frac{\omega_p^2}{\omega} \int dv \frac{v}{\omega - kv} \frac{\partial F(v)}{\partial v} \exp[i(\omega - kv)(t - t_0)].$$

Evaluate this integral for a one-dimensional Maxwellian distribution

$$F(v) = \frac{1}{(2\pi)^{1/2} V_\alpha} \exp[-v^2/2V_\alpha^2].$$

Specifically show that the result reduces to

$$K^{(TR)}(\omega, k) = -\sum \frac{\omega_{p\alpha}^2}{\omega^2} \frac{1}{\sqrt{2k}V_\alpha} \left\{ 2[1 - \phi(\bar{\omega})][\bar{\omega} - i\sqrt{2}kV_\alpha(t - t_0)] \right.$$

$$\left. + \frac{k^2 V_\alpha^2(t - t_0)^2}{\bar{\omega}} \right\} \exp[i\omega(t - t_0)] \exp\left[-\frac{k^2 V_\alpha^2(t - t_0)^2}{2} \right]$$

with

$$\bar{\omega} = \frac{\omega + ik^2 V_\alpha^2(t - t_0)}{\sqrt{2}kV_\alpha}.$$

6

Weak turbulence theory

6.1 The emission formula: Cerenkov emission

The inhomogeneous wave equation (1.28), viz.

$$\Lambda_{ij}(\omega, \mathbf{k})E_j(\omega, \mathbf{k}) = -\frac{i}{\varepsilon_0 \omega} J_i^{ext}(\omega, \mathbf{k}), \tag{6.1}$$

may be solved to find the power generated in waves in an arbitrary mode by the arbitrary source $\mathbf{J}^{ext}$. First let us solve (6.1) for the electric field generated by the source. This is achieved by contracting (6.1) with $\lambda_{li}(\omega, \mathbf{k})$ and using (2.54):

$$E_i(\omega, \mathbf{k}) = -\frac{i}{\varepsilon_0 \omega} \frac{\lambda_{ij}(\omega, \mathbf{k})J_j^{ext}(\omega, \mathbf{k})}{\Lambda(\omega, \mathbf{k})} \tag{6.2}$$

We shall be interested in the part of (6.2) which corresponds to waves in some mode M. The dispersion equation $\Lambda(\omega, \mathbf{k}) = 0$ has solutions at $\omega = \omega_M(\mathbf{k})$ and at $\omega = -\omega_M(-\mathbf{k})$. In the neighbourhood of either solution we have

$$\Lambda(\omega, \mathbf{k}) \approx (\omega \mp \omega_M(\pm \mathbf{k}) + i0) \left[\frac{\partial \Lambda(\omega, \mathbf{k})}{\partial \omega} \right]_{\omega = \pm \omega_M(\pm \mathbf{k})} \tag{6.3}$$

Then using (2.55) and (2.66) we find the electric field for waves in the mode M generated by the source term:

$$\mathbf{E}_M(\omega, \mathbf{k}) = -\frac{\pi}{\varepsilon_0} R_M(\mathbf{k})\mathbf{e}_M(\mathbf{k})[\mathbf{e}_M^*(\mathbf{k}) \cdot \mathbf{J}^{ext}(\omega, \mathbf{k})]\{\delta(\omega - \omega_M(\mathbf{k})) + \delta(\omega + \omega_M(-\mathbf{k}))\}. \tag{6.4}$$

The time-averaged rate work is done on the wave system has been evaluated in (2.60), and this may be equated to the rate of increase in the wave energy i.e. to $dW_M(\mathbf{k})/dt$ integrated over $\mathbf{k}$-space. This gives

$$\int \frac{d^3\mathbf{k}}{(2\pi)^3} \frac{dW_M(\mathbf{k})}{dt} = -\lim_{T \to \infty} \frac{1}{T} \int \frac{d\omega d^3\mathbf{k}}{(2\pi)^4} [\mathbf{J}^{ext}(\omega, \mathbf{k}) \cdot \mathbf{E}(\omega, \mathbf{k})]_M, \tag{6.5}$$

where the subscript M on the right hand side indicates that only the contribution to the mode M is to be retained. On substituting (6.4) in (6.5) one obtains

$$\frac{dW_M(\mathbf{k})}{dt} = \frac{1}{T\varepsilon_0} R_M(\mathbf{k})|\mathbf{e}_M^*(\mathbf{k}) \cdot \mathbf{J}^{ext}(\omega_M(\mathbf{k}), \mathbf{k})|^2. \tag{6.6}$$

We refer to (6.6) as the *emission formula*. It is implicit that we take the limit $T \to \infty$.

Emission by a single particle (sp) may be treated by identifying the extraneous current with that due to the single particle. Consider a particle of charge q and mass m. Let its orbit be described by (5.1), viz.

$$\mathbf{x} = \mathbf{X}(t), \tag{6.7}$$

so that its instantaneous velocity is $\mathbf{v}(t) = d\mathbf{X}(t)/dt$. The current (density) is

$$[\mathbf{J}(t, \mathbf{x})]^{sp} = q\mathbf{v}(t)\delta^3(\mathbf{x} - \mathbf{X}(t)), \tag{6.8}$$

and its Fourier transform is

$$[\mathbf{J}(\omega, \mathbf{k})]^{(sp)} = q \int_{-\infty}^{\infty} dt \, \mathbf{v}(t) \exp\left[i\{\omega t - \mathbf{k} \cdot \mathbf{X}(t)\}\right] \tag{6.9}$$

For the particular case of Cerenkov emission the particle is assumed to be in constant rectilinear motion. Let its velocity be $\mathbf{v}$ and its orbit be described by (6.7) with $\mathbf{X}(t) = \mathbf{x}_0 + \mathbf{v}t$; both $\mathbf{x}_0$ and $\mathbf{v}$ are constant. Then (6.9) gives

$$[\mathbf{J}(\omega, \mathbf{k})]^{sp} = 2\pi q\mathbf{v} \exp\left[-i\mathbf{k} \cdot \mathbf{x}_0\right]\delta(\omega - \mathbf{k} \cdot \mathbf{v}). \tag{6.10}$$

On inserting (6.10) in the emission formula (6.6), the square of the δ-function is rewritten using (5.28), and one obtains

$$\frac{dW_M(\mathbf{k})}{dt} = \frac{2\pi q^2}{\varepsilon_0} R_M(\mathbf{k}) |\mathbf{e}_M^*(\mathbf{k}) \cdot \mathbf{v}|^2 \delta\{\omega_M(\mathbf{k}) - \mathbf{k} \cdot \mathbf{v}\} \tag{6.11}$$

Later, in §6.3, we use a semiclassical description in which the waves are described as a collection of wave quanta with energy $\hbar|\omega_M(\mathbf{k})|$ momentum $\hbar\mathbf{k}$ and occupation number

$$N_M(\mathbf{k}) = \frac{W_M(\mathbf{k})}{\hbar|\omega_M(\mathbf{k})|V}. \tag{6.12}$$

where V is the volume of the system. Then we may write (6.11) in the form

$$\frac{dN_M(\mathbf{k})}{dt} = \frac{1}{V} w_M(\mathbf{k}, \mathbf{p}) \tag{6.13}$$

where

$$w_M(\mathbf{k}, \mathbf{p}) = \frac{2\pi q^2 R_M(\mathbf{k})}{\varepsilon_0 \hbar |\omega_M(\mathbf{k})|} |\mathbf{e}_M^*(\mathbf{k}) \cdot \mathbf{v}|^2 \delta\{\omega_M(\mathbf{k}) - \mathbf{k} \cdot \mathbf{v}\} \tag{6.14}$$

is the probability per unit time that a particle with momentum $\mathbf{p}$ emit a quantum in the mode M in the range $d^3\mathbf{k}/(2\pi)^3$. More simply we refer to (6.14) as the probability for Cerenkov emission. The appearance of the volume V in (6.13) may be interpreted in terms of the power radiated per unit volume

$$\int \frac{d^3\mathbf{k}}{(2\pi)^3} \hbar\omega_M(\mathbf{k}) \frac{dN_M(\mathbf{k})}{dt}$$

being proportional to the number density of radiating particles.

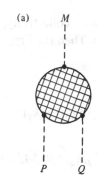

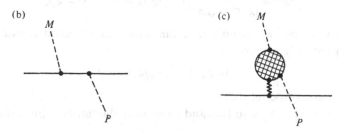

Fig. 6.1 Diagrammatic representation of three-wave coupling (Fig. 6.1(a)), Thomson scattering (Fig. 6.1(b)) and nonlinear scattering (Fig. 6.1(c)). The hatched circle represents the nonlinear response described by α_{ijl}, the solid line represents the scattering particle and the dashed lines represent waves in the modes as labelled. A crossed relation involves transferring one or more wave lines from the initial state (below) to the final state (above) or vice versa.

6.2 Probabilities for nonlinear processes

There is a hierarchy of nonlinear processes which can be ordered according to the degree of nonlinearity. Here we consider only second order processes explicitly. The specific processes are three-wave interactions and particle-wave scattering. These are represented diagrammatically in Figure 6.1. (A different second order process, called turbulent bremsstrahlung, was proposed by Tsytovich, Stenflo & Wilhelmsson (1975); this process has been shown not to exist by Kuijpers & Melrose (1985) and Melrose & Kuijpers (1984).)

Three-wave interactions may be treated by identifying the extraneous current in the emission formula as that due to the nonlinear response of the plasma to two fields $\mathbf{A}_P(\omega', \mathbf{k}')$ and $A_Q(\omega'', \mathbf{k}'')$ in wave modes P and Q. The relevant (quadratic) nonlinear response is the $n = 2$ term in the expansion (1.25). It is convenient to change our notation here in two ways. First, we describe the fields in terms of the vector potential $\mathbf{A}$ in the temporal gauge (electric potential $= 0$). One then has

$$\mathbf{E}(\omega, \mathbf{k}) = i\omega\mathbf{A}(\omega, \mathbf{k}), \quad \mathbf{B}(\omega, \mathbf{k}) = i\mathbf{k} \times \mathbf{A}(\omega, \mathbf{k}). \tag{6.15a, b}$$

Second, we denote the arguments $\omega, \mathbf{k}$ by k collectively; similarly ω', $\mathbf{k}'$ is denoted k', $d\omega' d^3 \mathbf{k}'$ as $d^4 k'$ and $\delta(\omega)\delta^3(\mathbf{k})$ as $\delta^4(k)$. Then the first few terms in the expansion (1.25) become

$$J_i^{(1)}(k) = \alpha_{ij}(k) A_j(k) \tag{6.16a}$$

$$J_i^{(2)}(k) = \int d\lambda^{(2)} \alpha_{ijl}(k, k_1, k_2) A_j(k_1) A_l(k_2). \tag{6.16b}$$

$$J_i^{(3)}(k) = \int d\lambda^{(3)} \alpha_{ijlm}(k, k_1, k_2, k_3) A_j(k_1) A_l(k_2) A_m(k_3) \tag{6.16c}$$

with

$$d\lambda^{(n)} = \frac{d^4 k_1}{(2\pi)^4} \cdots \frac{d^4 k_n}{(2\pi)^4} (2\pi)^4 \delta^4(k - k_1 - \cdots - k_n). \tag{6.17}$$

For formal purposes we assume that the nonlinear response tensors are symmetrized, specifically we assume

$$\alpha_{ijl}(k, k_1, k_2) = \alpha_{ilj}(k, k_2, k_1) \tag{6.18}$$

for the quadratic response tensor.

We write $\mathbf{A} = \mathbf{A}_P + \mathbf{A}_Q$ in (6.16b) and retain only the response involving the cross terms, i.e.

$$J_i^{(2)}(k) = 2 \int d\lambda^{(2)} \alpha_{ijl}(k, k_1, k_2) A_{Pj}(k_1) A_{Ql}(k_2) \tag{6.19}$$

where we use (6.18). On substituting (6.19) in (6.6) the amplitudes are rewritten as in (2.56), e.g.

$$\mathbf{A}_M(k) = -\frac{i e_M(\mathbf{k})}{\omega} E_M(\mathbf{k}) \exp\left[i\psi_M(\mathbf{k})\right] 2\pi \{\delta(\omega - \omega_M(\mathbf{k})) + \delta(\omega - \omega_M(-\mathbf{k}))\} \tag{6.20}$$

where an explicit phase $\psi_M(\mathbf{k})$ of the wave is now included. We now assume random phases and perform an average over phases (denoted by a bar). This average corresponds to

$$\overline{A_{Mi}(k) A_{Mj}^*(k')} = \frac{(2\pi)^4 \delta^4(k - k')}{TV} A_{Mi}(k) A_{Mj}^*(k), \tag{6.21}$$

where V is the volume of the system. The right hand side of (6.21) is independent of the phase. The square of the amplitude on the right hand side of (6.21) is rewritten using (2.59), (2.66) and (6.12):

$$|E_M(\mathbf{k})|^2 = \frac{V R_M(\mathbf{k})}{\varepsilon_0} \hbar |\omega_M(\mathbf{k})| N_M(\mathbf{k}). \tag{6.22}$$

The square of the 4-dimensional δ-function is reduced using an obvious generalization of (2.58):

$$[\delta^4(k)]^2 = \frac{TV}{(2\pi)^4} \delta^4(k). \tag{6.23}$$

Then, after dividing (6.6) by $\hbar|\omega_M(\mathbf{k})|$, one finds

$$\frac{dN_M(\mathbf{k})}{dt} = \int \frac{d^3k'}{(2\pi)^3}\frac{d^3k''}{(2\pi)^3} u_{MPQ}(\mathbf{k}, \mathbf{k}', \mathbf{k}'')N_P(\mathbf{k}')N_Q(\mathbf{k}'') \qquad (6.24)$$

with

$$u_{MPQ}(\mathbf{k}, \mathbf{k}', \mathbf{k}'') = \frac{4\hbar}{\varepsilon_0^3} \frac{R_M(\mathbf{k})R_P(\mathbf{k}')R_Q(\mathbf{k}'')}{|\omega_M(\mathbf{k})\omega_P(\mathbf{k}')\omega_Q(\mathbf{k}'')|}|\alpha_{MPQ}(\mathbf{k}, \mathbf{k}', \mathbf{k}'')|^2(2\pi)^4\delta^4(k_M - k_P' - k_Q'')$$

$$(6.25)$$

and with

$$\alpha_{MPQ}(\mathbf{k}, \mathbf{k}', \mathbf{k}'') = e_{Mi}^*(\mathbf{k})e_{Pj}(\mathbf{k}')e_{Ql}(\mathbf{k}'')\alpha_{ijl}(k_M, k_P', k_Q''), \qquad (6.26)$$

where k_M denotes $(\omega_M(\mathbf{k}), \mathbf{k})$, k_P' denotes $(\omega_P(\mathbf{k}'), \mathbf{k}')$ and k_Q'' denotes $(\omega_Q(\mathbf{k}''), \mathbf{k}'')$. In (6.25) we have combined the negative and positive frequency solutions, adopting the convention

$$\omega_M(-\mathbf{k}) = -\omega_M(\mathbf{k}), \quad \mathbf{e}_M(-\mathbf{k}) = e_M^*(\mathbf{k}), \quad R_M(-\mathbf{k}) = R_M(\mathbf{k}). \qquad (6.27)$$

(Technically, we also need to separate into forward and backward waves, e.g. denoted as two different modes $M \pm$ each of which satisfies (6.27); this has been done explicitly by Kuijpers (1980a).)

The other second order process we consider is scattering of waves by particles. There are two contributions to the current in this case. One is referred to as Thomson scattering. It involves the current due to the perturbed motion of the scattering particle in the field of the unscattered waves. Let the unscattered waves be in the mode P. The equation of motion for the particle.

$$\frac{d\mathbf{p}}{dt} = q[\mathbf{E} + \mathbf{v} \times \mathbf{B}]$$

may then be written in the form

$$\frac{d^2X_i(t)}{dt^2} = \frac{iq}{m\gamma}\left(\delta_{ij} - \frac{v_iv_j}{c^2}\right)\int\frac{d^4k'}{(2\pi)^4}g_{jr}(k', \mathbf{v})\exp\left[-i\{\omega't - \mathbf{k}'\cdot\mathbf{X}(t)\}\right]A_r(k')$$

$$(6.28)$$

with

$$g_{ij}(k, \mathbf{v}) = (\omega - \mathbf{k}\cdot\mathbf{v})\delta_{ij} + k_iv_j. \qquad (6.29)$$

The first order equation of motion is obtained by setting $\mathbf{X}(t) = \mathbf{X}^{(1)}(t)$ on the left hand side of (6.28) and writing $\mathbf{v} = $ constant, $\mathbf{X}(t) = \mathbf{x}_0 + \mathbf{v}t$ on the right hand side. One then identifies the relevant current as

$$J_i^{(1)}(k) = q\int dt[\dot{X}_i^{(1)}(t) - i\mathbf{k}\cdot\mathbf{X}^{(1)}(t)v_i]\exp\left[i(\omega - \mathbf{k}\cdot\mathbf{v})t - i\mathbf{k}\cdot\mathbf{x}_0\right]$$

$$= \frac{2\pi q^2}{m}\int\frac{d^4k'}{(2\pi)^4}a_{ij}(k, k', \mathbf{v})A_j(k')\delta\{(\omega - \mathbf{k}\cdot\mathbf{v}) - (\omega' - \mathbf{k}'\cdot\mathbf{v})\}$$

$$\times \exp\left[-i(\mathbf{k} - \mathbf{k}')\cdot\mathbf{x}_0\right] \qquad (6.30)$$

with

$$a_{ij}(k, k', \mathbf{v}) = \frac{1}{\gamma}\left[\delta_{ij} + \frac{k_i'v_j}{\omega' - \mathbf{k}'\cdot\mathbf{v}} + \frac{k_jv_i}{\omega - \mathbf{k}\cdot\mathbf{v}} + \frac{(\mathbf{k}\cdot\mathbf{k}' - \omega\omega'/c^2)v_iv_j}{(\omega' - \mathbf{k}'\cdot\mathbf{v})(\omega - \mathbf{k}\cdot\mathbf{v})}\right] \qquad (6.31)$$

The other contribution to the current is referred to as nonlinear scattering. It is due to the nonlinear response to the field $\mathbf{A}_P$ and the field $\mathbf{A}^{(q)}$ due to the unperturbed motion of the charge q:

$$A_i^q(k) = -\frac{1}{\varepsilon_0\omega^2}\frac{\lambda_{ij}(k)}{\Lambda(k)}[J_j(k)]^{sp}, \qquad (6.32)$$

with $[\mathbf{J}(k)]^{sp}$ given by (6.10). Thus the nonlinear scattering current is

$$J_i(k) = 2\int d\lambda^{(2)}\alpha_{ijl}(k,k_1,k_2)A_{Pj}(k_1)A_l^{(q)}(k_2). \qquad (6.33)$$

After adding (6.30) and (6.33) and inserting the sum in (6.6), we average over phases using (6.21), insert the expression (6.20) for $A_P(k_1)$ and use (6.22). The rate of increase of the occupation number $N_M(\mathbf{k})$ of the scattered waves is then given by a formula analogous to (6.13):

$$\frac{dN_M(\mathbf{k})}{dt} = \frac{1}{V}\int\frac{d^3k'}{(2\pi)^3}w_{MP}(\mathbf{k},\mathbf{k}',\mathbf{p})N_P(\mathbf{k}'), \qquad (6.34)$$

with

$$w_{MP}(\mathbf{k},\mathbf{k}',\mathbf{p}) = \frac{2\pi q^4}{\varepsilon_0^2 m^2}\frac{R_M(\mathbf{k})R_P(\mathbf{k}')}{|\omega_M(\mathbf{k})\omega_P(\mathbf{k}')|}|A_{MP}(\mathbf{k},\mathbf{k}',\mathbf{v})|^2\delta\{(\omega_M(\mathbf{k})-\mathbf{k}\cdot\mathbf{v})-(\omega_P(\mathbf{k}')-\mathbf{k}'\cdot\mathbf{v})\} \qquad (6.35)$$

and with

$$A_{MP}(\mathbf{k},\mathbf{k}',\mathbf{v}) = e_{Mi}^*(\mathbf{k})e_{Pj}(\mathbf{k}')\left[a_{ij}(k_M,k_P',\mathbf{v}) \right.$$
$$\left. +\frac{2m}{q\varepsilon_0}\frac{\alpha_{ijl}(k_M,k_P',k_M-k_P')\lambda_{lm}(k_M-k_P')v_m}{\{\omega_M(\mathbf{k})-\omega_P(\mathbf{k}')\}^2\Lambda(k_M-k_P')} \right]. \qquad (6.36)$$

Equation (6.35) gives the probability per unit time that a particle with momentum $\mathbf{p}$ scatter a quantum in the mode P in the range $d^3k'/(2\pi)^3$ into a quantum in the mode M in the range $d^3k/(2\pi)^3$.

6.3 Kinetic equations

Recall that in the semiclassical formalism introduced in §6.1, the waves are regarded as a collection of quanta with energy $\hbar|\omega_M(\mathbf{k})|$ momentum $\hbar\mathbf{k}$ and occupation number $N_M(\mathbf{k})$. Implicit in this description is the random phase approximation; the uncertainty principle implies that if the occupation number is specified then we have no information on the phase of the wave. More generally the uncertainties are related by $\Delta\psi_M\Delta N_M \sim 1$.

The major advantage of the semiclassical formalism is that it allows one to appeal to detailed balance or microscopic reversibility. This is an expression of the second law of thermodynamics on the microscopic level. In radiation theory, detailed balance is usually expressed in terms of the Einstein coefficients. Consider the probability $w_M(\mathbf{k},\mathbf{p})$, as given classically by (6.14), for the Cerenkov emission of a wave quantum by a particle. The state of the particle changes from $\mathbf{p}$ to $\mathbf{p} - \hbar\mathbf{k}$, where the components of $\mathbf{p}$ may be regarded as continuous quantum numbers. Let $N_M(\mathbf{k})$ be the occupation number of the wave quanta and $N(\mathbf{p})$ be the occupation number of the particles. Let us concentrate first on the wave quanta. The Einstein coefficients imply that the rate of emission can be stimulated by the presence of the emitted waves such that the total probability of emission is $w_M(\mathbf{k}, \mathbf{p})\{1 + N_M(\mathbf{k})\}$. The unit term describes spontaneous emission and the term $N_M(\mathbf{k})$ describes stimulated emission. Transitions from $\mathbf{p} - \hbar\mathbf{k}$ back to $\mathbf{p}$ correspond to absorption and are described by a probability $w_M(\mathbf{k}, \mathbf{p})N_M(\mathbf{k})$.

When including the particles one needs to know whether they are bosons or fermions. In these two cases the probability of emission $\mathbf{p} \to \mathbf{p} - \hbar\mathbf{k}$ is proportional to $N(\mathbf{p})\{1 \pm N(\mathbf{p} - \hbar\mathbf{k})\}$ and the probability of absorption $\mathbf{p} - \hbar\mathbf{k} \to \mathbf{p}$ is proportional to $N(\mathbf{p} - \hbar\mathbf{k})\{1 \pm N(\mathbf{p})\}$, with the upper sign for bosons and the lower sign for fermions. We treat the particles classically, which corresponds to the limit in which $N(\mathbf{p})$ is infinitesimal. The product $N(\mathbf{p})N(\mathbf{p} - \hbar\mathbf{k})$ is then negligible and there is no difference between bosons and fermions. We use a classical distribution function $f(\mathbf{p})$, normalized according to

$$\int d^3\mathbf{p}\, f(\mathbf{p}) = n \tag{6.37}$$

where n is the number density of the particles. To see how $f(\mathbf{p})$ is included consider the rate of increase of $N_M(\mathbf{k})$ due to Cerenkov emission: from (6.13) we have

$$\frac{dN_M(\mathbf{k})}{dt} = \frac{1}{V}w_M(\mathbf{k}, \mathbf{p}),$$

which applies for emission by one particle. When we sum over a set of particles, $1/V$ times this sum may be replaced by the integral $d^3\mathbf{p}\, f(\mathbf{p})$. Thus the net rate of emission per unit volume becomes

$$w_M(\mathbf{k}, \mathbf{p})\{1 + N_M(\mathbf{k})\}f(\mathbf{p})$$

and the net rate of absorption becomes

$$w_M(\mathbf{k}, \mathbf{p})N_M(\mathbf{k})f(\mathbf{p} - \hbar\mathbf{k}),$$

each of which is to be operated on with the integral $\int d^3\mathbf{p}$.

It is now elementary to derive a kinetic equation for the waves taking into account spontaneous emission, stimulated emission and absorption. Each time a quantum is emitted $N_M(\mathbf{k})$ increases by unity and each time a quantum is absorbed $N_M(\mathbf{k})$ decreases by unity. Hence we have

$$\frac{dN_M(\mathbf{k})}{dt} = \int d^3\mathbf{p}\, w_M(\mathbf{k}, \mathbf{p})[\{1 + N_M(\mathbf{k})\}f(\mathbf{p}) - N_M(\mathbf{k})f(\mathbf{p} - \hbar\mathbf{k})]. \tag{6.38}$$

We approximate $f(\mathbf{p} - \hbar\mathbf{k})$ by $f(\mathbf{p}) - \hbar\mathbf{k}\cdot\partial f(\mathbf{p})/\partial\mathbf{p}$ in the classical limit, and then (6.38) becomes

$$\frac{dN_M(\mathbf{k})}{dt} = \alpha_M(\mathbf{k}) - \gamma_M(\mathbf{k})N_M(\mathbf{k}), \tag{6.39}$$

where

$$\alpha_M(\mathbf{k}) = \int d^3p\, w_M(\mathbf{k}, \mathbf{p}) f(\mathbf{p}) \tag{6.40}$$

is an emission coefficient, and where

$$\gamma_M(\mathbf{k}) = - \int d^3p\, w_M(\mathbf{k}, \mathbf{p}) \hbar\mathbf{k} \cdot \frac{\partial f(\mathbf{p})}{\partial \mathbf{p}} \tag{6.41}$$

is the absorption coefficient. The expression implied by (6.41) and (6.14) is equivalent to that implied by (2.67) with (2.24).

The back-reaction of the emission and absorption on the particles may be treated in a similar way. In this case it is necessary to include transitions $\mathbf{p} + \hbar\mathbf{k} \leftrightarrow \mathbf{p}$ as well as transitions $\mathbf{p} \leftrightarrow \mathbf{p} - \hbar\mathbf{k}$. The rate of change of $f(\mathbf{p})$ is then determined by the increase due to emission $\mathbf{p} + \hbar\mathbf{k} \to \mathbf{p}$ and absorption $\mathbf{p} - \hbar\mathbf{k} \to \mathbf{p}$ and to the decrease due to the reverse processes. The expansion in $\hbar\mathbf{k}$ needs to be carried out to second order in this case (the first order terms cancel), i.e. one writes

$$f(\mathbf{p} \pm \hbar\mathbf{k}) = \left(1 \pm \hbar\mathbf{k} \cdot \frac{\partial}{\partial \mathbf{p}} + \tfrac{1}{2}\hbar^2 k_i k_j \frac{\partial^2}{\partial p_i \partial p_i} \right) f(\mathbf{p}).$$

In the classical limit one finds

$$\frac{df(\mathbf{p})}{dt} = \frac{\partial}{\partial p_i}[A_i(\mathbf{p}) f(\mathbf{p})] + \frac{\partial}{\partial p_i}\left[D_{ij}(\mathbf{p}) \frac{\partial f(\mathbf{p})}{\partial p_j} \right] \tag{6.42}$$

with

$$\begin{bmatrix} A_i(\mathbf{p}) \\ D_{ij}(\mathbf{p}) \end{bmatrix} = \int \frac{d^3k}{(2\pi)^3} w_M(\mathbf{k}, \mathbf{p}) \begin{bmatrix} \hbar k_i \\ \hbar^2 k_i k_j N_M(\mathbf{k}) \end{bmatrix}. \tag{6.43}$$

Together (6.39) and (6.43) are called the quasilinear equations. The term $\alpha_M(\mathbf{k})$ in (6.39) and the term involving $A_i(\mathbf{p})$ in (6.42) describe the effects of spontaneous emission. The term involving $\gamma_M(\mathbf{k})$ in (6.39) describes absorption and the corresponding term in (6.42) is the final term which describes a diffusion in momentum space due to the effects of the induced processes. A one-dimensional form of the quasilinear equations, without the spontaneous term, has been used in §4.2.

It is simple to derive the kinetic equations for the three wave processes $P + Q \leftrightarrow M$ in a similar way. The probability for the transition $P + Q \to M$ is $u_{MPQ}(\mathbf{k}, \mathbf{k}', \mathbf{k}'')$ $\times \{1 + N_M(\mathbf{k})\} N_P(\mathbf{k}') N_Q(\mathbf{k}'')$ and the probability of the transition $M \to P + Q$ is $u_{MPQ}(\mathbf{k}, \mathbf{k}', \mathbf{k}'') N_M(\mathbf{k})\{1 + N_P(\mathbf{k}')\}\{1 + N_Q(\mathbf{k}'')\}$. These transitions cause $N_M(\mathbf{k})$, $N_P(\mathbf{k}')$ and $N_Q(\mathbf{k}'')$ to change by ± 1, ∓ 1 and ∓ 1 respectively. Hence one finds

$$\frac{dN_M(\mathbf{k})}{dt} = \int \frac{d^3k'}{(2\pi)^3} \frac{d^3k''}{(2\pi)^3} u_{MPQ}(\mathbf{k}, \mathbf{k}', \mathbf{k}'')[N_P(\mathbf{k}')N_Q(\mathbf{k}'') - N_M(\mathbf{k})\{N_P(\mathbf{k}') + N_Q(\mathbf{k}'')\}]$$

$$\tag{6.44}$$

and

$$\frac{dN_P(\mathbf{k}')}{dt} = - \int \frac{d^3k}{(2\pi)^3} \frac{d^3k''}{(2\pi)^3} u_{MPQ}(\mathbf{k}, \mathbf{k}', \mathbf{k}'')[N_P(\mathbf{k}')N_Q(\mathbf{k}'') - N_M(\mathbf{k})\{N_P(\mathbf{k}') + N_Q(\mathbf{k}'')\}]$$

$$\tag{6.45}$$

plus another equation obtained from (6.45) by interchanging primed and doubly primed quantities. In (6.44) and (6.45) a term $N_M(\mathbf{k})$ has been omitted from inside the square brackets; this term describes "photon splitting" in a fully quantum treatment but it has no classical counterpart and so is not included here.

The kinetic equations for scattering may be derived in a similar manner. They are

$$\frac{dN_M(\mathbf{k})}{dt} = \int d^3\mathbf{p} \int \frac{d^3\mathbf{k}'}{(2\pi)^3} w_{MP}(\mathbf{k}, \mathbf{k}', \mathbf{p}) \left[\{N_P(\mathbf{k}') - N_M(\mathbf{k})\} f(\mathbf{p}) \right.$$
$$\left. - N_P(\mathbf{k}')N_M(\mathbf{k})\hbar(\mathbf{k}' - \mathbf{k}) \cdot \frac{\partial f(\mathbf{p})}{\partial \mathbf{p}} \right] \tag{6.46a}$$

$$\frac{dN_P(\mathbf{k}')}{dt} = \int d^3\mathbf{p} \int \frac{d^3\mathbf{k}}{(2\pi)^3} u_{MP}(\mathbf{k}, \mathbf{k}', \mathbf{p}) \left[\{N_M(\mathbf{k}) - N_P(\mathbf{k}')\} f(\mathbf{p}) \right.$$
$$\left. - N_M(\mathbf{k})N_P(\mathbf{k}')\hbar(\mathbf{k} - \mathbf{k}') \cdot \frac{\partial f(\mathbf{p})}{\partial \mathbf{p}} \right] \tag{6.46b}$$

and

$$\frac{df(\mathbf{p})}{dt} = \int \frac{d^3\mathbf{k}}{(2\pi)^3} \frac{d^3\mathbf{k}'}{(2\pi)^3} \hbar(\mathbf{k} - \mathbf{k}') \cdot \frac{\partial}{\partial \mathbf{p}} \left[w_{MP}(\mathbf{k}, \mathbf{k}', \mathbf{p}) \left[\{N_P(\mathbf{k}') - N_M(\mathbf{k})\} f(\mathbf{p}) \right. \right.$$
$$\left. \left. - N_P(\mathbf{k}')N_M(\mathbf{k})\hbar(\mathbf{k}' - \mathbf{k}) \cdot \frac{\partial f(\mathbf{p})}{\partial \mathbf{p}} \right] \right] \tag{6.46c}$$

The terms proportional to N_P and N_M on the right hand sides describe the effects of 'spontaneous' scattering $P \to M$ and $M \to P$ respectively, and the terms proportional to $N_P N_M$ describe the effects of induced scattering.

The probabilities satisfy crossing symmetries. For example the probability for the three wave processes $M + P \leftrightarrow Q$ is $u_{MPQ}(\mathbf{k}, -\mathbf{k}', \mathbf{k}'')$ and the probability for double emission or double absorption (both wave quanta emitted or absorbed) is $w_{MP}(\mathbf{k}, -\mathbf{k}', \mathbf{p})$. In these cases the "crossed" processes are related to the original processes by the wave in the mode P being transferred from the initial state to the final state (or vice versa). Formally one does this by replacing $\mathbf{k}'$ by $-\mathbf{k}'$ and using the conventions (6.27). More generally, if a crossing symmetry relates two processes by moving a quantum from the initial state to the final state (or vice versa) then the relevant probabilities are related by interchanging the sign of the relevant wavevector. In particular the probability of the process $M \to P + Q$ is $u_{MPQ}(-\mathbf{k}, -\mathbf{k}', -\mathbf{k}'')$, which is readily shown to be equal to the probability $u_{MPQ}(\mathbf{k}, \mathbf{k}', \mathbf{k}'')$ for the process $P + Q \to M$.

6.4 Scattering of Langmuir waves off thermal ions

Langmuir turbulence generated by a stream is confined to a narrow region in k-space, with $\mathbf{k}$ peaked around $k = \omega_p/v_b$ and around the direction parallel to $\mathbf{v}_b$. Scattering of Langmuir waves can transfer them from one region of k-space to another, changing both their k-spectrum and their angular distribution. There are two weak-turbulence processes which can be important in scattering Langmuir waves. One of these is induced scattering off thermal ions. This is also sometimes called nonlinear Landau damping because it can lead to removal at an exponential rate of Langmuir waves from the region of **k**-space where they are being generated.

The scattering is attributed to ions when the contribution from nonlinear scattering dominates that from Thomson scattering. This occurs when the inverse wavenumber of the waves (actually $1/|\mathbf{k} - \mathbf{k}'|$) is much longer than a Debye length. Physically, nonlinear scattering may be attributed to scattering by the shielding field around an individual particle. For a charge q at rest at $\mathbf{x} = \mathbf{x}_0$ in a thermal plasma the net field has electrostatic potential

$$\Phi(\mathbf{x}) = \frac{q}{4\pi\varepsilon_0|\mathbf{x} - \mathbf{x}_0|}\exp[-|\mathbf{x} - \mathbf{x}_0|/\lambda_D], \tag{6.47}$$

and this may be regarded as composed of the 'bare' Coulomb field $\Phi(\mathbf{x}) = q/4\pi\varepsilon_0|\mathbf{x} - \mathbf{x}_0|$ plus the 'shielding' or 'self-consistent' field. The particle with its shielding field is said to be 'dressed'. At short wavelengths $k^{-1} \ll \lambda_D$ the wave does not feel the shielding field which is smeared out over a distance $\approx \lambda_D$; the wave feels only the bare field of the particle and the scattering is then Thomson scattering. The cross section for Thomson scattering is

$$\sigma_T = \frac{8\pi}{3}r_0^2 \tag{6.48}$$

where $r_0 = q^2/4\pi\varepsilon_0 mc^2$ is the classical radius for the particle. This cross-section is much greater for electrons than for ions, and hence for $k^{-1} \ll \lambda_D$ the scattering is Thomson scattering by electrons.

At long wavelengths $k^{-1} \gg \lambda_D$ the wave sees the combination of bare field and shielding field. The shielding field is due primarily to an inhomogeneity in the field electrons. For scattering by an electron the scattering field acts like that of a positive electron and the contributions from Thomson scattering and nonlinear scattering nearly cancel each other. For scattering by an ion, Thomson scattering is negligible and for $k^{-1} \gg \lambda_D$ nonlinear scattering is like Thomson scattering by a particle with the mass of the electron and charge equal and opposite to that of the ion.

To treat scattering by ions we need first to find an appropriate approximation to the probability (6.35). Important approximations are to α_{ijl} and to λ_{ij}/Λ. If we assume that the shielding field is predominantly electrostatic, this corresponds to making the longitudinal approximation

$$\frac{\lambda_{ij}(\omega, \mathbf{k})}{\Lambda(\omega, \mathbf{k})} = \frac{k_i k_j}{|\mathbf{k}|^2}\frac{1}{K^L(\omega, \mathbf{k})}. \tag{6.49}$$

(N.B. We use the shorthand notation k for ω, $\mathbf{k}$ in this Section only in connection with the nonlinear responses, cf. (6.16) et seq.) Before making approximations to $\alpha_{ijl}(k, k_1, k_2)$ it is appropriate to write down general expressions for it.

Extending the kinetic theory calculation of §2.2 to the lowest order of nonlinearity leads to an expression

$$\alpha_{ijl}(k, k_1, k_2) = q^3 \int d^3\mathbf{p} \frac{v_i g_{rj}(k_1, \mathbf{v})}{\omega - \mathbf{k} \cdot \mathbf{v}} \frac{\partial}{\partial p_r} \left[\frac{g_{sl}(k_2, \mathbf{v})}{\omega_2 - \mathbf{k}_2 \cdot \mathbf{v}} \frac{\partial f(\mathbf{p})}{\partial p_s} \right], \tag{6.50}$$

with $g_{ij}(k, \mathbf{v})$ given by (6.29). After partially integrating twice and imposing the symmetry (6.18), one obtains an expression analogous to (2.25). An identical expression may be obtained by extending the cold plasma calculation of §2.1 to the lowest order of nonlinearity. This expression is

$$\alpha_{ijl}(k, k_1, k_2) = \frac{q^3}{2m^2} \int d^3\mathbf{p} \frac{f(\mathbf{p})}{\gamma} [a_{ij}(k, k_1, \mathbf{v}) d_l(k_2, \mathbf{v})$$
$$+ a_{il}(k, k_2, \mathbf{v}) d_j(k_1, \mathbf{v}) + a_{jl}(k_1, k_2, \mathbf{v}) d_i(k, \mathbf{v})] \tag{6.51}$$

with

$$d_i(k, \mathbf{v}) = \frac{1}{\omega - \mathbf{k} \cdot \mathbf{v}} \left[k_i + \frac{(|\mathbf{k}|^2 - \omega^2/c^2) v_i}{\omega - \mathbf{k} \cdot \mathbf{v}} \right] \tag{6.52}$$

and with $a_{jl}(k_1, k_2, \mathbf{v})$ given by (6.31), viz.

$$a_{jl}(k_1, k_2, \mathbf{v}) = \frac{1}{\gamma} \left[\delta_{jl} + \frac{k_{1l} v_j}{\omega_1 - \mathbf{k}_1 \cdot \mathbf{v}} + \frac{k_{2j} v_l}{\omega_2 - \mathbf{k}_2 \cdot \mathbf{v}} + \frac{(\mathbf{k}_1 \cdot \mathbf{k}_2 - \omega_1 \omega_2/c^2) v_j v_l}{(\omega_1 - \mathbf{k}_1 \cdot \mathbf{v})(\omega_2 - \mathbf{k}_2 \cdot \mathbf{v})} \right] \tag{6.53}$$

In (6.51) only the contribution from one species of particle is retained. Due to the dependence as m^{-2} the electronic contribution is the dominant one, and henceforth only the electronic contribution ($q = -e$, $m = m_e$) is retained.

We consider an approximation in which one field ($\omega_2, \mathbf{k}_2$) is electrostatic with a large wavenumber and low phase speed and the other two fields ($\omega, \mathbf{k}$ and $\omega_1, \mathbf{k}_1$) have high phase speeds. Then with $|\omega_2| \ll |\omega_1|$, $|\omega|$ we have $\omega_1 \approx \omega$ and in (6.50) we may make the approximation

$$\frac{v_i g_{rj}(k_1, \mathbf{v})}{\omega - \mathbf{k} \cdot \mathbf{v}} \approx v_i \delta_{rj}.$$

Then partially integrating with respect to p_r and making the nonrelativistic approximation, the electronic contribution becomes

$$\alpha_{ijl}(k, k_1, k_2) \approx \frac{e^3}{m_e} \int d^3\mathbf{p} \frac{\delta_{ij} g_{sl}(k_2, \mathbf{v})}{\omega_2 - \mathbf{k}_2 \cdot \mathbf{v}} \frac{\partial f(\mathbf{p})}{\partial p_s}.$$

Now contracting with k_{2l}, which corresponds to assuming the field $\omega_2, \mathbf{k}_2$ to be longitudinal, one finds

$$\alpha_{ijl}(k, k_1, k_2) \approx \frac{e \varepsilon_0 \omega_2}{m_e} \delta_{ij} k_{2l} \{ \chi^{L(e)}(\omega_2, \mathbf{k}_2) \} \tag{6.54}$$

where $\chi^{L(e)}(\omega_2, \mathbf{k}_2)$ is the electronic contribution to $K^L(\omega_2, \mathbf{k}_2)$. An analogous result to (6.54) may be derived from the form (6.51), but the derivation is complicated by the symmetry imposed. The symmetry prevents us from assuming that $\omega_2, \mathbf{k}_2$ is qualitatively different from $\omega_1, \mathbf{k}_1$; one needs to artificially break the symmetry by omitting the factor $\frac{1}{2}$ in (6.51) and then assuming that $\omega_2, \mathbf{k}_2$ and $\omega_1, \mathbf{k}_1$ are different.

Now when (6.54) with (6.49) is inserted in (6.36) and the nonrelativistic approximation to $a_{ij}(\approx \delta_{ij})$ is made, one finds (omitting the factor 2 in the final term because of our use of the unsymmetrized form (6.54))

$$A_{MP}(\mathbf{k}, \mathbf{k}', \mathbf{v}) \approx e_M^*(\mathbf{k}) \cdot e_P(\mathbf{k}') \left[1 + \frac{em}{qm_e} \frac{\chi^{L(e)}(\omega - \omega', \mathbf{k} - \mathbf{k}')}{K^L(\omega - \omega', \mathbf{k} - \mathbf{k}')} \right]. \tag{6.55}$$

For electrons ($q = -e$, $m = m_e$) a strong cancellation between the two terms in square brackets results in view of

$$K^L(\omega, \mathbf{k}) = 1 + \chi^{L(e)}(\omega, \mathbf{k}) + \chi^{L(i)}(\omega, \mathbf{k}) \approx 1 + \frac{1}{k^2 \lambda_{De}^2} - \omega_{pi}^2/\omega^2 \tag{6.56}$$

for $\sqrt{2}kV_i \lesssim \omega \ll \sqrt{2}kV_e$. For ions the units term in square brackets in (6.55) is negligible. More generally we have

$$\frac{\chi^{L(e)}(\omega, \mathbf{k})}{K^L(\omega, \mathbf{k})} \approx \begin{cases} -\dfrac{\omega^2}{\omega_p^2} & \text{for} \quad \omega \gg \sqrt{2}kV_e \\[2mm] 1 & \text{for} \quad \sqrt{2}kV_i \ll \omega \ll \sqrt{2}kV_e \\[2mm] \dfrac{T_i}{T_e + T_i} & \text{for} \quad \omega \ll \sqrt{2}kV_e \end{cases} \tag{6.57}$$

When treating scattering by thermal ions it is convenient to average over a Maxwellian distribution of ions. The averaged probability obtained from (6.35) is then

$$\bar{w}_{MP}(\mathbf{k}, \mathbf{k}') = \frac{(2\pi)^{1/2} e^4}{\varepsilon_0^2 m_e^2 |\mathbf{k} - \mathbf{k}'| V_i} \left[\frac{\chi^{L(e)}(\omega - \omega', \mathbf{k} - \mathbf{k}')}{K^L(\omega - \omega', \mathbf{k} - \mathbf{k}')} \right]^2$$

$$\times |e_M^*(\mathbf{k}) \cdot e_P(\mathbf{k}')|^2 \exp\left[-\frac{(\omega - \omega')^2}{2|\mathbf{k} - \mathbf{k}'|^2 V_i^2} \right], \tag{6.58}$$

with $\omega = \omega_M(\mathbf{k})$ and $\omega' = \omega_P(\mathbf{k}')$. The kinetic equation (6.46a) then becomes

$$\frac{dN_M(\mathbf{k})}{dt} = n_i \int \frac{d^3 k'}{(2\pi)^3} \bar{w}_{MP}(\mathbf{k}, \mathbf{k}') \left[N_P(\mathbf{k}') - N_M(\mathbf{k}) \right.$$

$$\left. + N_P(\mathbf{k}') N_M(\mathbf{k}) \frac{\hbar \{\omega_P(\mathbf{k}') - \omega_M(\mathbf{k})\}}{T_i} \right] \tag{6.59}$$

and (6.49b) reduces to (6.59) with primed and unprimed quantities interchanged.

Induced scattering is described by the final term in (6.59). An important property is that $N_M(\mathbf{k})$ decreases for $\omega_P(\mathbf{k}') < \omega_M(\mathbf{k})$; thus induced scattering transfers wave quanta from higher to lower frequencies. The energy difference $\hbar\{\omega_M(\mathbf{k}) - \omega_P(\mathbf{k}')\}$ is absorbed by the thermal plasma just as though a fluctuation $\omega - \omega'$, $\mathbf{k} - \mathbf{k}'$ were being Landau damped. This is the basis for the name 'nonlinear Landau damping'.

For Langmuir waves with, cf. (1.3),

$$\omega_L(\mathbf{k}) \approx \omega_P + \frac{3k^2 V_e^2}{2\omega_p}, \tag{6.60}$$

a decrease in frequency implies a decrease in wavenumber. Hence induced scattering tends to pump Langmuir waves from larger to smaller k, or equivalently from smaller to larger phase speeds $v_\phi \approx \omega_p/k$. The maximum effective change Δk in k can be inferred from the exponential factor in (6.58) with (6.60). Let us write

$$k* = \frac{2\sqrt{2}\omega_p V_i}{3V_e^2}.$$

Then the exponential in (6.60) implies that induced scattering is weak except for

$$k^2 - k'^2 \lesssim |\mathbf{k} - \mathbf{k'}|k*. \tag{6.61}$$

For $k < k*$ we may satisfy (6.61) for virtually any $k' < k$; in this case the Langmuir waves may be scattered from k to $k' \ll k$ in one step. However, for $k \gg k*$ (6.61) can only be satisfied for $\mathbf{k'} \approx -\mathbf{k}$ with $k - k'$ of order $k*$. Thus for $k \gg k*$ the Langmuir waves tend to be scattered into the backward direction with a small change in k of order $k*$. The waves can be scattered to small k' and to a wider range of angles in a step-wise manner.

The important qualitative property of induced scattering of Langmuir waves is that it transfers Langmuir turbulence towards small ks, either in one step or in a sequence of steps. One says that a 'condensate' tends to form at small k as the Langmuir turbulence collects there.

Induced scattering can lead to a nonlinear saturation of a beam instability. The nonlinear damping rate of the beam modes, i.e. of waves with $\mathbf{k} \approx \omega_p \mathbf{v}_b/v_b^2$, is proportional to the level of turbulence at $\mathbf{k'}$. As this level builds up, the rate of damping of the beam modes also builds up. In principle the nonlinear damping rate can reach a value at which it balances the growth rate, and then the instability is suppressed. There are several other nonlinear processes which can suppress a beam instability in a similar way, cf. §6.5 and §7.2.

6.5 Scattering of Langmuir waves by ion sound waves

Ion sound turbulence can lead to a scattering of Langmuir waves through the processes $L + s \to L'$ and $L \to L' + s$, where L refers to the initial Langmuir waves, L' to the scattered Langmuir waves and s to the ion sound waves. Here we assume that relevant ion sound waves are present (as they appear always to be in the solar wind) and explore the consequences of the scattering.

The kinematic conditions for a three-wave process are expressed in the δ-function in the probability (6.25). For the processes $L \pm s \to L'$ these conditions are

$$\omega_L(\mathbf{k}_L) \pm \omega_s(\mathbf{k}_s) = \omega_L(\mathbf{k}_{L'}) \tag{6.62a}$$

$$\mathbf{k}_L \pm \mathbf{k}_s = \mathbf{k}_{L'}. \tag{6.62b}$$

After inserting the dispersion relation (6.60) for the Langmuir waves and the dispersion relation $\omega_s(\mathbf{k}) = kv_s$ for ion sound waves, (6.62) implies

$$k_L^2 \pm k_0 k_s = k_L^2, \tag{6.63}$$

with

$$k_0 = 2\omega_p v_s/3V_e^2. \tag{6.64}$$

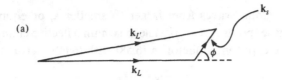

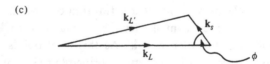

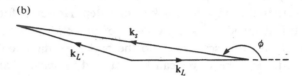

Fig. 6.2 Schematic vector diagrams for the coalescence process $L + s \rightarrow L'$ for $k_s \ll k_L$ (Fig. 6.1(a)) and for $k_s \lesssim k_0 + 2k_L$ (Fig. 6.1(b)) and for the process $L - s \rightarrow L'$ (Fig. 6.1(c)). The angle ϕ is equal to arc cos $\{(k_0 \mp k_s)/2k_L\}$.

Let κ_L and κ_s be unit vectors along $\mathbf{k}_L$ and $\mathbf{k}_s$ respectively. Then the square of (6.62b) with (6.63) implies

$$\kappa_L \cdot \kappa_s = \frac{k_0 \mp k_s}{2k_L}. \tag{6.65}$$

Suppose we refer to the cases with the upper and lower signs in (6.62a, b) as up-conversion (u) and down-conversion (d) respectively. Let us consider these processes starting with arbitrarily small k_s and then increasing k_s. For very low frequency ion sound waves ($k_s \ll k_0$), (6.65) gives $\kappa_L \cdot \kappa_s = k_0/2k_L$, which implies that the process is allowed only for $k_L > k_0/2$, when the angle between $\mathbf{k}_L$ and $\mathbf{k}_s$ is arc cos $(k_0/2k_L)$; we also have $k_{L'}^2 = k_L^2$ from (6.63). As k_s is increased, for the u-case the angle between $\mathbf{k}_L$ and $\mathbf{k}_s$ increases and reaches 180° for $k_s = k_0 + 2k_L$ when we have $k_{L'} = k_L + k_0$; for the d-case the angle between $\mathbf{k}_L$ and $\mathbf{k}_s$ decreases and reaches 0° for $k_s = 2k_L - k_0$ when we have $k_{L'} = k_L - k_0$. These kinematic conditions are illustrated in Figure 6.2.

The probability (6.25) for the cases $P \pm s \rightarrow M$ with the two modes M and P having phase speeds $\gg V_e$ may be evaluated using the approximation (6.54) to α_{ijl}.

(Note that the factor 4 in (6.25) is omitted when the unsymmetrized form (6.54) is used.) We have

$$\chi^{L(e)}(\omega, \mathbf{k}) = \frac{1}{k^2 \lambda_{De}^2} = \frac{\omega_{pi}^2}{\omega^2};$$ (6.66)

the latter identity follows from $\omega = k v_s$. Then using $R_s = \omega_s^2/2\omega_{pi}^2$ from (2.50), one finds

$$u_{MPs}(\mathbf{k}, \mathbf{k}', \pm \mathbf{k}'') = \frac{(2\pi)^4 e^2 \hbar \omega_p^2}{2\varepsilon_0 m_e^2 V_e^2} \left| \frac{\omega_s(\mathbf{k}'')}{\omega_M(\mathbf{k})\omega_P(\mathbf{k}')} \right| R_M(\mathbf{k}) R_P(\mathbf{k}')$$
$$\times |\mathbf{e}_M^*(\mathbf{k}) \cdot \mathbf{e}_P(\mathbf{k}')|^2 \delta\{\omega_M(\mathbf{k}) - \omega_P(\mathbf{k}') \mp \omega_s(\mathbf{k}'')\} \delta^3(\mathbf{k} - \mathbf{k}' \mp \mathbf{k}'').$$
(6.67)

For Langmuir waves ($M = P = L$) we have the expression (2.48) for R_L, the dispersion relation (6.60) and $\mathbf{e}_M = \kappa = \mathbf{k}/k$. The evaluation of the factor $|\mathbf{e}_M^* \cdot \mathbf{e}_P|^2$ for transverse waves is treated in Exercise 6.8.

In discussing the effects of the three-wave interaction from a quantitative viewpoint a major simplification occurs if the interaction saturates. Saturation corresponds to the processes $P + Q \to M$ and $M \to P + Q$ being in balance. According to (6.44) or (6.45) saturation occurs for

$$N_P(\mathbf{k}') N_Q(\mathbf{k}'') = N_M(\mathbf{k})\{N_P(\mathbf{k}') + N_Q(\mathbf{k}'')\}.$$ (6.68)

It is useful to rewrite (6.68) in terms of the *effective temperature* of the waves:

$$T_M(\mathbf{k}) = \hbar \omega_M(\mathbf{k}) N_M(\mathbf{k}),$$ (6.69)

where we now assume $\omega_M(\mathbf{k}) > 0$ whenever no ambiguity should arise. Then

$$W_M = \int \frac{d^3 \mathbf{k}}{(2\pi)^3} T_M(\mathbf{k})$$ (6.70)

is the energy density in the waves. In terms of effective temperatures, (6.68) gives

$$\frac{T_P(\mathbf{k}') T_Q(\mathbf{k}'')}{\omega_P(\mathbf{k}')\omega_Q(\mathbf{k}'')} = \frac{T_M(\mathbf{k})}{\omega_M(\mathbf{k})} \left\{ \frac{T_P(\mathbf{k}')}{\omega_P(\mathbf{k}')} + \frac{T_Q(\mathbf{k}'')}{\omega_Q(\mathbf{k}'')} \right\}$$ (6.71)

A relatively simple case occurs when we have

$$\frac{T_s}{\omega_s} \gg \frac{T_L}{\omega_L}, \frac{T_{L'}}{\omega_{L'}}.$$ (6.72)

Then (6.71), with $Q = s$, $M = L$ and $P = L'$, implies $T_L(\mathbf{k}) = T_{L'}(\mathbf{k}')$. That is the effective temperature of the Langmuir waves tends to a uniform value throughout the accessible region of $\mathbf{k}$-space, i.e. throughout the region to which Langmuir waves may be transferred consistent with the kinematic restrictions (6.62a, b).

This simple case requires both that (6.72) be satisfied and that the three-wave interaction saturate. This latter condition is more difficult to quantify. In effect the timescale t_s, given by setting $N_M(\mathbf{k}) = N_P(\mathbf{k}')$ in (6.44), i.e.

$$t_s = \left[\int \frac{d^3 \mathbf{k}'}{(2\pi)^3} \frac{d^3 \mathbf{k}''}{(2\pi)^3} u_{LLs}(\mathbf{k}, \mathbf{k}', \mathbf{k}'') N_s(\mathbf{k}'') \right]^{-1},$$ (6.73)

must be shorter than other relevant timescales affecting the evolution of the Langmuir turbulence. However, if the waves are being generated by a beam instability, then an obvious difficulty arises if t_s^{-1} is greater than the growth rate for the instability: the ion sound waves would then suppress the instability. Thus we require that t_s^{-1} be less than the growth rate, so that the Langmuir waves grow; only then is any Langmuir turbulence generated. It is possible to have the evolution of the Langmuir turbulence dominated by growth at an early stage (with $t_s^{-1} \ll$ growth rate) and then to have the subsequent evolution determined by the effect of the ion sound turbulence after the stream has passed. The simple case applies only if the evolution in this latter stage leads to saturation of the processes $L \pm s \rightarrow L'$.

6.6 Plasma emission

Emission at the fundamental ($\omega \approx \omega_p$) and the second harmonic ($\omega \approx 2\omega_p$) is a characteristic feature of certain solar radio emissions at metre and longer wavelengths (e.g. Wild, Smerd & Weiss 1963, Kundu 1965). The first attempt at a quantitative theory for these processes was by Ginzburg & Zheleznyakov (1958). Their theory, specifically for so-called type III solar radio bursts, involves a stream of electrons generating Langmuir waves, with the Langmuir waves being scattered into fundamental transverse waves by thermal ions, and with the Langmuir waves also coalescing with thermal Langmuir waves to produce second harmonic transverse waves. In detail these two mechanisms are now outdated, but in broad outline current theories for these so-called *plasma emission processes* are qualitatively similar to Ginzburg & Zheleznyakov's theory. The deficiencies in Ginzburg & Zheleznyakov's theory are quantitative: the scattering process $L \rightarrow T$ (T denotes a transverse wave) is too slow to account for the observed emission and the coalescence $L + L' \rightarrow T$ with L' a thermal Langmuir wave restricts the brightness temperature of the second harmonic radiation to twice the electron temperature, i.e. $T_T(\mathbf{k}) \lesssim 2T_e$. Numerous modifications to and alternatives for the detailed processes involved have been suggested (cf. Melrose 1980, Goldman 1983). At present there are many different detailed theories and there is no strong consensus on which of them actually operate in type III bursts.

The most important modification for second harmonic emission is that some of the Langmuir waves generated by the beam be scattered into the backward direction. These scattered waves can then coalesce nearly 'head on' with the initial Langmuir waves to produce an adequate level of second harmonic emission. Satisfactory modifications for fundamental emission have proved more difficult to identify. One suggestion is that induced scattering $L \rightarrow T$ by thermal ions is important. This scattering may be described by (6.59) with $M' = L$ and $M = T$; the dispersion relation for transverse waves is

$$\omega_T(\mathbf{k}) = (\omega_p^2 + k^2 c^2)^{1/2}. \tag{6.74}$$

Induced scattering causes transverse waves to grow for $\omega_T(\mathbf{k}) < \omega_L(\mathbf{k}')$. However induced scattering is only important when the final term in the square brackets in (6.59) dominates the other term, and this requires

$$T_T(\mathbf{k}) = \hbar\omega_T(\mathbf{k})N_T(\mathbf{k}) > \frac{T_i\omega_T(\mathbf{k})}{\omega_L(\mathbf{k}') - \omega_T(\mathbf{k})}. \tag{6.75}$$

The exponential function in (6.59) requires $|\omega_L(\mathbf{k}') - \omega_T(\mathbf{k})| \lesssim \sqrt{2}k'V_i$, where we assume $k \ll k'$, which is well satisfied for $L \to T$. Then (6.75) gives

$$T_T \gtrsim T_i \frac{v_\phi}{\sqrt{2}V_i} \tag{6.75'}$$

as the condition for induced scattering to be important. Type III electron streams have $v_b \lesssim c/3$, and $v_\phi = v_b$ then implies $T_T \gtrsim 10^9\,K$. Thus once T_T reaches $\approx 10^9$ K it can then start to increase exponentially due to induced scattering. One might then expect $T_T \gg 10^9$ K to result. In fact type III bursts can have brightness temperatures as high as about 10^{13} K. However if induced scattering does become important then it can lead to absorption as well as amplification. Along the ray path of the escaping transverse wave ω_p decreases and hence $\omega_L(\mathbf{k}') - \omega_T(\mathbf{k})$ changes sign; as a consequence amplification of the transverse waves then changes into absorption. In a simple model the growth for $\omega_T(\mathbf{k}) < \omega_L(\mathbf{k}')$ is balanced by absorption for $\omega_T(\mathbf{k}) > \omega_L(\mathbf{k}')$ so that the net effect of induced scattering $L \to T$ leaves $T_T \approx 10^9$ K. Thus an additional assumption, e.g. an inhomogeneous distribution of Langmuir turbulence, is required before induced scattering can be effective in producing fundamental type III emission with $T_T \gg 10^9$ K.

A relatively simple theory for plasma emission is one based on the assumption that ion sound turbulence is pesent. The fundamental emission is then due to the processes $L \pm s \to T$, which requires $\mathbf{k}_s \approx \mp \mathbf{k}_L$. This process is closely analogous to the processes $L \pm s \to L'$ disussed in §6.5. A notable difference is that the time required for saturation can be determined by propagation effects, rather than as in (6.73). The second harmonic emission then results either from coalescence between initial and scattered Langmuir waves or between scattered Langmuir waves. As discussed in §6.5, when the scattering is due to the processes $L \pm s \to L'$, with a sufficiently high level of ion sound turbulence, T_L tends to become uniform throughout the available k-space. An associated argument implies that T_{T1} and T_{T2}, for the fundamental and second harmonic respectively, both also approach this value of T_L. This is one of the most attractive features of this simple theory: T_{T1} and T_{T2} are often comparable for type III bursts and no other theory leads naturally to an equality $T_{T1} \approx T_{T2}$. Of course the equality is not required in this theory because either the fundamental or the second harmonic or both may not saturate, leaving T_{T1} and/or T_{T2} less than T_L and determined by the detailed structure of the source.

Our understanding of plasma emission is in an unsatisfactory state. It seems that the problems with our understanding of plasma emission are of an astrophysical nature and will eventually be solved through new observational data. There are several different possible mechanisms which can lead to fundamental plasma emission, and it is still not clear which is the relevant one in practice. Although this leaves the theory of fundamental plasma emission in a somewhat uncertain state, the theory for second harmonic emission is well understood; there seems to be no reasonable alternative to the coalescence process $L + L \to T$. To complete this discussion of plasma emission let us summarize the theory for second harmonic emission briefly.

In the coalescence process $L + L \to T$ all three waves have phase speed $\gg V_e$. The relevant form of the nonlinear response tensor is then obtained from (6.51) by assuming $\omega \gg \mathbf{k} \cdot \mathbf{v}$, $\omega_1 \gg \mathbf{k}_1 \cdot \mathbf{v}$, $\omega_2 \gg \mathbf{k}_2 \cdot \mathbf{v}$. The electronic contribution becomes

$$\alpha_{ijl}(k, k', k'') = -\frac{e^3 n_e}{2m_e^2} \left\{ \delta_{jl} \frac{k_i}{\omega} + \delta_{il} \frac{k_j'}{\omega'} + \delta_{ij} \frac{k_l''}{\omega''} \right\} \tag{6.76}$$

This is to be projected onto the transverse polarization $\mathbf{e}$ and the longitudinal polarizations $\mathbf{e}' = \boldsymbol{\kappa}'$ and $\mathbf{e}'' = \boldsymbol{\kappa}''$, cf. (6.26). One finds

$$\alpha_{TLL}(\mathbf{k}, \mathbf{k}', \mathbf{k}'') = -\frac{e^3 n_e}{2m_e^2} \mathbf{e} \cdot \left[\frac{k' \boldsymbol{\kappa}''}{\omega_L(\mathbf{k}')} + \frac{k'' \boldsymbol{\kappa}'}{\omega_L(\mathbf{k}'')} \right]. \tag{6.77}$$

As we are not interested in the polarization of the radiation in this case, we sum over the two states of transverse polarization, using

$$\sum_{\text{pol}} |\mathbf{e} \cdot \mathbf{A}|^2 = |\boldsymbol{\kappa} \times \mathbf{A}|^2. \tag{6.78}$$

Then with $\mathbf{k} = \mathbf{k}' + \mathbf{k}''$ from the δ-function in the probability (6.25) one finds

$$\sum_{\text{pol}} |\alpha_{TLL}(\mathbf{k}, \mathbf{k}', \mathbf{k}'')|^2 = \frac{e^6 n_e^2}{4m_e^4 \omega_p^2} \frac{(k'^2 - k''^2)^2}{k^2} |\boldsymbol{\kappa}' \times \boldsymbol{\kappa}''|^2, \tag{6.79}$$

where we also set $\omega_L(\mathbf{k}') \approx \omega_L(\mathbf{k}'') \approx \omega_p$. One then inserts (6.79) into the expression probability (6.25) for the probability, assumes $R_T \approx R_L \approx \frac{1}{2}$ and replaces the factor 4 in (6.25) by 2 to take account of the identity of the two modes $M' = M'' = L$. The result is

$$u_{TLL}(\mathbf{k}, \mathbf{k}', \mathbf{k}'') = \frac{(2\pi)^4 \hbar e^2}{32\varepsilon_0 m_e^2 \omega_p} \frac{(k'^2 - k''^2)^2}{k^2} |\boldsymbol{\kappa}' \times \boldsymbol{\kappa}''|^2$$
$$\times \delta\{\omega_T(\mathbf{k}) - \omega_L(\mathbf{k}') - \omega_L(\mathbf{k}'')\} \delta^3(\mathbf{k} - \mathbf{k}' - \mathbf{k}''). \tag{6.80}$$

Further simplification occurs in the 'head on' approximation. For transverse waves with $\omega = 2\omega_p$, the dispersion relation (6.74) implies $k = \sqrt{3}\omega_p/c$. For Langmuir waves with phase speed $v_\phi \ll c/\sqrt{3}$ we have $k', k'' \gg k$ and then $\mathbf{k} = \mathbf{k}' + \mathbf{k}''$ requires $\mathbf{k}'' \approx -\mathbf{k}'$, i.e. that the coalescence occur nearly head on. In this limit we have

$$\delta\{\omega_T(\mathbf{k}) - \omega_L(\mathbf{k}') - \omega_L(\mathbf{k}'')\} \approx \frac{1}{\sqrt{3}c} \delta\left(k - \sqrt{3}\frac{\omega_p}{c} \right) \tag{6.81}$$

and

$$\frac{(k'^2 - k''^2)^2}{k^2} |\boldsymbol{\kappa}' \times \boldsymbol{\kappa}''| \approx \frac{3\omega_p^2}{c^2} |\boldsymbol{\kappa} \cdot \boldsymbol{\kappa}'|^2 |\boldsymbol{\kappa} \times \boldsymbol{\kappa}'|^2, \tag{6.82}$$

and (6.80) simplifies accordingly.

Exercise set 6

6.1 Derive the formula

$$P = \frac{q^2 v}{4\pi\varepsilon_0 c^2} \int d\omega\, \omega \left(1 - \frac{c^2}{v^2 N^2(\omega)}\right)$$

for the power radiated P in Cerenkov emission into transverse waves in an isotropic dielectric with the refractive index $N(\omega)$. The integral is over the region $v/c > 1/N(\omega)$.

Remark: Integrate (6.14) over $d^3k/(2\pi)^3$ using $k = N(\omega)\omega/c$, $R_M(\mathbf{k}) = [2N(\omega)d\{\omega N(\omega)\}/d\omega]^{-1}$, and sum over the two states of transverse polarization using (6.78).

6.2 Show that the power radiated by a particle due to Cerenkov emission of Langmuir waves is given by

$$P = \frac{q^2 \omega_p^2}{4\pi\varepsilon_0 v} \ln \frac{v}{V_e},$$

where the Langmuir waves are assumed to exist for $v_\phi > V_e$ and to have properties $\omega_L(\mathbf{k}) = \omega_p$, $R_L(\mathbf{k}) = \frac{1}{2}$.

6.3 When the quasilinear equation (6.42) is written in terms of an orthogonal coordinate system α, β, γ, with

$$|d\mathbf{p}|^2 = (h_\alpha d\alpha)^2 + (h_\beta d\beta)^2 + (h_\gamma d\gamma)^2,$$

the result is

$$\frac{df}{dt} = \sum_\lambda \frac{1}{\sqrt{g}} \frac{\partial}{\partial\lambda}(\sqrt{g}A_\lambda f) + \sum_{\lambda,\mu} \frac{1}{\sqrt{g}} \frac{\partial}{\partial\lambda}\left(\sqrt{g}D_{\lambda\mu}\frac{\partial}{\partial\mu}\right)$$

with λ, μ running over α, β, γ and with $\sqrt{g} = h_\alpha h_\beta h_\gamma$.
Show that the coefficients A_λ and $D_{\lambda\mu}$ are given by

$$\begin{bmatrix} A_\lambda \\ D_{\lambda\mu} \end{bmatrix} = \int \frac{d^3k}{(2\pi)^3} w_M(\mathbf{k},\mathbf{p}) \begin{bmatrix} \Delta\lambda \\ \Delta\lambda\Delta\mu N_M(\mathbf{k}) \end{bmatrix}$$

with

$$\Delta\lambda = \hbar\mathbf{k}\cdot\frac{\partial\lambda}{\partial\mathbf{p}}.$$

Evaluate A_λ and $D_{\lambda\mu}$ explicitly for
(a) cylindrical polar coordinates $p_\perp$, ϕ, $p_\parallel$ with

$$h_\perp = 1, \quad h_\phi = p_\perp, \quad h_\parallel = 1$$

and
(b) spherical polar coordinates, p, θ, ϕ with

$$h_p = 1, \quad h_\theta = p, \quad h_\phi = p\sin\theta.$$

6.4 Write down the kinetic equations in the classical limit for the following processes:-
(a) Double emission in which two quanta in modes M and P are emitted by a

particle with momentum $\mathbf{p}$; the relevant probability is given by $w_{MP}(\mathbf{k}, -\mathbf{k}', \mathbf{p})$ with $w_{MP}(\mathbf{k}, \mathbf{k}', \mathbf{p})$ given by (6.35).

(b) The four-wave process in which two wave quanta in modes M & P interact and produce two other wave quanta in modes Q & R.

6.5   Show that the probability for the four-wave process in which waves in modes P, Q & R coalesce into a wave in the mode M is given by

$$u_{MPQR}(\mathbf{k}, \mathbf{k}_1, \mathbf{k}_2, \mathbf{k}_3) = \frac{36\hbar^2}{\varepsilon_0^4} \frac{R_M(\mathbf{k}) R_P(\mathbf{k}_1) R_Q(\mathbf{k}_2) R_R(\mathbf{k}_3)}{|\omega_M(\mathbf{k})\omega_P(\mathbf{k}_1)\omega_Q(\mathbf{k}_2)\omega_R(\mathbf{k}_3)|}$$
$$\times |\alpha_{MPQR}(\mathbf{k}, \mathbf{k}_1, \mathbf{k}_2, \mathbf{k}_3)|^2 (2\pi)^4 \delta^4(k_M - k_P - k_Q - k_R)$$

where the notation is the same as in (6.25) and (6.26) and where the cubic nonlinear response has been written in the symmetrized form

$$J_i^{(3)}(k) = \int d\lambda^{(3)} \alpha_{ijlm}(k, k_1, k_2, k_3) A_j(k_1) A_l(k_2) A_m(k_3)$$

with

$$\alpha_{ijlm}(k, k_1, k_2, k_3) = \alpha_{ijml}(k, k_1, k_3, k_2)$$
$$= \alpha_{iljm}(k, k_2, k_1, k_3).$$

6.6   Let $w(\mathbf{k}, \mathbf{k}', \mathbf{p}, \mathbf{p}')$ be the probability that an electron ($\mathbf{p}$) and a positron ($\mathbf{p}'$) annihilate to produce two photons $\mathbf{k}$ and $\mathbf{k}'$. Let $N_+(\mathbf{p})$ and $N_-(\mathbf{p})$ be the occupation numbers of the electrons and positrons, both of which are fermions.

(a) Show that the kinetic equation for the photons is

$$\frac{dN(\mathbf{k})}{dt} = \int \frac{d^3\mathbf{p}}{(2\pi\hbar)^3} \int \frac{d^3\mathbf{p}'}{(2\pi\hbar)^3} \int \frac{d^3\mathbf{k}'}{(2\pi)^3} w(\mathbf{k}, \mathbf{k}', \mathbf{p}, \mathbf{p}')$$
$$\times [\{1 + N(\mathbf{k})\}\{1 + N(\mathbf{k}')\} N_+(\mathbf{p}) N_-(\mathbf{p}')$$
$$- N(\mathbf{k}) N(\mathbf{k}') \{1 - N_+(\mathbf{p})\} \{1 - N_-(\mathbf{p}')\}]$$

(b) Derive the kinetic equation for the electrons.

(c) It was pointed out by Ramaty, McKinley & Jones (1982) that the kinetic equation in part (a) allows the possibility of a GRASAR (Gamma Ray Amplification by Stimulated Annihilation Radiation). Determine conditions under which GRASAR action is possible.

6.7   Show that the brightness temperature $T_T(\mathbf{k})$ of second harmonic plasma emission is restricted by $T_T(\mathbf{k}) < 2T_e$ in the case considered by Ginzburg & Zheleznyakov (1958), i.e. for coalescence of two Langmuir waves one of which is from a one-dimensional distribution with $T_L(\mathbf{k}) \gg T_e$ and the other of which is from a thermal distribution (which is isotropic with $T_L(\mathbf{k}) = T_e$).

6.8   Consider the factor $|\mathbf{e}_M^* \cdot \mathbf{e}_P|^2$ in the probabilities (6.58) and (6.67) when the waves are either longitudinal or transverse. Sum over the states of transverse polarization for transverse waves in the final state, and average over the states of polarization for transverse waves in the initial state. Hence show that one has

$$|e_M^* \cdot e_P|^2 = \begin{cases} |\kappa_L \cdot \kappa_{L'}|^2 & M = L, \quad P = L' \\ |\kappa_L \times \kappa_T|^2 & M = L, \quad P = T \\ \frac{1}{2}|\kappa_T \times \kappa_L|^2 & M = T, \quad P = L \\ \frac{1}{2}(1 + |\kappa_T \cdot \kappa_{T'}|^2) & M = T, \quad P = T'. \end{cases}$$

6.9 Show that conservation of energy

$$\varepsilon' = \varepsilon - \hbar\omega$$

and momentum

$$\mathbf{p}' = \mathbf{p} - \hbar\mathbf{k},$$

with $\varepsilon = (m^2c^4 + p^2c^2)^{1/2}$, $\varepsilon' = (m^2c^4 + p'^2c^2)^{1/2}$, imply the relation

$$\omega - \mathbf{k} \cdot \mathbf{v} + \frac{\hbar}{2\varepsilon}\{k^2c^2 - (\mathbf{k} \cdot \mathbf{v})^2\} + \cdots = 0,$$

where...indicates an expansion in powers of $\hbar$.

Remarks. (i) Thus energy–momentum conservation on the microscopic level implies the classical resonance condition $\omega - \mathbf{k} \cdot \mathbf{v} = 0$. (ii) The term proportional to $\hbar$ is the first quantum recoil and it needs to be retained when treating the back-reaction of the emission on the particle, i.e. it contributes in (6.42). (iii) If one assumes $\varepsilon = p^2/2m$ the term k^2c^2 in the recoil does not appear, cf. comment at the end of §2.2.

6.10 In the case of a three wave process $M \pm Q \to M$ with $k'' \ll k$ and $\omega_Q \ll \omega_M$, it is appropriate to expand in powers of k''/k and ω_Q/ω_M. Show that on making such an expansion the kinetic equations, cf. (6.44) and (6.45), may be approximated by

$$\frac{dN_Q(\mathbf{k}'')}{dt} = \left[\int \frac{d^3\mathbf{k}}{(2\pi)^3} w_{MQ}(\mathbf{k}, \mathbf{k}'') \mathbf{k}'' \cdot \frac{\partial N_M(\mathbf{k})}{\partial \mathbf{k}} \right] N_Q(\mathbf{k}'')$$

$$\frac{dN_M(\mathbf{k})}{dt} = \frac{\partial}{\partial k_i} \left[D_{ij}^{MQ}(\mathbf{k}) \frac{\partial N_M(\mathbf{k})}{\partial k_j} \right]$$

with

$$D_{ij}^{MQ}(\mathbf{k}) = \int \frac{d^3\mathbf{k}''}{(2\pi)^3} k_i'' k_j'' w_{MQ}(\mathbf{k}, \mathbf{k}'') N_Q(\mathbf{k}''),$$

and where the relevant probability follows from (6.25) through

$$w_{MQ}(\mathbf{k}, \mathbf{k}'') = \int \frac{d^3\mathbf{k}'}{(2\pi)^3} u_{MMQ}(\mathbf{k}, \mathbf{k}', \mathbf{k}'')$$

with

$$\delta^4(k_M - k_M' - k_Q'') \approx \delta^3(\mathbf{k} - \mathbf{k}' - \mathbf{k}'') \delta(\omega_Q(\mathbf{k}'') - \mathbf{k}'' \cdot \mathbf{v}_{gM}(\mathbf{k})).$$

Remarks: (i) The wave–wave interaction is then directly analogous to the wave–particle interaction, cf. (6.39) to (6.43), with the low-frequency (Q) wave being emitted by the high-frequency (M) wave. (ii) Here no analogue of spontaneous emission appears. Such an analogue exists; it is an intrinsically quantum mechanical effect referred to as 'photon splitting'.

7

Nonlinear instabilities and strong turbulence

7.1 Nonlinear correction to the dispersion equation

The growth of waves due to induced scattering or to the effects of a three-wave interaction may be regarded as a nonlinear instability. To see this, note that there exists terms in (6.46a) and (6.44) which may be written in the form

$$\frac{dN_M(\mathbf{k})}{dt} = -\gamma_M^{NL}(\mathbf{k})N_M(\mathbf{k}).\tag{7.1}$$

The specific contributions to the nonlinear absorption coefficient from (6.46a) and (6.44) are

$$\gamma_M^{NL}(\mathbf{k}) = \int\frac{d^3\mathbf{k}'}{(2\pi)^3}\int d^3\mathbf{p}\,w_{MP}(\mathbf{k},\mathbf{k}',\mathbf{p})\hbar(\mathbf{k}-\mathbf{k}')\cdot\frac{\partial f(\mathbf{p})}{\partial\mathbf{p}}N_P(\mathbf{k}')$$

$$+\int\frac{d^3\mathbf{k}'}{(2\pi)^3}\int\frac{d^3\mathbf{k}''}{(2\pi)^3}u_{MPQ}(\mathbf{k},\mathbf{k}'\mathbf{k}'')\{N_P(\mathbf{k}')+N_Q(\mathbf{k}'')\}.\tag{7.2}$$

As already indicated $\gamma_M^{NL}(\mathbf{k})$ can be negative. The associated wave growth may be treated as a kinetic instability in the following sense. Suppose we identify a nonlinear correction $\alpha_{ij}^{NL}(k)$ to the linear response tensor. (We use the shorthand notation k for ω, $\mathbf{k}$ in this Section.) The homogeneous wave equation is then, cf. (1.28),

$$\left[\Lambda_{ij}(k) + \frac{1}{\varepsilon_0\omega^2}\alpha_{ij}^{NL}(k)\right]A_j(k) = 0,\tag{7.3}$$

where $\alpha_{ij}^{NL}(k)$ is assumed proportional to the energy density in microturbulence. Now if we construct a nonlinear absorption coefficient by analogy with (2.67), then we have

$$\gamma_M^{NL}(\mathbf{k}) = \frac{2R_M(\mathbf{k})}{\varepsilon_0\omega_M(\mathbf{k})}\,\text{Im}\,[e_{Mi}^*(\mathbf{k})e_{Mj}(\mathbf{k})\alpha_{ij}^{NL}(k_M)]\tag{7.4}$$

where k_M denotes $\omega_M(\mathbf{k})$, $\mathbf{k}$. In the sense that the nonlinear instabilities implied by (7.1) may be derived from (7.4), they are kinetic instabilities.

There are also reactive versions of these and other nonlinear kinetic instabilities; indeed there is a wide class of reactive nonlinear instabilities. Many of these instabilities may be treated using (7.3) or a related equation. Several examples are discussed in §7.2.

The nonlinear correction to the linear response tensor in (7.3) has a variety of other implications. One is self-focussing. This is most familiar for transverse waves in nonlinear optics. Suppose we apply (7.3) to transverse waves and assume that the dominant correction to $\alpha_{ij}^{NL}(k)$ is from the energy density in these waves. When we solve (7.3) for the refractive index for transverse waves, it includes a term proportional to the energy density in the transverse waves. Suppose this correction term is positive. Waves are refracted toward increasing refractive index, and hence a local enhancement in the wave energy density tends to cause more waves to refract into that region. This effect is called self-focussing, and it leads to a fila-mentation instability in which a beam of radiation breaks up into filaments. Similar types of instability occur for Langmuir waves, e.g. the modulational instability and Langmuir collapse; these are mentioned further in §7.4.

A further implication of (7.3) is the existence of a nonlinear frequency shift in the dispersion relation for waves in any particular mode. The nonlinear frequency shift for waves in the mode M is

$$\Delta\omega_M^{NL}(\mathbf{k}) = -\frac{R_M(\mathbf{k})}{\varepsilon_0\omega_M(\mathbf{k})}e_{Mi}^*(\mathbf{k})e_{Mj}(\mathbf{k})\alpha_{ij}^{NL}(k_M). \qquad (7.5)$$

(Minus twice the imaginary part of this is equal to the nonlinear absorption coefficient (7.4).) Such a nonlinear correction to the dispersion relation for ion sound waves has been invoked to allow the three-wave interaction $s' + s'' \leftrightarrow s$, which is kinematically forbidden with the unmodified dispersion relation for ion sound waves. Tsytovich (1972) referred to this effect as resonance broadening due to turbulent collisions.

The calculation of $\alpha_{ij}^{NL}(k)$ itself is relatively simple from a formal viewpoint. However, the general form of the result is not particularly useful, and approxi-mations are required to treat specific instabilities. For completeness the general results and the nature of the approximations made are outlined here. The resulting approximate expressions are usually derived in quite different ways, involving physical models for particular classes of instabilities. The most widely used of these models is that implicit in the Zakharov equations (§7.4).

In the remainder of this Section we present some of the technical details required to treat kinetic and reactive nonlinear instabilities. It is not necessary to follow the detailed derivations in order to use the resulting formulas, e.g. (7.19) to (7.21), in the subsequent sections.

Let there be waves in some mode P present in the plasma. Our aim is to derive a nonlinear current which is proportional to both $|A_P(k')|^2$ and to the amplitude $A(k)$ of a test field. There are two contributions, one from the quadratic response (6.16b) operating twice, and one from the cubic response (6.16c). The latter leads to

$$\alpha_{ij}^{NL(3)}(k) = \frac{1}{VT}\int\frac{d^4k'}{(2\pi)^4}\,3\alpha_{irjs}(k, k', k, -k')A_{Pr}(k')A_{Ps}(-k') \qquad (7.6)$$

where the factor $1/VT$ comes from an integral over $d^4k''/(2\pi)^4$ with $k'' = k'$, and where the factor 3 arises from the relevant cross terms in $(\mathbf{A}_P + \mathbf{A})^3$. The other contribution arises as follows. The nonlinear response to $\mathbf{A}_P$ and $\mathbf{A}$ is

$$J_i^{(2)}(k) = \int d\lambda^{(2)} \, 2\alpha_{ijl}(k, k', k'') A_j(k) A_{Pl}(k'). \tag{7.7}$$

When this is inserted as a source term in the wave equation

$$\Lambda_{ij}(k) A_j(k) = -\frac{1}{\varepsilon_0 \omega^2} J_i(k) \tag{7.8}$$

it gives rise to a beat field

$$A_i^{(2)}(k) = -\frac{1}{\varepsilon_0 \omega^2} \frac{\lambda_{ij}(k)}{\Lambda(k)} J_j^{(2)}(k), \tag{7.9}$$

where (2.54) has been used. The quadratic response to $\mathbf{A}_P$ and $\mathbf{A}^{(2)}$ leads to another current proportional to $|\mathbf{A}_P|^2$ and to $\mathbf{A}$. This is the other contribution to $\alpha_{ij}^{NL}(k)$:

$$\alpha_{ij}^{NL(2)}(k) = -\frac{1}{VT\varepsilon_0} \int \frac{d^4k'}{(2\pi)^4} 2\alpha_{ira}(k, k', k - k')$$

$$\times \frac{\lambda_{ab}(k - k')}{(\omega - \omega')^2 \Lambda(k - k')} 2\alpha_{bjs}(k - k', k, -k') A_{Pr}(k') A_{Ps}(-k'). \tag{7.10}$$

Explicit expressions for the quadratic response tensor are given by (6.50), (6.51) and (6.54). The expression (6.50) is unsymmetrized. The analogous unsymmetrized form for the cubic response tensor is

$$\alpha_{ijlm}(k, k_1, k_2, k_3) = q^4 \int d^3\mathbf{p} \frac{v_i g_{rj}(k_1, \mathbf{v})}{\omega - \mathbf{k}\cdot\mathbf{v}} \frac{\partial}{\partial p_r}$$

$$\times \left[\frac{g_{sl}(k_2, \mathbf{v})}{\omega_2 + \omega_3 - (\mathbf{k}_2 + \mathbf{k}_3)\cdot\mathbf{v}} \frac{\partial}{\partial p_s} \left\{ \frac{g_{tm}(k_3, \mathbf{v})}{\omega_3 - \mathbf{k}_3\cdot\mathbf{v}} \frac{\partial f(\mathbf{p})}{\partial p_t} \right\} \right]. \tag{7.11}$$

As with (6.50), the form (7.11) needs to be symmetrized (over j, k_1; l, k_2; m, k_3) before being inserted in (7.6). Then (6.50) and (7.11) may be used to rederive the nonlinear absorption coefficient (7.1) with (7.2), cf. Exercise 7.1. However, simplified forms of the nonlinear response tensors are required to discuss the reactive versions of these instabilities.

Let us assume that k and k' describe high frequency ($\omega, \omega' > \omega_p$) waves with phase speed $\gg V_e$ and that $k - k'$ is of low frequency with phase speed $\ll V_e$. We neglect relativistic effects and neglect v in comparison with $\omega/|\mathbf{k}|$ and $\omega'/|\mathbf{k}'|$. In evaluating (7.11), we partially integrate over p_r and operate with the p_s-derivative only on $g_{tm}(k_3, \mathbf{v})/(\omega_3 - \mathbf{k}_3\cdot\mathbf{v})$, then neglect v in comparison with the phase speeds, set $k_1 = k'$, $k_2 = k$ and $k_3 = -k'$, and write, cf. (2.24),

$$q_\alpha^2 \int d^3\mathbf{p} \frac{1}{\omega - \mathbf{k}\cdot\mathbf{v}} \mathbf{k} \cdot \frac{\partial f^{(\alpha)}(\mathbf{p})}{\partial \mathbf{p}} = \frac{|\mathbf{k}|^2}{\omega^2} \varepsilon_0 \chi^{L(\alpha)}(k). \tag{7.12}$$

It is conventional in this context to use the susceptibility

$$\chi^{L(\alpha)}(k) = K^{L(\alpha)}(k) - 1 \tag{7.13}$$

for each species. Only the electronic contribution is important. Thus one finds

$$\alpha_{irjs}(k, k', k, -k') = \frac{e^2}{m_e^2} \varepsilon_0 \delta_{ir} \delta_{js} |\mathbf{k} - \mathbf{k}'|^2 \chi^{L(e)}(k - k'). \tag{7.14}$$

The symmetry properties implicit in (7.6) are such that the factor 3 is to be omitted when the form (7.14) is used.

The analogous approximation to (7.14) for the quadratic response tensor is given by (6.54), viz.

$$\alpha_{iji}(k, k', k - k') = \frac{e\varepsilon_0}{m_e} (\omega - \omega') \delta_{ij} (\mathbf{k} - \mathbf{k}')_i \chi^{L(e)}(k - k'). \tag{7.15}$$

When (7.15) is used the factors 2 in (7.10) are to be omitted. The factor λ_{ab}/Λ in (7.10) may be evaluated for an isotropic plasma using (2.34) and (2.35) in the definition (2.54) of λ_{ij}:

$$\frac{\lambda_{ij}(k)}{\Lambda(k)} = \frac{\kappa_i \kappa_j}{K^L(k)} + \frac{\delta_{ij} - \kappa_i \kappa_j}{K^T(k) - N^2}. \tag{7.16}$$

Combining these results, one finds

$$\frac{1}{\varepsilon_0 \omega^2} \alpha_{ij}^{NL}(k) = \frac{e^2}{m_e^2 \omega^2} \int \frac{d^4 k'}{(2\pi)^4} \frac{|\mathbf{k} - \mathbf{k}'|^2 \chi^{L(e)}(k - k')}{K^L(k - k')} \{1 + \chi^{L(i)}(k - k')\} A_{Pi}(k') A_{Pj}^*(k'), \tag{7.17}$$

where we write

$$K^L(k) = 1 + \chi^{L(e)}(k) + \chi^{L(i)}(k). \tag{7.18}$$

Let us now assume that all waves are longitudinal, and average over the phase of the waves in the mode M', now relabelled as L', using (6.20) to (6.22). One finds

$$K^L(k) + \frac{e^2 \hbar}{m_e^2 \omega^2 \varepsilon_0} \int \frac{d^3 k'}{(2\pi)^3} \frac{R_L(k')}{\omega_L(k')} N_L(k') |\boldsymbol{\kappa} \cdot \boldsymbol{\kappa}'|^2$$

$$\times \left[\frac{|\mathbf{k} - \mathbf{k}'|^2}{K^L(k - k')} \chi^{L(e)}(k - k') \{1 + \chi^{L(i)}(k - k')\} \right.$$

$$\left. + \frac{|\mathbf{k} + \mathbf{k}'|^2}{K^L(k + k')} \chi^{L(e)}(k + k') \{1 + \chi^{L(i)}(k + k')\} \right] = 0. \tag{7.19}$$

An analogous result may be derived for transverse waves (Exercise 7.2).

Equation (7.19) is still not quite in the form required to treat reactive nonlinear instabilities. Suppose the spectrum of L waves is narrow, e.g. $N_L(\mathbf{k}') \propto \delta^3(\mathbf{k}' - \mathbf{k}_0)$, and that k describes another Langmuir wave with $\omega \approx \omega' = \omega_0$. Further, let us approximate $R_L(k_0)$ by $\frac{1}{2}$ and denote the energy density in the waves as W_{L0}. Finally, let us relabel k in (7.19) in two ways, firstly as $k - k_0$, neglecting the contribution at $k - 2k_0$, and secondly as $k + k_0$, neglecting the contribution at $k + 2k_0$. In this way we 'invert' (7.19) to find

$$\frac{1}{\chi^{L(e)}(k)} + \frac{1}{1 + \chi^{L(i)}(k)} + \frac{e^2 W_{LO}}{2m_e^2 \varepsilon_0} \frac{|\mathbf{k}|^2}{\omega_0^4} \left[\frac{\mu_+^2}{K^L(k + k_0)} + \frac{\mu_-^2}{K^L(k - k_0)} \right] = 0 \quad (7.20)$$

with

$$\mu_\pm = \frac{\mathbf{k}_0 \cdot (\mathbf{k} \pm \mathbf{k}_0)}{|\mathbf{k}_0||\mathbf{k} \pm \mathbf{k}_0|}. \quad (7.21)$$

Again there is an analogous formula for transverse waves [cf. Exercise 7.3].

Formula (7.20) can be derived more directly by evaluating $\alpha_{ir\,js}(k, k', k, -k')$ for the case where k describes the low-frequency and k' the high-frequency disturbance. One finds that the dominant contribution in this case comes from the combination of two quadratic responses, i.e. from (7.10). In this case one uses the approximation (7.15) and the symmetry properties of α_{ijl} to obtain the relevant approximate form (for k low-frequency and k' high-frequency and $k - k'$ also high-frequency).

$$\alpha_{ijl}(k, k', k - k') = \frac{e\varepsilon_0}{m_e} \omega k_i \chi^{L(e)}(k) \delta_{jl}.$$

Then (7.20) follows after assuming $K^L(k) \approx 0$.

7.2 Parametric instabilities

The terminology relating to parametric instabilities arises from the theory of driven oscillators. Let the oscillator have a natural frequency ω, and perhaps a second natural frequency $\omega_i \ll \omega$ called the *idler*. The oscillations are driven by a *pump* with frequency ω_0 such that there is a *frequency mismatch* $\omega - \omega_0$. In some applications the frequency mismatch coincides with an idler frequency, and in others it does not. If the pump is at low frequency $\omega_0 \ll \omega$, then the high frequency oscillation may develop *sidebands* at $\omega \pm \omega_0$, $\omega \pm 2\omega_0, \ldots$. The application of such terminology to nonlinear plasma instabilities is natural and obvious. One notable difference from laboratory oscillators is that in plasmas the relevant oscillations are waves with both a wavevector and a frequency. A pump then has frequency ω_0 and wavevector $\mathbf{k}_0$. Suppose the pump drives a wave ω, $\mathbf{k}$ with a small frequency mismatch $|\omega - \omega_0| \ll \omega_0$; it does not necessarily follow that $\mathbf{k}$ is approximately equal to $\mathbf{k}_0$. For example, in some cases $\mathbf{k}$ is nearly opposite to $\mathbf{k}_0$ (a backscatter) and in others $|\mathbf{k}|$ is much larger than $|\mathbf{k}_0|$.

There is a vast literature on parametric instabilities in plasmas (e.g. the reviews by Liu & Kaw 1976 and Cap 1978). The discussion of them here is aimed at identifying parametric instabilities related to the instabilities discussed in Chapters 3, 4, & 6. For this purpose the dispersion equation (7.20) suffices. Note that this equation applies only to longitudinal waves and to a monochromatic pump. It is convenient to assume that the electric field in the pump is of the form

$$\mathbf{E}(t, \mathbf{x}) = [\mathbf{E}_0 \exp\{-i(\omega_0 t - \mathbf{k}_0 \cdot \mathbf{x})\} + \text{c.c.}] \quad (7.22)$$

It is also sometimes convenient to introduce the *quiver velocity*

$$\mathbf{v}_0 = \frac{e\mathbf{E}_0}{m_e\omega_0}, \tag{7.23}$$

which is characteristic of the perturbed motion of an electron in the pump field. Then (7.20) may be rewritten as

$$\frac{1}{\chi^{L(e)}(k)} + \frac{1}{1 + \chi^{L(i)}(k)} + \frac{|\mathbf{k}|^2}{\omega_0^2}\left[\frac{|(\mathbf{k}+\mathbf{k}_0)\cdot\mathbf{v}_0|^2}{|\mathbf{k}+\mathbf{k}_0|^2}\frac{1}{K^L(k+k_0)}\right.$$

$$\left. + \frac{|(\mathbf{k}-\mathbf{k}_0)\cdot\mathbf{v}_0|^2}{|\mathbf{k}-\mathbf{k}_0|^2}\frac{1}{K^L(k-k_0)}\right] = 0. \tag{7.24}$$

Explicit expressions for the susceptibilities follow from the expression (2.31a) for the longitudinal response of a thermal plasma. For $\sqrt{2}kV_e \gg \omega \gg \sqrt{2}kV_i$ we have [cf. Exercise 7.3]

$$\frac{1}{\chi^{L(e)}(k)} + \frac{1}{1 + \chi^{L(i)}(k)} \approx \frac{\omega^2 - \omega_s^2(\mathbf{k}) + i\omega\gamma_s(\mathbf{k})}{\omega_{pi}^2} \tag{7.25}$$

where the imaginary parts combine into the absorption coefficient $\gamma_s(\mathbf{k})$ for ion sound waves. For a pump in the Langmuir mode, and for $\omega \ll \omega_p$, $\omega_0 \pm \omega$ are close to the plasma frequency, and then we may make the approximation

$$K^L(k \pm k_0) \approx \pm \frac{2\{\omega \pm \delta_\pm + i\gamma_L(\mathbf{k}_0 \pm \mathbf{k})/2\}}{\omega_p} \tag{7.26}$$

with $\omega_0 = \omega_L(\mathbf{k}_0)$ and

$$\delta_\pm = \omega_L(\mathbf{k}_0) - \omega_L(\mathbf{k}_0 \pm \mathbf{k}). \tag{7.27}$$

(Here it is understood that the frequencies $\omega_L(\mathbf{k})$, $\omega_s(\mathbf{k})$, etc. are strictly positive.)

Now let us briefly outline some particular parametric instabilities.

Decay instability

Let the pump be a Langmuir wave, and suppose that ω is close to the ion sound frequency $\omega_s(\mathbf{k})$. We retain only one of the terms $k \pm k_0$ in (7.24), which then reduces to

$$(\omega^2 + i\omega\gamma_s - \omega_s^2)(\omega + i\gamma_L/2 \pm \delta_\pm) \mp \frac{\omega_{pi}^2}{2\omega_p}\frac{|\mathbf{k}|^2}{|\mathbf{k}\pm\mathbf{k}_0|^2}|(\mathbf{k}\pm\mathbf{k}_0)\cdot\mathbf{v}_0|^2 = 0, \tag{7.28}$$

where arguments $\mathbf{k}$ etc. are omitted. Now if ω is close to ω_s, then (7.28) may be further simplified to

$$(\omega + i\gamma_s/2 - \omega_s)(\omega + i\gamma_L/2 \pm \delta_\pm) \mp \omega_\pm^2 = 0, \tag{7.29}$$

with

$$\omega_\pm^2 = \frac{\omega_{pi}^2}{4\omega_p\omega_s}\frac{|\mathbf{k}|^2}{|\mathbf{k}\pm\mathbf{k}_0|^2}|(\mathbf{k}\pm\mathbf{k}_0)\cdot\mathbf{v}_0|^2. \tag{7.30}$$

For $\gamma_s = 0$, $\gamma_L = 0$ one readily confirms that (7.29) has complex solutions only for the lower sign and then only for $4\omega_-^2 > (\omega_s - \delta_-)^2$. Inclusion of $\gamma_s \neq 0$, $\gamma_L \neq 0$ does not affect the qualitative conclusion that growth is possible only for the lower sign. The growing solutions occur when the imaginary part of

$$\omega = \tfrac{1}{2}\{\omega_s - i\gamma_s/2 + \delta_- - i\gamma_L/2 \pm [\{(\omega_s - i\gamma_s/2) - (\delta_- - i\gamma_L/2)\}^2 - 4\omega_-^2]^{1/2}\} \tag{7.31}$$

is positive [cf. Exercise 7.4].

The growing solution of (7.31) describes a reactive version of the three-wave decay process $L \to L' + s$, where L denotes the pump. The coalescence process $L + s \to L'$, with L the pump, leads to a kinetic nonlinear instability in which the waves L' grow, but there is no reactive counterpart of this.

Decay into a reactive quasimode

Now suppose that the growth is so fast that we have $|\omega| \gg \omega_s$. If we also assume $|\omega| \gg \gamma_s/2, \gamma_L/2, |\delta_\pm|$ then (7.28) reduces to a cubic equation

$$\omega^3 \mp 2\omega_s\omega_\pm^2 = 0, \tag{7.32}$$

with $\omega_\pm^2$ given by (7.30). There are complex solutions to (7.32) for either sign.

In this case the low-frequency mode cannot be regarded as an ion sound wave, and is called a quasimode. This instability may be regarded as a reactive version of induced scattering $L \to L'$.

There is a related instability in which the reactive quasimode is replaced by a resistive quasimode. That is the low frequency disturbance is so heavily Landau damped by thermal ions that it cannot be regarded as a propagating mode.

Oscillating two-stream instability

Consider a long-wavelength pump ($k_0 \ll k$) and suppose that the frequency ω is $\ll \sqrt{2}kV_i$ in magnitude. Then (7.24) with (7.26) may be approximated by

$$k^2\lambda_{De}^2\left(1 + \frac{T_i}{T_e}\right) + \frac{(\mathbf{k}\cdot\mathbf{v}_0)^2}{2\omega_p}\left[\frac{1}{\omega + i\gamma_L/2 + \delta_+} - \frac{1}{\omega + i\gamma_L/2 - \delta_-}\right] = 0. \tag{7.33}$$

Algebraically (7.33) is similar in form to the dispersion equation for a two-stream instability, and this is partly the basis for the peculiar name "oscillating two-stream instability". Actually (7.33) is simpler than the dispersion equations for streaming instabilities (cf. §§3.1 & 3.2) in that (7.33) is a quadratic equation for ω. In the limit $k_0 = 0$, (7.27) implies $\delta_+ = \delta_- (= \delta$ say), and the solutions of (7.33) are

$$\omega = -i\gamma_L/2 \pm \left[\frac{(\mathbf{k}\cdot\mathbf{v}_0)^2\delta}{\omega_p k^2\lambda_{De}^2(1 + T_i/T_e)} + \delta^2\right]^{1/2}. \tag{7.34}$$

The conditions for growth are readily deduced from (7.34) [cf. Exercise 7.5].

A notable feature of the oscillating two-stream instability is that it transfers Langmuir waves from small k_0 to larger k, i.e. from lower to higher frequencies. This is opposite to the direction of transfer in induced scattering (§6.4). A similar transfer from lower to higher k occurs in Langmuir collapse (§7.4).

The instability mechanism for the oscillating two-stream instability involves the ponderomotive force (§5.6). Here this force has Fourier components ω and $\mathbf{k} \pm \mathbf{k}_0 \approx \mathbf{k}$, and it gives rise to an electric field pressure which opposes the electron pressure in the usual dispersion relation for the ion sound waves. As a consequence ω^2 becomes less than $k^2 v_s^2$, and beyond a threshold value ω^2 becomes negative leading to a purely imaginary ω. Thus, as implied by (7.34), this instability involves a purely growing low-frequency quasimode.

The foregoing examples suggest that there are reactive versions of all the weak turbulence processes which can lead to wave growth. This seems to be the case. The reactive version for the processes $L \pm s \leftrightarrow T$ is closely analogous to that for $L \pm s \leftrightarrow L'$, and there is also a reactive version of $L + L \leftrightarrow T$ (Kamilov, Khakimov, Stenflo & Tsytovich 1974). The processes $T \pm s \leftrightarrow T$ and $T \pm L \leftrightarrow T$, called Brillouin and Raman scattering respectively, also have weak turbulence an reactive versions.

Note that in a nonlinear instability in weak turbulence theory at least two wave distributions need to be excited to drive a third. This follows from the saturation conditions (6.44) or (6.45) for a three wave process: if only waves in a mode P are excited ($N_P(\mathbf{k}') \gg N_Q(\mathbf{k}''), N_M(\mathbf{k})$) then saturation of the process $P + Q \rightarrow M$ occurs for $N_M(\mathbf{k}) = N_Q(\mathbf{k}'')$, i.e. the third waves ($M$) cannot be excited beyond the level of the second waves (Q). The situation is different for parametric instabilities. A pump at P can excite both waves Q and M. Only the pump needs to be excited for the parametric instability to cause the other waves to grow.

There are further examples of parametric instabilities which have no counterpart in weak turbulence theory. These include filamentation instabilities, Langmuir collapse (§7.4), and various modulational instabilities, e.g. Bardwell & Goldman (1976) and the review articles cited in §7.4.

7.3 The free electron maser: plasma maser

The free electron maser (or laser) and the cyclotron maser or gyrotron (§11.2) are the basis for laboratory devices which produce high power coherent radiation at centimetre to millimetre wavelengths. In a free electron maser a beam of energetic electrons is subjected to a periodic pump field; in practice this is usually a spatially periodic magnetic field called a 'wiggler'. In the rest frame of the electron beam this 'wiggler' field has a frequency $\gamma|\mathbf{k}_0 \cdot \mathbf{v}|$, where γ and $\mathbf{v}$ refer to the electron beam and where $\mathbf{k}_0$ is the wavevector of the 'wiggler'. The electrons subject to such a perturbation radiate at this same frequency in their rest frame, giving radiation at ω, $\mathbf{k}$ satisfying

$$\omega - \mathbf{k} \cdot \mathbf{v} = \pm (\omega_0 - \mathbf{k}_0 \cdot \mathbf{v}) \tag{7.35}$$

in the laboratory frame. In (7.35) we have allowed for the possibility that the 'wiggler' may have a natural frequency ω_0 as well as a spatial structure described by k_0. Either sign is possible in (7.35) in principle, but for given ω_0, k_0 and v only the one implying $\omega > 0$ is allowed.

The free electron maser can operate in three different regimes in the laboratory (e.g. Sprangle & Coffey 1984). There is a 'Compton regime' which is essentially the weak turbulence regime where the nonlinear growth may be treated as induced scattering. In the 'Raman regime' the growth is reactive, involving space-charge bunching, and in the highest-gain regime the instability involves the ponderomotive force. There is an obvious analogy with plasma instabilities, and this analogy also applies to gyrotrons.

Another feature of some laboratory free electron masers is that the electron beam is relativistic. The result (7.35) applies in the relativistic case. For $\gamma \gg 1$ the emission is strongly beamed along the forward direction (k is nearly parallel to v). Then for $\omega \gg \omega_p$, $k \approx \omega/c$ implies that the left hand side of (7.35) is of order ω/γ^2. Thus emission in the ultrarelativistic case (for $k_0 \gg \omega_0/c$) is near the characteristic frequency

$$\omega_c = \gamma^2 k_0 c. \tag{7.36}$$

A process analogous to the free electron maser can occur in a plasma and is called a plasma maser or laser (Lin, Kaw & Dawson 1973). Consider an electron beam passing through a region where ion sound waves are excited. Then radiation (transverse waves) can be produced at ω, k by scattering of the ion sound waves. Assuming $|k \cdot v| \ll \omega$ for the transverse waves, and $|k_s \cdot v| \gg \omega_s$ for the ion sound waves, the radiation is at $\omega \approx |k_s \cdot v|$ according to (7.35). The maximum frequency is $\omega = \omega_p (k_s \lambda_{De})(v/V_e)$. Under quite mild conditions (e.g. for type III electron beams in the solar wind) such as $k_s \lambda_{De} \approx 0.1$ to 0.3 and $v/V_e \approx 10$ to 30 this process is allowed in that it gives rise to emission above the plasma frequency. The levels of ion sound turbulence in the solar wind can be high enough (a) to suppress the bump-in-tail growth of Langmuir waves due to the nonlinear damping $L \pm s \rightarrow L'$, and (b) to allow the plasma maser to operate. Although these facts suggest that the plasma maser should be considered in this context (Melrose 1982, Goldman & DuBois 1982) it is by no means clear that it is important in practice.

Here we discuss two applications of the free electron or plasma maser to astrophysical situations. One is for electron streams in the solar wind, and the other is to radio emission from pulsars.

Scattering of ion sound waves into transverse waves involves both the scattering $s \rightarrow T$ and also double emission in which the electron emits at $\omega_T + \omega_s$, $k_T + k_s$. The probabilities for scattering and double emission are related by a crossing symmetry. Labelling them as $\pm$ respectively, the probabilities follow from (6.35) and (6.36) with the approximations (6.49) and (6.54):

$$w_{\pm}^{Ts}(k, k', p) = \frac{e^4}{4\varepsilon_0^2 m_e^2} \frac{\omega_p^4}{\omega_{pi}^2 V_e^4 |k'|^4} \left(\frac{\omega_s(k')}{\omega_T(k)} \right)^3$$

$$\times \frac{|e \cdot \kappa'|^2}{|K^L(k_T \mp k_s')|^2} 2\pi \delta \{ \omega_T(k) \pm \omega_s(k') - (k \mp k') \cdot v \} \tag{7.37}$$

where **e** denotes the polarization vector of the transverse waves. For $\omega(=\omega_T(\mathbf{k})) \gg \omega_p$ we may approximate $K^L(k_T \mp k'_s)$ by unity. However for ω close to ω_p, $K^L(k_T \mp k'_s)$ has a zero at $k_T \mp k'_s = k''_L$ where $\omega_T \mp \omega_s$, $\mathbf{k} \mp \mathbf{k}'$ satisfies the dispersion relation for Langmuir waves, i.e. one has $\omega_L(\mathbf{k}'') = \omega_T(\mathbf{k}) \mp \omega_s(\mathbf{k}')$, $\mathbf{k}'' = \mathbf{k} \mp \mathbf{k}'$. This leads to a singular contribution in (7.37). One is to regard the processes as scattering or double emission if the frequency mismatch $\omega_T(\mathbf{k}) \pm \omega_s(\mathbf{k}') - \omega_L(\mathbf{k}'')$ is much larger in magnitude than the relevant absorption coefficient; in the neighbourhood of the singularity, where the mismatch is less than the absorption coefficient, these processes are to be reinterpreted as three-wave processes $T \pm s \leftrightarrow L$, and treated accordingly. Clearly the plasma maser favours emission at $\omega \approx \omega_p$ which is where $K^L(k_T - k'_s)$ is much less than unity.

The kinetic equations for the two processes may be combined in the form, cf. (6.44),

$$\frac{dN_T(\mathbf{k})}{dt} = \int d^3\mathbf{p} \int \frac{d^3\mathbf{k}'}{(2\pi)^3} \sum_{\pm} w_{\pm}^{Ts}(\mathbf{k}, \mathbf{k}', \mathbf{p}) \left[\{N_s(\mathbf{k}') \mp N_T(\mathbf{k})\} f(\mathbf{p}) \right.$$
$$\left. + N_T(\mathbf{k}) N_s(\mathbf{k}') \hbar(\mathbf{k} \mp \mathbf{k}') \cdot \frac{\partial f(\mathbf{p})}{\partial \mathbf{p}} \right]. \tag{7.38}$$

Now the δ-function in (7.37) requires that $(\mathbf{k} \mp \mathbf{k}') \cdot \mathbf{v}$ be positive, and hence for $k \ll k'$ one requires $\mathbf{k}' \cdot \mathbf{v} < 0$ for the scattering process and $\mathbf{k}' \cdot \mathbf{v} > 0$ for the double emission process. A bump-in-tail instability can occur for the lower sign in (7.38). Formally the growth is analogous to the bump-in-tail instability for Langmuir waves (§4.1) with the Langmuir wave being replaced by the combination of a transverse wave and an ion sound wave. This is the plasma maser.

The plasma maser is related to the sequential processes of growth of L waves due to the stream followed by the three-wave processes $L \pm s \rightarrow T$. In the plasma maser the counterpart of the Langmuir wave is non-resonant, and is simply the beat between the transverse wave and ion sound wave. This beat disturbance is radiated directly in the instability, and the growth is induced by the presence of the ion sound turbulence. Note however that it is only the double emission which leads to growth; the scattering process $s \rightarrow T$ leads to damping and not to growth of the transverse waves. If the ion sound waves were isotropic the growth and damping due to the two mechanisms would balance. Hence growth is possible only when the ion sound waves are anisotropic favouring the streaming direction.

A potentially serious difficulty with this mechanism is that the net growth rate is the sum of the growth rate due to the instability itself minus the natural damping rates of the ion sound and transverse waves (Goldman & DuBois 1982); that is one is to sum the absorption coefficients and the sum must be negative for instability to result. In practice the damping rate of ion sound waves is so large that it could not be overcome by the growth for a stream. It seems that the ion sound waves would have to be marginally stable, rather than damped, for growth to be possible. Alternatively the reactive counterpart of this instability could grow fast enough to overcome the damping of the ion sound waves.

Now let us turn to the other suggested application of free electron laser emission: to the radio emission from pulsars. Pulsar radio remission is extremely bright (brightness temperature perhaps as high as 10^{30} K) and so involves some maser or other coherent emission mechanisms, cf. the book by Manchester & Taylor (1977). Despite an extensive literature (e.g. Sieber & Wielebinski 1980), the detailed mechanisms involved are not understood. Until recently the most widely accepted mechanisms has been coherent curvature emission, in which electron bunches radiate coherently due to the perpendicular acceleration as they move along a curved magnetic field line. However, detailed theories for this process can be criticized in a variety of ways; suffice it to say that the subject has stagnated over the past few years due in large part to the unsatisfactory features of this mechanism and to the difficulty in identifying definitive tests for it and for alternative mechanism. Once we reject coherent curvature radiation the most plausible mechanism is a form of free electron maser emission. The 'wiggler' is identified as an electric field parallel to the magnetic field oscillating at around the plasma frequency, i.e. a large amplitude Langmuir wave. The argument for this is somewhat speculative: in a favoured model for pulsars (Ruderman & Sutherland 1975) the pulsar magnetosphere is populated by highly relativistic electron-positron pairs formed as a result of a 'vacuum discharge' involving a potential $\approx 10^{12}$ V; it seems unlikely that this discharge can neutralize the electric field to less than potential variations of order ε/e, where $\varepsilon = \gamma m c^2$ is the energy of a typical electron or positron.

Suppose then that we have a highly relativistic (electron-positron) pair plasma with an electric field whose potential $\Phi \approx \gamma m c^2 / e$ is oscillating at about the plasma frequency $\omega_p = (n_e e^2 / \gamma \varepsilon_0 m_e)^{1/2}$. The wavenumber k_0 for these oscillations is likely to be of order ω_p / c, giving a natural frequency $\omega_c \approx \gamma^2 \omega_p$, cf. (7.36). Due to the superstrong magnetic field the electrons and positrons radiate away their perpendicular energy very rapidly, and it is reasonable to assume that all particles have zero perpendicular energy. The pair plasma is then one dimensional, and one may describe the electrons $(+)$ and positrons $(-)$ in terms of one-dimensional distribution function $F_{\pm}(\gamma)$ normalized according to

$$n_{\pm} = \int_1^{\infty} d\gamma F_{\pm}(\gamma)$$

where $n_{\pm}$ are the number densities.

A derivation of the absorption coefficient for transverse waves (averaged over polarization and direction of emission) was given by Melrose (1978) for the idealized case $\mathbf{E}(t) = \mathbf{E}_0 \cos \omega_0 t$ with $\mathbf{E}_0$ directed along the magnetic field. The result is

$$\gamma(\omega) = -\frac{3\pi e^4 E_0^2}{4\varepsilon_0 m_e^3 c^2 \omega_0^3} \sum_{\pm} \int_{\gamma_0}^{\infty} d\gamma \frac{F_{\pm}(\gamma)}{\gamma^7} \left(1 - \frac{2\omega}{3\omega_0 \gamma^2}\right) \tag{7.39}$$

with $\gamma_0 = (2\omega_0/\omega - \omega_p^2/\omega^2)^{-1/2}$. Growth occurs for $\omega \lesssim 3\omega_0 \gamma^2/2$, i.e. only below about the characteristic frequency (7.36). With $eE_0 \approx \gamma m_e c^2 k_0$, $k_0 \approx \omega_0/c$, and $\omega_0 \approx \omega_p$, one finds that $\gamma(\omega)$ can be of order $-\omega/\gamma^6$ where γ is some characteristic

Lorentz factor. This mechanism requires $\gamma \lesssim 10$ for the growth rate to be fast enough to account for the observed emission.

As already stated, these ideas are speculative. However at present the free electron maser mechanism seems the most favourable basis for explaining the radio emission from pulsars.

7.4 The Zakharov equations and strong Langmuir turbulence

The most general useful model for strong Langmuir turbulence currently available is based on the Zakharov equations (Zakharov 1972). These are a pair of coupled equations for the envelope of the Langmuir waves and for the density fluctuations in the electron gas. The derivation of these equations is heuristic rather than rigorous. That is, unlike say (7.19), they are not derived within the framework of a more general theory. They are simply written down as the analytic expression of a physical model.

The model involves two important ideas. First, the dispersive properties of Langmuir waves can be altered by a nonlinear response which causes a local perturbation in the plasma frequency. Let $\delta n_e (\ll n_e)$ be a variation in the electron density, causing a variation $(\delta n_e/2n_e)\omega_p$ in the plasma frequency. The dispersion relation for Langmuir waves is then approximated by

$$\omega = \omega_p + 3k^2 V_e^2/2\omega_p + (\delta n_e/2n_e)\omega_p - i\gamma_L/2, \qquad (7.40)$$

where a contribution γ_L from Landau damping is retained. The term $(\delta n_e/2n_e)\omega_p$ couples the Langmuir waves to any processes which cause localized variations in δn_e.

The electric field of the Langmuir waves is assumed to be of the form

$$\mathbf{E}(t, \mathbf{x}) = [\mathscr{E}(t, \mathbf{x})e^{-i\omega_p t} + \text{c.c.}], \qquad (7.41)$$

where $\mathbf{E}(t, \mathbf{x})$ is the envelope. The variation of the envelope is assumed to be governed by an equation obtained from (7.40) by replacing $\omega - \omega_p$ and $\mathbf{k}$ by the operators $i\partial/\partial t$ and $-i\partial/\partial \mathbf{x}$ respectively. That is, $\mathscr{E}(t, \mathbf{x})$ is assumed to evolve according to

$$[i\partial/\partial t + (3V_e^2/2\omega_p)\nabla^2 + i\gamma_L/2]\mathscr{E}(t, \mathbf{x}) = \frac{\omega_p \delta n_e(t, \mathbf{x})\mathscr{E}(t, \mathbf{x})}{2n_e}. \qquad (7.42)$$

The evolution of $\delta n_e(t, \mathbf{x})$ is governed by a dispersion relation related to that for ion sound waves, but with the effects of the ponderomotive force taken into account. According to (5.63) the ponderomotive force acting on the electrons is

$$\mathbf{F}_P = -\frac{e^2}{m_e \omega_p^2} \text{grad} |\mathscr{E}(t, \mathbf{x})|^2. \qquad (7.43)$$

The electrons cannot move relative to the ions without setting up an ambipolar electric field, and this leads to the electrons dragging along the ions. The force $\mathbf{F}_P$ on the electrons therefore leads to an acceleration $\dot{\mathbf{v}} = \mathbf{F}_P/m_i$ of the entire fluid, where m_i is the effective mass of electron plus ion. The equation of continuity $\partial n/\partial t + \text{div}(n\mathbf{v}) = 0$ then implies

$$\frac{\partial^2}{\partial t^2} \delta n_e(t, \mathbf{x}) = \frac{e^2 n_e}{m_e m_i \omega_p^2} \nabla^2 |\mathscr{E}(t, \mathbf{x})|^2. \tag{7.44}$$

In the absence of the ponderomotive force $\delta n_e(t, x)$ oscillates as for an ion sound wave with dispersion relation, cf. (7.28)

$$\omega^2 + i\omega\gamma_s - |\mathbf{k}|^2 v_s^2 = 0.$$

Replacing ω and $\mathbf{k}$ by $i\partial/\partial t$ and $-i\partial/\partial \mathbf{x}$, respectively, and including (7.44) as the driving term, one finds

$$\left[\frac{\partial^2}{\partial t^2} + \gamma_s \frac{\partial}{\partial t} - v_s^2 \nabla^2 \right] \delta n_e(t, \mathbf{x}) = \frac{\varepsilon_0}{m_i} \nabla^2 |\mathscr{E}(t, \mathbf{x})|^2. \tag{7.45}$$

Equations (7.42) and (7.45) are the Zakharov equations.

The Zakharov equations may be used to derive the results of §7.2. In this case one separates $\mathscr{E}$ into a Langmuir pump field $\mathscr{E}^{(0)}$ and another Langmuir field $\mathscr{E}^{(1)}$ and one linearizes the Zakharov equations neglecting terms $|\mathscr{E}^{(1)}|^2$ and $\delta n_e \mathscr{E}^{(1)}$. The equations may then be reduced to a single equation or to a set of coupled linear equations. An example is given in Exercise 7.6.

The Zakharov equations are the basis for a theory of strong Langmuir turbulence currently under development. An important effect in this theory is called 'Langmuir collapse'. Langmuir collapse is related to the modulational instability, identified by Nishikawa (1968) in an early discussion of parametric instabilities, and to the oscillating two-stream instability which is a particular modulational instability. The common feature is the role of the ponderomotive force. In a modulational instability an initially spatially uniform envelope of Langmuir turbulence breaks up into envelopes with smaller natural lengths. In Langmuir collapse this process continues to a stage where the Langmuir wave energy is confined to a number of localized regions several Debye lengths across called *cavitons*. One describes the Langmuir turbulence as collapsing (in coordinate space) into cavitons and being pumped (in k-space) to larger k. A favourable starting condition for collapse is the Langmuir condensate which forms at small k due to the effects of induced scattering (§6.4).

The mechanism involved in Langmuir collapse is analogous to that involved in self-focussing in nonlinear optics. Specifically, a localized increase in the energy density in Langmuir turbulence leads to a ponderomotive force which drives plasma out of the localized region. This decreases the plasma frequency locally. The dispersion relation $\omega = \omega_p + 3k^2 V_e^2/2\omega_p$ implies that k^2 increases when ω_p decreases along a ray path. Waves refract towards the direction of increasing refractive index, that is in the direction of increasing k^2. Hence Langmuir waves refract into the localized region thereby enhancing the electric energy there and providing a feedback mechanism to drive the instability.

In Langmuir collapse, and also in the oscillating two-stream instability, the

Langmuir turbulence is driven from smaller to larger k by the ponderomotive force. The turbulence can then be Landau damped. (Note that the Landau damping coefficient (2.49) varies as $\exp(-\omega_p^2/2k^2 V_e^2)$ and so becomes important only for ω_p/k less than several times V_e.) This damping accelerates electrons in the tail of the Maxwellian distribution, leading to a nonthermal electron tail which modifies the Landau damping. As a consequence, detailed studies of strong turbulence require that the evolution of the electron distribution and of the Langmuir turbulence be taken into account simultaneously. It also follows that Landau damping is the saturation mechanism for the collapse instability.

As already mentioned, the theory of strong Langmuir turbulence is still under development. Significant progress has been made e.g. the reviews by Porkolab (1976), Rudakov & Tsytovich (1978), Thornhill & ter Haar (1978), Sagdeev (1979) and Goldman (1984), but much remains to be done, especially in connection with statistical theories for an ensemble of cavitons.

Note that 'strong' turbulence really means 'non-weak' turbulence, and 'strong' does not mean that the energy density in the plasma is predominantly in the turbulence. Indeed even 'strong' turbulence theory applies only if the energy density in the Langmuir waves is much less than the thermal energy density in the plasma. The threshold at which strong turbulence effects become important depends on the details of the processes involved. However, estimates made under a wide variety of conditions leads to a condition

$$\frac{W_L}{n_e T_e} \lesssim 10^{-4} \tag{7.46}$$

for weak turbulence theory to apply. Conversely, strong turbulence effects need to be taken into account when the energy density W_L in the Langmuir waves exceeds about 10^{-4} times the thermal energy density.

7.5 Solitons

When the effects of the ponderomotive force due to Langmuir waves are retained in the wave equation for the waves themselves, one obtains a nonlinear wave equation. Certain specific nonlinear wave equations have exact solutions, called solitary waves or *solitons*, which have remarkable properties. Solitons propagate without changing their profile. This may be attributed to the effects of nonlinearity and of dispersion balancing exactly in the solution. Two solitons can pass through each other also without changing their profiles.

The nonlinear equation for Langmuir waves may be derived from the Zakharov equations. For this purpose we neglect the wave damping, setting $\gamma_L = 0$ in (7.42) and $\gamma_s = 0$ in (7.45). The second time-derivative in (7.45) may be attributed to the inertia of the ions, and this term is also neglected. From another viewpoint, this approximation corresponds to assuming $\omega^2 \ll k^2 v_s^2$ for the density perturbations; the variations in the density are then "subsonic" which means they are adiabatic.

Then integrating (7.45) twice, and using $m_i v_s^2 = m_e V_e^2 = T_e$ we find

$$\delta n_e = -\frac{\varepsilon_0 |\mathscr{E}|^2}{T_e} \tag{7.47}$$

where the constants of integration are set equal to zero. Substituting (7.47) into (7.42), with $\gamma_L = 0$, gives

$$\left[i\partial/\partial t + (3V_e^2/2\omega_p)\nabla^2 + \omega_p \frac{\varepsilon_0 |\mathscr{E}|^2/2}{n_e T_e} \right] \mathscr{E} = 0. \tag{7.48a}$$

This is called the *nonlinear Schrödinger equation*.

The nonlinear Schrodinger equation has soliton solutions in one dimension. The soliton solution of

$$[i\partial/\partial t + (3V_e^2/2\omega_p)\partial^2/\partial x^2 + \omega_p \varepsilon_0 |\mathscr{E}|^2/2n_e T_e]\mathscr{E} = 0 \tag{7.48b}$$

is [Exercise 7.7]

$$\mathscr{E}(t, x) = 4\left(\frac{\Omega n_e T_e}{\varepsilon_0 \omega_p}\right)^{1/2} \exp\left[i\Omega t\right] \mathrm{sech}\left[\left(\frac{2\Omega \omega_p}{3 V_e^2}\right)^{1/2} x\right]. \tag{7.49}$$

The frequency Ω is a free parameter.

An early theory for strong Langmuir turbulence (Kingsep, Rudakov & Sudan 1973) was based on solitons. Other theories are now favoured. The role, if any, of Langmuir solitons in turbulence theory is uncertain.

There are also soliton solutions for ion sound waves. In these waves the electrons drag along the ions so that the plasma moves with a fluid velocity v. The fluid equation for the ions (assumed singly charged) is

$$\frac{\partial v}{\partial t} + v\frac{\partial v}{\partial x} = \frac{eE}{m_i}, \tag{7.50}$$

where only one-dimensional motions are considered. The term $v\partial v/\partial x$ is the non-linear term, and we may use the properties for a linear ion sound wave to relate E to v on the right hand side of (7.50). The relevant relation is

$$E = -\frac{im_i}{en_e} \frac{kv_s}{(1 + k^2 \lambda_{De}^2)^{1/2}} v. \tag{7.51}$$

Now expanding $(1 + k^2 \lambda_{De}^2)^{-1/2} \approx 1 - k^2 \lambda_{De}^2/2$ and then interpreting ik as $\partial/\partial x$, (7.51) and (7.50) give

$$\frac{\partial v}{\partial t} + (v + v_s)\frac{\partial v}{\partial x} + \frac{\lambda_{De}^2 v_s}{2}\frac{\partial^3 v}{\partial x^3} = 0, \tag{7.52}$$

which is the (modified) *Korteweg–de Vries equation*.

The soliton solution of the Korteweg-de Vries equation is well known. It involves a free parameter v_0 which plays the role of a phase velocity. The solution is

$$v = 3(v_0 - v_s)\mathrm{sech}^2\left[\left(\frac{v_0 - v_s}{2v_s}\right)^{1/2}\frac{x - v_0 t}{\lambda_{De}}\right], \tag{7.53}$$

as may be confirmed by substitution into (7.52).

The soliton solution for ion sound waves is the basis for a model of a *collisionless shock wave* in an unmagnetized plasma. The idea is to start from a soliton and to introduce some dissipation. Such shocks are referred to as *laminar shocks*; collisionless shocks which are not based on a soliton solution are called *turbulent shocks* (e.g. Tidman & Krall 1971).

Exercise set 7

7.1 Consider the contribution to the imaginary part of $\alpha_{ij}^{NL(2)}(k)$, as given by (7.10), from the pole in $1/\Lambda(k - k')$ at $\omega - \omega' = \omega''$ with $\mathbf{k}'' = \mathbf{k} - \mathbf{k}'$. Evaluate the singular contribution by writing

$$\Lambda(k - k') = (\omega - \omega' - \omega'' + i0)\left[\frac{\partial \Lambda(k - k')}{\partial(\omega - \omega')}\right]_{\omega - \omega' = \omega''}$$

with $\omega'' = \omega_Q(\mathbf{k}'')$, and using the Plemelj formula (1.32).

Show that the nonlinear absorption coefficient (7.4) arising from this contribution reproduces (7.1) when only the contribution from the $N_Q(\mathbf{k}'')$ in (7.2) is retained, and with the probability $u_{MPQ}(\mathbf{k}, \mathbf{k}', \mathbf{k}'')$ given by (6.25).

7.2 Show that (7.19) is modified for transverse waves as follows:
 (i) $K^L(k)$ is replaced by $K^T(k) - |\mathbf{k}|^2 c^2/\omega^2$,
 (ii) $|\mathbf{\kappa} \cdot \mathbf{\kappa}'|^2$ is replaced by $|\mathbf{\kappa} \times \mathbf{\kappa}'|^2$.

7.3 Derive (7.25) as follows. Assume

$$\chi^{L(e)}(k) = \frac{1}{k^2 \lambda_{De}^2} + i \operatorname{Im}[K^{L(e)}(k)],$$

$$\chi^{L(i)}(k) = -\frac{\omega_{pi}^2}{\omega^2} + i \operatorname{Im}[K^{L(i)}(k)],$$

supposing the imaginary parts to be small. Then use the dispersion relation

$$\omega = \frac{kv_s}{(1 + k^2 \lambda_{De}^2)^{1/2}}, \qquad v_s = \omega_{pi} \lambda_{De}$$

for ion sound waves, and the expression (2.51) for the absorption coefficient γ_s.

7.4 (a) Show that (7.31) implies that for $\omega_s = \delta_-$ the threshold condition for growth of the decay instability is

$$\omega_-^2 > \gamma_s \gamma_L/4.$$

 (b) Estimate the growth rate just above this threshold.

7.5 Using (7.34) derive conditions for the oscillating two-stream instability to grow. Specifically show that $(v_0/V_e)^2 > (3/2)k^2 \lambda_{De}^2$ is required.

7.6 Starting from the Zakharov equations, specifically (7.42), its complex conjugate and (7.45), assume that $\mathscr{E}$ is the sum of a pump field $\mathscr{E}^{(0)}$ with $k_0 = 0$ and another field $\mathscr{E}^{(1)}$ with Fourier components ω and $\mathbf{k}$, and linearize the equations. Hence deduce the dispersion equation

$$\omega^2 + i\omega\gamma_s - \omega_s^2 - \frac{\omega_{pi}^2}{2\omega_p}|\mathbf{k}\cdot\mathbf{v}_0|^2\left[\frac{1}{\omega + i\gamma_L/2 + \delta_+} - \frac{1}{\omega + i\gamma_L/2 - \delta_-}\right] = 0.$$

7.7 Solve the nonlinear Schrödinger equation (7.48b) by writing

$$\mathscr{E} = e^{i\Omega t}f(x),$$

deriving the equation

$$\frac{d^2 f}{dx^2} - \frac{2\omega_p\Omega}{3V_e^2}f + \frac{\omega_p^2\varepsilon_0}{3V_e^2 n_e T_e}f^3 = 0$$

and integrating it twice to find $f(x)$.

Hint: Multiply by df/dx.

PART III
Collision-dominated magnetized plasmas

8
Magnetohydrodynamics

8.1 The MHD equations

At frequencies below relevant collision frequencies plasmas act like fluids. Almost all plasmas have magnetic fields and are highly electrically conducting. The magnetic field is then frozen into the fluid and moves with the fluid. A distinction is made between high-β and low-β plasmas, where

$$\beta = \frac{P}{B^2/2\mu_0} \tag{8.1}$$

is the ratio of the plasma pressure P to the magnetic pressure $B^2/2\mu_0$. In high-β plasmas the magnetic field is drawn along with the plasma and in low-β plasmas the plasma is drawn along by the magnetic field. Put another way, for $\beta \gg 1$ the stresses in the plasma are predominantly gas-like and are transferred by sound waves, and for $\beta \ll 1$ the stresses are predominantly magnetic and are transferred by Alfvén waves. To within a factor of order unity β is the ratio of the square of the sound speed c_s to the square of the Alfvén speed v_A, with

$$v_A = \frac{B}{(\mu_0\eta)^{1/2}}, \tag{8.2}$$

where η is the mass density.

The theory which describes the motion of a collision-dominated magnetized plasma is called magnetohydrodynamics, or MHD theory. The MHD equations consist of three of Maxwell's equations

$$\mathrm{curl}\,\mathbf{E} = -\frac{\partial}{\partial t}\mathbf{B} \tag{8.3}$$

$$\mathrm{curl}\,\mathbf{B} = \mu_0\mathbf{J} + \frac{1}{c^2}\frac{\partial \mathbf{E}}{\partial t} \tag{8.4}$$

$$\mathrm{div}\,\mathbf{B} = 0, \tag{8.5}$$

plus equations which describe the plasma as a fluid. It is conventional to omit the displacement current $(1/c^2)\,\partial\mathbf{E}/\partial t$ in (8.4), but it is important to retain it for low density plasmas with $v_A \gtrsim c$. We retain it here for the purpose of deriving the wave properties (§8.2). The fluid equations are the equations of (mass) continuity

$$\frac{\partial \eta}{\partial t} + \text{div}\,(\eta \mathbf{v}) = 0 \tag{8.6}$$

where $\mathbf{v}$ is the fluid velocity, the equation of fluid motion

$$\eta \frac{d\mathbf{v}}{dt} = -\,\text{grad}\,P + \mathbf{J} \times \mathbf{B} + \eta \mathbf{f} \tag{8.7}$$

where $\mathbf{f}$ is an external force per unit mass, Ohm's law, which in its simplest form is

$$\mathbf{E} + \mathbf{v} \times \mathbf{B} = \mathbf{J}/\sigma, \tag{8.8}$$

where σ is the electrical conductivity, and an equation of state. Three different equations of state may be relevant in different contexts. For most purposes the adiabatic equation of state applies:

$$\frac{d}{dt}(P\eta^{-\Gamma}) = 0 \tag{8.9a}$$

where Γ is the adiabatic index. Another equation of state (when the thermal conductivity is large) is the isothermal equation

$$\frac{d}{dt}\,T = 0, \tag{8.9b}$$

where T is the plasma temperature. Finally for an incompressible fluid the equation of state is

$$\text{div}\,\mathbf{v} = 0. \tag{8.9c}$$

The MHD equations with infinite conductivity, i.e. with $\mathbf{J}/\sigma$ in (8.8) replaced by zero, imply an equation which may be interpreted as expressing conservation of energy. For simplicity let us neglect the external force (it may be included simply if it is a potential force). Then using

$$\frac{d}{dt} = \frac{\partial}{\partial t} + \mathbf{v} \cdot \text{grad}, \tag{8.10}$$

the scalar product of (8.8) with $\mathbf{v}$ may be reduced as follows:

$$\eta \mathbf{v} \cdot \frac{d\mathbf{v}}{dt} = \frac{\partial}{\partial t}(\tfrac{1}{2}\eta v^2) + \text{div}\,(\tfrac{1}{2}\eta v^2 \mathbf{v})$$

where (8.6) is used,

$$\mathbf{v} \cdot \text{grad}\,P = \frac{\partial}{\partial t}\left(\frac{P}{\Gamma - 1}\right) + \text{div}\left(\frac{\Gamma}{\Gamma - 1}\,P\mathbf{v}\right)$$

where (8.6), (8.9a) and (8.10) are used, and

$$\mathbf{v} \cdot \mathbf{J} \times \mathbf{B} = -\frac{\partial}{\partial t}\left[\frac{\varepsilon_0 E^2}{2} + \frac{B^2}{2\mu_0}\right] - \text{div}\left[\frac{\mathbf{E} \times \mathbf{B}}{\mu_0}\right],$$

where (8.3), (8.4) and (8.8) with $\sigma = \infty$ are used. Combining these, the equation for energy continuity becomes

$$\frac{\partial}{\partial t}\left[\tfrac{1}{2}\eta v^2 + \frac{P}{\Gamma - 1} + \frac{\varepsilon_0 E^2}{2} + \frac{B^2}{2\mu_0}\right] + \mathrm{div}\left[\tfrac{1}{2}\eta v^2 \mathbf{v} + \frac{\Gamma}{\Gamma - 1}P\mathbf{v} + \frac{\mathbf{E} \times \mathbf{B}}{\mu_0}\right] = 0.$$

(8.11)

The four contributions to the energy density are interpreted as the kinetic energy density, the thermal energy density, the electric energy density and the magnetic energy density respectively. The three contributions to the energy flux are the kinetic energy flux, the thermal energy flux and the Poynting vector respectively.

Another equation of continuity is that for the ith component momentum:

$$\frac{\partial}{\partial t}\left[\eta v_i + \frac{1}{\mu_0 c^2}(\mathbf{E} \times \mathbf{B})_i\right] + \frac{\partial}{\partial x_j}\left[\eta v_i v_j + P\delta_{ij}\right.$$
$$\left. + \left(\frac{\varepsilon_0 E^2}{2} + \frac{B^2}{2\mu_0}\right)\delta_{ij} - \varepsilon_0 E_i E_j - \frac{B_i B_j}{\mu_0}\right] = 0.$$

(8.12)

Here $\eta\mathbf{v}$ and $(\mathbf{E} \times \mathbf{B})/\mu_0 c^2$ are the fluid and electromagnetic momentum densities, the terms $\eta v_i v_j$, $P\delta_{ij}$ and the remaining three terms are the stress tensors associated with the fluid motions, the thermal motions and the electromagnetic field respectively.

8.2 Small amplitude MHD waves

We now seek solutions of the MHD equations for small amplitude waves. We set the conductivity equal to infinity in (8.8) and ignore the external force in (8.7). It is convenient to introduce the displacement $\xi(t, \mathbf{x})$ of a fluid element from its equilibrium position. The fluid displacement is related to the fluid velocity by

$$\frac{d\xi}{dt} = \mathbf{v}.$$

(8.13)

In the absence of any wave, we assume that the fluid is at rest and that there is no electric field. Then (8.7) requires

$$-\mathrm{grad}\, P_0 + \mathbf{J}_0 \times \mathbf{B}_0 = 0,$$

(8.14)

where subscripts 0 describe the unperturbed state. Let η_1, P_1, $\mathbf{J}_1$, $\mathbf{B}_1$, $\mathbf{E}$ and ξ be perturbations associated with the wave motion. Using the MHD equations and (8.13) these are related by

$$\eta_1 = -\eta_0 \,\mathrm{div}\,\xi, \quad P_1 = -\Gamma P_0 \,\mathrm{div}\,\xi + \xi\cdot\mathrm{grad}\,P_0, \qquad \text{(8.15a, b)}$$

$$\mathbf{J}_1 = \frac{1}{\mu_0}\,\mathrm{curl}\,\mathbf{B}_1 + \frac{1}{\mu_0 c^2}\frac{d^2\xi}{dt^2} \times \mathbf{B}_0, \qquad \text{(8.15c)}$$

$$\mathbf{B}_1 = \mathrm{curl}\,(\xi \times \mathbf{B}_0), \quad \mathbf{E} = -\frac{d\xi}{dt} \times \mathbf{B}_0. \qquad \text{(8.15d, e)}$$

The linearized equation of fluid motion (8.7) is

$$\eta_0 \frac{d^2\xi}{dt^2} = - \operatorname{grad} P_1 + \mathbf{J}_1 \times \mathbf{B}_0 + \mathbf{J}_0 \times \mathbf{B}_1. \tag{8.16}$$

We assume that the wavelength of the waves is much shorter than the characteristic lengths over which the unperturbed quantities change and hence neglect grad P_0 and $\mathbf{J}_0$. Then (8.16) with (8.15) reduces to

$$\frac{d^2\xi}{dt^2} = c_s^2 \operatorname{grad} \operatorname{div} \xi + v_A^2 \{\operatorname{curl} \operatorname{curl} (\xi \times \mathbf{b})\} \times \mathbf{b} + \frac{v_A^2}{c^2} \left(\frac{d^2\xi}{dt^2} \times \mathbf{b} \right) \times \mathbf{b} \tag{8.17}$$

where $\mathbf{b}$ is a unit vector along $\mathbf{B}_0$ and where

$$c_s = \left(\frac{\Gamma P_0}{\eta_0} \right)^{1/2}, \quad v_A = \frac{B_0}{(\mu_0 \eta_0)^{1/2}} \tag{8.18a, b}$$

are the adiabatic sound speed and the Alfvén speed respectively.

Let us Fourier transform (8.17) in space and time. Then we obtain a set of three equations for the three vector components of $\xi(\omega, \mathbf{k})$. This may be written in the form

$$\Gamma_{ij}(\omega, \mathbf{k}) \xi_j(\omega, \mathbf{k}) = 0 \tag{8.19}$$

with

$$\Gamma_{ij}(\omega, \mathbf{k}) = \omega^2 \left[\delta_{ij} + \frac{v_A^2}{c^2} (\delta_{ij} - b_i b_j) \right]$$
$$- k^2 c_s^2 \kappa_i \kappa_j - k^2 v_A^2 [\kappa_i \kappa_j - \boldsymbol{\kappa} \cdot \mathbf{b} (\kappa_i b_j + \kappa_j b_i) + (\boldsymbol{\kappa} \cdot \mathbf{b})^2 \delta_{ij}], \tag{8.20}$$

where $\boldsymbol{\kappa}$ is a unit vector along $\mathbf{k}$.

The condition for a solution to (8.19) to exist is that the determinant of the 3×3 matrix Γ_{ij} vanish. This condition leads to the dispersion equation (cf. Exercise 8.2)

$$\left[\omega^2 \left(1 + \frac{v_A^2}{c^2} \right) - k^2 v_A^2 \cos^2 \theta \right] \left[\omega^4 \left(1 + \frac{v_A^2}{c^2} \right) \right.$$
$$\left. - \omega^2 k^2 (v_A^2 + c_s^2) + k^4 v_A^2 c_s^2 \cos^2 \theta \right] = 0 \tag{8.21}$$

where θ is the angle between $\boldsymbol{\kappa}$ and $\mathbf{b}$, i.e. with $\boldsymbol{\kappa} \cdot \mathbf{b} = \cos \theta$. There are three solutions to (8.21) corresponding to three MHD wave modes. We refer to these as the Alfvén (A) mode, the fast (F) mode and the slow (S) mode. The dispersion relation $\omega = \omega_A(\mathbf{k})$ for the Alfvén mode arises from setting the first factor in (8.21) equal to zero, implying

$$\omega_A(\mathbf{k}) = \frac{k v_A |\cos \theta|}{(1 + v_A^2/c^2)^{1/2}}. \tag{8.22}$$

The other two modes have

$$\omega_F(\mathbf{k}) = k v_+(\theta), \quad \omega_S(\mathbf{k}) = k v_-(\theta), \tag{8.23a, b}$$

with $v_{\pm}^2(\theta)$ found by setting the second factor in (8.21) equal to zero and solving for $v_\phi^2 = \omega^2/k^2$:

$$v_{\pm}^2(\theta) = \frac{1}{2(1 + v_A^2/c^2)} \{v_A^2 + c_s^2 \pm [(v_A^2 + c_s^2)^2 - 4v_A^2 c_s^2 \cos^2 \theta (1 + v_A^2/c^2)]^{1/2}\}$$

$$(8.24)$$

The wave equation may be solved for the direction of the fluid displacement $\boldsymbol{\xi}_M$ in each mode M. We introduce the following unit vectors and coordinate system:

$$\mathbf{b} = \mathbf{B}_0/B_0 = (0, 0, 1) \tag{8.25a}$$

$$\boldsymbol{\kappa} = \mathbf{k}/k = (\sin \theta, 0, \cos \theta), \tag{8.25b}$$

$$\mathbf{a} = -\mathbf{k} \times \mathbf{B}_0/|\mathbf{k} \times \mathbf{B}_0| = (0, 1, 0) \tag{8.25c}$$

$$\mathbf{t} = \mathbf{a} \times \boldsymbol{\kappa} = (\cos \theta, 0, -\sin \theta) \tag{8.25d}$$

The component of $\boldsymbol{\xi}_M$ may be found from the matrix of cofactors of $\Gamma_{ij}(\omega, \mathbf{k})$: one substitutes the relevant dispersion relation in any column of this matrix and normalizes to unity. One finds (cf. Exercise 8.1)

$$\boldsymbol{\xi}_A = \mathbf{a} = (0, 1, 0) \tag{8.26a}$$

$$\boldsymbol{\xi}_F = \boldsymbol{\xi}_+, \quad \boldsymbol{\xi}_S = \boldsymbol{\xi}_- \tag{8.26b, c}$$

with

$$\boldsymbol{\xi}_{\pm} = \boldsymbol{\kappa} \cos(\Psi_{\pm} - \theta) + \mathbf{t} \sin(\Psi_{\pm} - \theta) = (\sin \Psi_{\pm}, \ 0, \ \cos \Psi_{\pm}), \tag{8.27}$$

$$\tan \Psi_{\pm} = \frac{v_{\pm}^2(\theta) - c_s^2 \cos^2 \theta}{c_s^2 \sin \theta \cos \theta} = \frac{c_s^2 \sin \theta \cos \theta}{v_{\pm}^2(\theta)(1 + v_A^2/c^2) - c_s^2 \sin^2 \theta - v_A^2}$$

The direction or form of the other perturbations in the waves follow by inserting (8.26) in (8.15).

Alfvén waves have $\mathbf{B}_1$ along $\mathbf{a}$, $\mathbf{E}$ along $(1, 0, 0)$, with $\eta_1 = 0$ and $P_1 = 0$. Thus they involve no compression of the plasma. To first order in the amplitude we have $B^2 = (\mathbf{B}_0 + \mathbf{B}_1)^2 = B_0^2 + 2\mathbf{B}_1 \cdot \mathbf{B}_0$, and $\mathbf{B}_1 \cdot \mathbf{B}_0 = 0$ for Alfvén waves implies that there is also no compression of the magnetic field. Thus Alfvén waves are non-compressional shear waves in the magnetic field.

For the fast and slow modes $\mathbf{B}_1$ is along $\mathbf{t}$ and $\mathbf{E}$ is along $\mathbf{a}$. There are non-zero contributions η_1, P_1 and $\mathbf{B}_1 \cdot \mathbf{B}_0$ implying that these waves involve compression of both the plasma and the magnetic field. For $c_s^2 \gg v_A^2$ the fast mode is essentially a sound wave in the gas, with minor modification due to the magnetic field, and for $v_A^2 \gg c_s^2$ the fast mode is a magneto-acoustic mode slightly modified by the gas. In these two cases the slow mode is a modified magneto-acoustic mode and a modified sound wave in the gas respectively; the important modification is that the wave properties become strongly anisotropic, with $\omega_S(k) \approx \min[v_A, c_s]k|\cos \theta|$

The energy in small amplitude MHD waves may be regarded as composed of four contributions, corresponding to the four contributions to the energy density in the MHD equation for continuity of energy, cf. (8.11). These are the kinetic energy (K), the thermal energy (T), the electric energy (E) and the magnetic energy (M). The thermal energy is given by

$$W_T = \frac{1}{2} \frac{P_1 \eta_1}{\eta_0^2} \tag{8.28}$$

with $P_1/P_0 = \Gamma\eta_1/\eta_0$ and $\eta_1/\eta_0 = -\operatorname{div}\xi$. (Note that the thermal energy in (8.11) is derived under the assumption that the volume is fixed.) Then using (8.15) one finds

$$W_K : W_T : W_E : W_M = \tfrac{1}{2}\eta_0 |\mathbf{v}|^2 : \frac{1}{2}\frac{P_1\eta_1}{\eta_0^2} : \frac{\varepsilon_0}{2}|\mathbf{E}|^2 : \frac{|\mathbf{B}_1|^2}{2\mu_0}$$

$$= |\xi|^2 : \frac{c_s^2}{v_\phi^2}|\boldsymbol{\kappa}\cdot\xi|^2 : \frac{v_A^2}{c^2}|\xi\times\mathbf{b}|^2 : \frac{v_A^2}{v_\phi^2}|\boldsymbol{\kappa}\times(\xi\times\mathbf{b})|^2. \qquad (8.29)$$

An equipartition relation follows from the wave equation in the form $\xi_i^*\Gamma_{ij}\xi_j = 0$:

$$W_E + W_K = W_M + W_T. \qquad (8.30)$$

For the Alfvén mode W_T is zero, and for the other two modes all four contributions are non-zero in general.

8.3 Surface waves

So far we have been concerned with waves in a homogeneous system. Suppose our plasma has a gradient in the x-direction. Then pressure balance requires

$$\frac{\partial}{\partial x}\left(P_0 + \frac{B_0^2}{2\mu_0}\right) = 0 \qquad (8.31)$$

A zeroth order current

$$\mathbf{J}_0 = \left(0, \frac{1}{\mu_0}\frac{\partial B_0}{\partial x}, 0\right) \qquad (8.32)$$

must flow. We now seek wave solutions in such a system.

The wave equation follows from (8.16), with P_1, $\mathbf{J}_1$ and $\mathbf{B}_1$ given in terms of ξ by (8.15). We may Fourier transform in time and in space over y and z, but ξ must remain a function of x. These Fourier transforms are effectively equivalent to seeking a solution of the form

$$\xi(t, \mathbf{x}) = \xi(x)\exp[-i(\omega t - k_y y - k_z z)]. \qquad (8.33)$$

On substituting (8.33) into (8.15) and (8.16) one may eliminate all variables other than ξ_x. After some lengthy algebra one obtains the following differential equation for ξ_x when the displacement current is neglected (e.g. Ionson 1978, Wentzel 1979):

$$\frac{1}{\eta_0}\frac{d}{dx}\left\{\frac{\eta_0(\omega^2 - k_z^2 v_A^2)\{(v_A^2 + c_s^2)\omega^2 - k_z^2 v_A^2 c_s^2\}}{\omega^4 - \omega^2(k_y^2 + k_z^2)(v_A^2 + c_s^2) + k_z^2(k_y^2 + k_z^2)v_A^2 c_s^2}\frac{d\xi_x}{dx}\right\} = -(\omega^2 - k_z^2 v_A^2)\xi_x$$

$$(8.34)$$

where η_0, v_A, c_s and ξ_x are functions of x.

In the limit of a homogeneous plasma, when η_0, v_A and c_s do not depend on x, we may assume $\xi_x(x) \propto \exp[ik_x x]$. Then (8.34) reproduces the dispersion equation (8.21) (without the terms v_A^2/c^2) as required.

We may rewrite (8.34) in the form

$$\frac{d}{dx}\left[\frac{\varepsilon(x)}{q^2(x)}\frac{d\xi_x}{dx}\right] = \varepsilon(x)\xi_x \qquad (8.35)$$

with

$$\varepsilon(x) = \eta_0(x)\{\omega^2 - k_z^2 v_A^2(x)\} \tag{8.36}$$

and

$$q^2(x) = k_y^2 - \frac{\{\omega^2 - k_z^2 v_A^2(x)\}\{\omega^2 - k_z^2 c_s^2(x)\}}{\omega^2\{v_A^2(x) + c_s^2(x)\} - k_z^2 v_A^2(x)c_s^2(x)} \tag{8.37}$$

where the dependences on x are shown explicitly.

Two important implications of (8.35) are the existence of surface waves at a sharp discontinuity, and the existence of a dissipation mechanism for magnetoacoustic waves. Let us consider these briefly.

Suppose there is a sharp boundary between two homogeneous media labelled 1 and 2, and let subscripts 1 and 2 refer to the values of ε, q, v_A, c_s etc., in the two media. The characteristic feature of a *surface wave* is that it is localized to the boundary layer, with the amplitude decaying exponentially with x away from the surface. It is apparent from (8.35) that decaying solutions must vary as $\exp[-|qx|]$, with $q = q_1$ and $q = q_2$ on sides 1 and 2 of the surface respectively. Wentzel (1979) discussed the boundary conditions; he concluded from the fact that ξ_x and P_1 must be continuous and dP_1/dx must change sign at the surface, that ε must change sign and $|\varepsilon/q|$ must be continuous. Then, writing $r = \varepsilon_1/\varepsilon_2$, it follows that the solution for r of

$$\frac{q_1^2}{\varepsilon_1^2} = \frac{q_2^2}{\varepsilon_2^2} \tag{8.38}$$

must be negative. The resulting dispersion relation for the surface waves is

$$\frac{\omega^2}{k_z^2} = \frac{B_1^2 - rB_2^2}{\mu_0(\eta_1 - r\eta_2)} \tag{8.39}$$

with $r = \varepsilon_1/\varepsilon_2 < 0$.

A detailed discussion of the properties of surface waves is not appropriate here; the interested reader is referred to the literature, e.q. as cited by Roberts (1981). Surface waves can be subject to instabilities. Indeed two of the most familiar fluid instabilities, the Rayleigh-Taylor instability and the Kelvin-Helmholtz instability both involve surface waves. We consider surface waves further when discussing the latter instability in §9.2.

Now suppose that the boundary is not sharp but involves a gradient in x over a finite boundary layer between two semi-infinite media (Figure 8.1). Surface waves can still exist, provided the thickness of the layer is much less than $|q_1|^{-1}$ or $|q_2|^{-1}$. Another important effect was pointed out by Chen & Hasegawa (1974) and was interpreted in accord with the following discussion by Ott, Wersinger & Bonoli (1978). This effect is a type of collisionless damping, whose existence had earlier been identified by Sedlácek (1971).

Consider a magnetoacoustic wave propagating towards the boundary layer from the side with the smaller value of v_A^2 (we assume $v_A^2 \gg c_s^2$ here). Before the wave reaches the boundary layer it has $k^2 = k_x^2 + k_y^2 + k_z^2 = \omega^2/v_+^2(\theta)$, and in the boundary layer k_y and k_z remain constant. The value of $q^2(x)$, which is equal

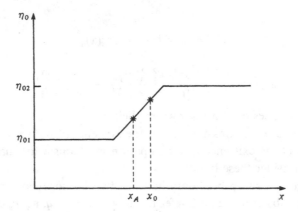

Fig. 8.1 A boundary between two homogeneous semi-infinite media. The density in region 2 (η_{02}) is assumed to be greater than the density in region 1 (η_{01}) implying $v_{A1} < v_{A2}$. The points $x = x_A$ ($\varepsilon(x_A) = 0$, cf. (8.36)) and the magnetoacoustic resonance point $x = x_0$ ($q^2(x_0) = k_y^2$) are indicated schematically.

to $-k_x^2$ outside the boundary layer, decreases in magnitude and approaches zero as x approaches the magnetoacoustic reflection point. The physical effect of interest here occurs near the resonance point $x = x_0$ where $q^2(x)$ equals k_y^2. The energy loss is localized around $x = x_0$.

In the immediate vicinity of $x = x_0$ we may write $q^2(x) \approx [q^2(x_0)]'(x - x_0)$, where $[q^2(x_0)]'$ denotes $d\{q^2(x)\}/dx$ evaluated at $x = x_0$. Then for x sufficiently close to x_0, (8.35) may be approximated by

$$\frac{d^2\xi_x}{dx^2} - \frac{1}{x - x_0}\frac{d\xi_x}{dx} - \xi_x = 0. \tag{8.40}$$

Sufficiently near $x = x_0$ the solution of (8.40) is dominated by a logarithmic singularity $\xi_x \propto \ln(x - x_0)$. As in the case of the resonant singularity at $\omega - \mathbf{k}\cdot\mathbf{v} = 0$ in kinetic theory, we must decide how to integrate around the logarithmic singularity at $x = x_0$. The logarithmic function changes by $i\pi$ as one circles the singularity, and this singular contribution leads to a non-zero value for the rate of energy dissipation (Chen & Hasegawa 1974).

As in the case of Landau damping, the calculation of this damping provides little insight into the mechanism causing the damping. The damping is due to mode coupling in which some of the energy in the magnetoacoustic mode is converted into energy in another mode (the kinetic Alfvén mode) at $x \approx x_0$. The fate of these other waves is not relevant to the calculation of the damping: the mode coupling itself causes the damping of the magnetoacoustic waves. In practice the waves produced can propagate further into the plasma and be Landau damped there. This is a possible mechanism for heating some laboratory plasmas (e.q. Ott et al., 1978) and for the solar corona (Kuperus, Ionson & Spicer 1981).

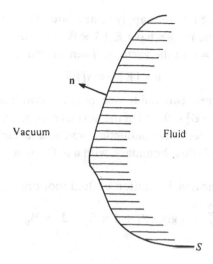

Fig. 8.2 The boundary S separates a fluid (hatched region) from a vacuum; the outward normal to S is denoted by the unit vector **n**.

8.4 The MHD energy principle

A standard method for determining the stability of a macroscopic system is to find the change in the energy when small variations are made to the system. If the change in the energy is positive then one needs to do work to make the change, and the system is stable. However, if the change in energy is negative then free energy is available to drive the system in the direction of this change, and the system is unstable. This leads to the concept of an energy principle in which one examines the sign of the change δW of the energy of the system subject to an arbitrary perturbation.

We consider a magnetized fluid confined to a volume V with a surface S. Confinement is due to the presence of an external magnetic field in the volume $\bar{V}$ outside the fluid. There are three contributions to δW: one, δW_F say, is from the body of the fluid, another, δW_S say, is associated with changes in the surface, and the third, δW_V say, is associated with the external magnetic field.

Consider the boundary between the fluid and the vacuum, with **n** the unit outward normal (Figure 8.2). Let [A] denote the jump in value in any physical quantity A across the boundary. The normal component of **B** must be continuous, requiring

$$\mathbf{n} \cdot [\mathbf{B}] = 0. \tag{8.41}$$

Pressure balance requires

$$[P + B^2/2\mu_0] = 0. \tag{8.42}$$

If there is a surface current $\mathbf{J}_S$, then curl $\mathbf{B} = \mu_0 \mathbf{J}$ implies

$$\mathbf{n} \times [\mathbf{B}] = \mu_0 \mathbf{J}_S. \tag{8.43}$$

To avoid an infinite force density tangential to the surface from $\mathbf{J}_S \times \mathbf{B}$ we require

$\mathbf{n} \cdot \mathbf{B} = 0$ for $\mathbf{J}_S \neq 0$. These relations apply to any boundary. For the particular case of a fluid-vacuum boundary we have $\mathbf{E} + \mathbf{v} \times \mathbf{B} = 0$ with $\mathbf{E} \neq 0$ and $\mathbf{v} \neq 0$ in the fluid and with $\mathbf{E} = 0$, $\mathbf{v} = 0$ in the vacuum. Then we require

$$\mathbf{n} \times [\mathbf{E}] = \mathbf{n} \cdot \mathbf{v}[\mathbf{B}]. \tag{8.44}$$

For an interface between two fluids one requires continuity of the tangential component of $\mathbf{E}$, i.e. $\mathbf{n} \times [\mathbf{E}] = 0$, and of the normal component of $\mathbf{v}$, i.e. $= \mathbf{n} \cdot [\mathbf{v}] = 0$. An interface between a perfectly conducting wall and a fluid may be regarded as a special case of the fluid–fluid boundary, with $\mathbf{n} \times \mathbf{E} = 0$, $\mathbf{n} \cdot \mathbf{v} = 0$ implied by $\mathbf{E} = 0$, $\mathbf{v} = 0$ at the wall.

Now consider the linearized equation of fluid motion:

$$\eta_0 \frac{d^2 \boldsymbol{\xi}}{dt^2} = -\operatorname{grad} P_1 + \mathbf{J} \times \mathbf{B}_1 + \mathbf{J}_1 \times \mathbf{B}_0 = \mathbf{F}(\boldsymbol{\xi}), \tag{8.45}$$

where (8.15) implies

$$\mathbf{F}(\boldsymbol{\xi}) = \operatorname{grad}[\Gamma P_0 \operatorname{div} \boldsymbol{\xi} + \boldsymbol{\xi} \cdot \operatorname{grad} P_0] + \mathbf{J}_0 \times \mathbf{Q} + \mu_0 (\operatorname{curl} \mathbf{Q}) \times \mathbf{B}_0 \tag{8.46}$$

with

$$\mathbf{Q} = \mathbf{B}_1 = \operatorname{curl} \boldsymbol{\xi} \times \mathbf{B}_0. \tag{8.47}$$

We now regard $\mathbf{F}(\boldsymbol{\xi})$ as a force per unit volume derived from a "potential" U by $\mathbf{F} = -\partial U / \partial \boldsymbol{\xi}$. Our energy principle involves a δW which is a quadratic function of $\boldsymbol{\xi}$, and we identify δW as the volume integral over the plasma of U:

$$\delta W = -\frac{1}{2} \int_V d^3 \mathbf{x}\, \boldsymbol{\xi} \cdot \mathbf{F}(\boldsymbol{\xi}). \tag{8.48}$$

On substituting (8.46) into (8.48) we use the standard identities

$$\operatorname{div}(\boldsymbol{\xi} \phi) = \phi \operatorname{div} \boldsymbol{\xi} + \boldsymbol{\xi} \cdot \operatorname{grad} \phi,$$

with

$$\phi = \boldsymbol{\xi} \cdot \operatorname{grad} P_0 + \Gamma P_0 \operatorname{div} \boldsymbol{\xi},$$

and

$$\operatorname{div}(\mathbf{A} \times \mathbf{C}) = \mathbf{C} \cdot \operatorname{curl} \mathbf{A} - \mathbf{A} \cdot \operatorname{curl} \mathbf{C},$$

with $\mathbf{A} = \boldsymbol{\xi} \times \mathbf{B}_0$ and $\mathbf{C} = \operatorname{curl} \boldsymbol{\xi} \times \mathbf{B}_0$. Then we use Gauss' theorem to find

$$\delta W = \frac{1}{2} \int_V d^3 \mathbf{x} \left\{ \frac{Q^2}{\mu_0} + \frac{1}{\mu_0} (\operatorname{curl} \mathbf{B}_0) \cdot (\boldsymbol{\xi} \times \mathbf{Q}) + (\operatorname{div} \boldsymbol{\xi})(\boldsymbol{\xi} \cdot \operatorname{grad}) P_0 + \Gamma P_0 (\operatorname{div} \boldsymbol{\xi})^2 \right\}$$

$$- \frac{1}{2} \int_S d^2 \mathbf{x}\, \mathbf{n} \cdot \left\{ \frac{1}{\mu_0} (\boldsymbol{\xi} \times \mathbf{B}_0) \times \mathbf{Q} + \boldsymbol{\xi}(\boldsymbol{\xi} \cdot \operatorname{grad}) P_0 + \boldsymbol{\xi} \Gamma P_0 \operatorname{div} \boldsymbol{\xi} \right\}. \tag{8.49}$$

Henceforth we omit subscripts 0.

For a vacuum plasma interface (8.42) implies

$$\left(P + \frac{B^2}{2\mu_0} \right)_{\text{plasma}} = \frac{B_V^2}{2\mu_0} \tag{8.50}$$

where $\mathbf{B}_V$ is the vacuum field. Let $(\mathbf{B})_{eq}$ and $(\mathbf{B}_V)_{eq}$ be the values at the equilibrium position of the interface. Then when the position of the interface is perturbed we have

$$\mathbf{B} - (\mathbf{B})_{eq} = \mathbf{Q} + (\boldsymbol{\xi} \cdot \text{grad})\mathbf{B} \qquad (8.51a)$$

inside the plasma and

$$\mathbf{B}_V - (\mathbf{B}_V)_{eq} = \delta\mathbf{B}_V + (\boldsymbol{\xi} \cdot \text{grad})\mathbf{B}_V \qquad (8.51b)$$

in the vacuum, where $\delta\mathbf{B}_V$ is the change in the vacuum field. Then (8.50) implies

$$-\Gamma P \operatorname{div} \boldsymbol{\xi} + \frac{\mathbf{B}}{\mu_0} \cdot \{\mathbf{Q} + (\boldsymbol{\xi} \cdot \text{grad})\mathbf{B}\} = \frac{\mathbf{B}_V}{\mu_0} \cdot \{\delta\mathbf{B}_V + (\boldsymbol{\xi} \cdot \text{grad})\mathbf{B}_V\}, \qquad (8.52)$$

where we use $P - P_{eq} = -\Gamma P \operatorname{div} \boldsymbol{\xi}$. We require $\mathbf{n} \cdot [\mathbf{B}] = 0$ on the surface and hence

$$\int_S d^2\mathbf{x}\, \mathbf{n} \cdot \left\{ \frac{1}{\mu_0} (\boldsymbol{\xi} \times \mathbf{B}) \times \mathbf{Q} + \boldsymbol{\xi}(\boldsymbol{\xi} \cdot \text{grad})P + \boldsymbol{\xi}\Gamma P \operatorname{div} \boldsymbol{\xi} \right\}$$

$$= -\int_S d^2\mathbf{x}\, \mathbf{n} \cdot \left[\text{grad}\left(P + \frac{B^2}{2\mu_0} \right) \right] (\boldsymbol{\xi} \cdot \mathbf{n})^2 - \int d^2\mathbf{x}\, \mathbf{n} \cdot \boldsymbol{\xi} \frac{\mathbf{B}_V \cdot \delta\mathbf{B}_V}{\mu_0}, \qquad (8.53)$$

where we use the fact that $P + B^2/2\mu_0$ is continuous along the interface. In the vacuum the magnetic field must be a potential field, and this allows one to rewrite the final term in (8.53) as a volume integral over $(\delta B_V)^2/2\mu_0$.

The final result is

$$\delta W = \delta W_F + \delta W_S + \delta W_V \qquad (8.54)$$

with

$$\delta W_F = \frac{1}{2} \int_V d^3\mathbf{x} \left\{ \frac{Q^2}{\mu_0} + \frac{1}{\mu_0}(\text{curl}\,\mathbf{B}) \cdot (\boldsymbol{\xi} \times \mathbf{Q}) \right.$$

$$\left. + (\text{div}\,\boldsymbol{\xi})(\boldsymbol{\xi} \cdot \text{grad})P + \Gamma P(\text{div}\,\boldsymbol{\xi})^2 \right\} \qquad (8.55a)$$

$$\delta W_S = \frac{1}{2} \int_S d^2\mathbf{x}\, \mathbf{n} \cdot \left[\text{grad}\left(P + \frac{B^2}{2\mu_0} \right) \right] (\mathbf{n} \cdot \boldsymbol{\xi})^2 \qquad (8.55b)$$

$$\delta W_V = \frac{1}{2} \int_V d^3\mathbf{x} \left(\frac{\delta B_V}{2\mu_0} \right)^2 \qquad (8.55c)$$

8.5 Jump conditions at a discontinuity

Another important application of MHD theory is to shock waves and related phenomena. Shock waves can transport large amounts of energy and are known to cause acceleration of particles and excitation of microturbulence. Although shocks in space plasmas are usually collisionless, it is desirable to understand the MHD shocks before considering collisionless shocks. In an MHD shock there is a boundary (where collisions transfer stresses from one side to the other) separating two regions which may be assumed homogeneous. We refer to such a boundary as a surface of discontinuity. Such surfaces differ from those considered in §8.3 in that there are steady flows on either side of them.

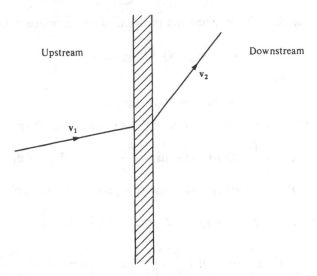

Fig. 8.3 A surface of discontinuity (shaded region) separates the preshock upstream plasma flowing into the surface from the postshock downstream plasma flowing out of the surface.

The boundary conditions at a surface of discontinuity relate fluid parameters on either side of the surface. These relations are called jump conditions, or Rankine-Hugoniot conditions.

Suppose the surface of discontinuity is a plane. The fluid in the frame in which the surface is at rest must flow into the surface on one side and out of the surface on the other side (or the component of flow velocity normal to the surface may be zero). Let the inflow region, called the *upstream* region, and the outflow region, called the *downstream* region, be labelled 1 and 2 respectively, as in Figure 8.3. The fluid properties in the two regions are η_1, P_1, $\mathbf{v}_1$, $\mathbf{B}_1,\ldots$ and η_2, P_2, $\mathbf{v}_2$, $\mathbf{B}_2,\ldots$ respectively. The jump in a quantity A is denoted

$$[A] = A_1 - A_2 \tag{8.56}$$

as in §8.4.

The jump conditions are derived from the equations of continuity. Let Q and $\mathbf{F}$ be the density and flux, respectively, of some conserved physical quantity. The continuity equation for this quantity is

$$\frac{\partial Q}{\partial t} + \operatorname{div} \mathbf{F} = 0 \tag{8.57}$$

In a steady flow the volume integral of Q must be constant, and thus (8.57) implies

$$\int_S d^2\mathbf{x}\, \mathbf{n} \cdot \mathbf{F} = 0, \tag{8.58}$$

where the integral is over any surface S. Choose a box as the surface, as illustrated in Figure 8.4, and let the thickness of the box across the surface shrink to zero.

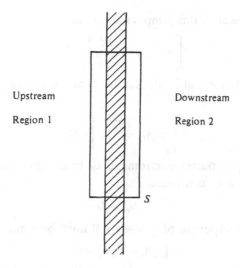

Upstream

Region 1

Downstream

Region 2

S

Fig. 8.4 The surface *S* over which the integral (8.58) is performed is a box extending across the surface of discontinuity with two sides normal to it.

Then (8.58) implies $(\mathbf{F}_1 - \mathbf{F}_2)\cdot\mathbf{n} = 0$. This is written as the jump condition

$$[F_n] = 0, \tag{8.59}$$

where subscript n denotes the normal component.

The relevant continuity equations are for mass, energy and momentum. These require

$$\operatorname{div}\{\eta\mathbf{v}\} = 0, \tag{8.60}$$

$$\operatorname{div}\left\{\tfrac{1}{2}\eta v^2\mathbf{v} + \frac{\Gamma}{\Gamma - 1}P\mathbf{v} + \frac{1}{\mu_0}(\mathbf{E} \times \mathbf{B})\right\} = 0, \tag{8.61}$$

and

$$\frac{\partial}{\partial x_j}\left\{\eta v_i v_j + P\delta_{ij} + \left(\frac{\varepsilon_0}{2}E^2 + \frac{B^2}{2\mu_0}\right)\delta_{ij} - \varepsilon_0 E_i E_j - \frac{B_i B_j}{\mu_0}\right\} = 0, \tag{8.62}$$

where we use (8.6), (8.11) and (8.12) respectively.

The jump condition implied by (8.50) is

$$[\eta v_n] = 0 \tag{8.63}$$

In (8.61) we first write $\mathbf{E} = -\mathbf{v} \times \mathbf{B}$, assuming infinite conductivity, and then we require

$$\left[\tfrac{1}{2}\eta v^2 v_n + \frac{\Gamma}{\Gamma - 1}Pv_n + \frac{B^2}{\mu_0}v_n - \frac{\mathbf{v}\cdot\mathbf{B}}{\mu_0}B_n\right] = 0 \tag{8.64}$$

In (8.62) the electric stress (for $\mathbf{E} = -\mathbf{v} \times \mathbf{B}$) is smaller than the magnetic stress by a factor v^2/c^2 and hence the electric stress is to be regarded as a relativistic correction. We do not include relativistic effects here and hence must ignore the electric stress. Equation (8.62) leads to jump conditions for a vectorial quantity.

The normal component of this jump condition is

$$\left[\eta v_n^2 + P + \frac{B_t^2}{2\mu_0} \right] = 0, \tag{8.65}$$

where t denotes the tangential (to the surface of discontinuity) part. The tangential component is

$$\left[\eta v_n \mathbf{v}_t - \frac{B_n}{\mu_0} \mathbf{B}_t \right] = 0. \tag{8.66}$$

In addition there are purely electromagnetic boundary conditions. The normal component of $\mathbf{B}$ must be continuous

$$[B_n] = 0 \tag{8.67}$$

and the tangential component of $\mathbf{E} = -\mathbf{v} \times \mathbf{B}$ must be continuous

$$[v_n \mathbf{B}_t - B_n \mathbf{v}_t] = 0. \tag{8.68}$$

In total we have eight jump conditions, four being scalar ones and the other four being the two tangential components of (8.66) and (8.68).

The admissible solutions of this set of jump conditions are classified as follows:

(i)   A *contact discontinuity or tangential discontinuity* involves no component of the mass flux into or out of the surface ($v_{1n} = 0$, $v_{2n} = 0$).
(ii) A *non-compressive shock* involves a non-zero mass flux ($\eta_1 v_{1n} = \eta_2 v_{2n} \neq 0$) and no density jump ($r = \eta_2/\eta_1 = 1$).
(iii) A *compressive shock* involves a non-zero mass flux and a density jump.

For a compressive shock the mass flux

$$j = \eta_1 v_{1n} = \eta_2 v_{2n} \tag{8.69}$$

is non-zero and then (8.66) and (8.68) imply

$$[\mathbf{B}_t] = \frac{\mu_0 j}{B_n^2} [v_n \mathbf{B}_t], \tag{8.70}$$

where $B_n = B_{1n} = B_{2n}$ is implied by (8.67). One may solve (8.70) for the two tangential components of $\mathbf{B}_2$ in terms of the tangential components of $\mathbf{B}_1$, and then it is straightforward to establish the identity

$$\mathbf{n} \cdot \mathbf{B}_1 \times \mathbf{B}_2 = 0. \tag{8.71}$$

This result is known as the *coplanarity theorem*: in a compressive shock $\mathbf{n}$, $\mathbf{B}_1$ and $\mathbf{B}_2$ are coplanar. The proof is invalid for $v_{1n} = v_{2n}$, and hence $\mathbf{n}$, $\mathbf{B}_1$ and $\mathbf{B}_2$ need not be coplanar in a contact discontinuity.

8.6 MHD shock waves

A shock wave is an interface separating an upstream flow from a downstream flow such that there is a mass flux across the interface and generation of entropy there. Shocks can be driven by a 'piston'; examples include a shock preceding a

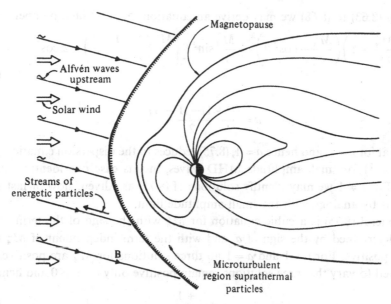

Fig. 8.5 The bowshock of the Earth; the solar wind impinges from the left and there is a turbulent region behind the shock and separated by a boundary (the magnetopause) from the Earth's magnetosphere.

mass flow and the bow shock of the Earth (Figure 8.5). A 'blast wave' is a non-driven shock initiated by an explosion at some point. Shocks can also form when finite amplitude waves 'break' as with surf on a beach. In a compressive wave c_s and v_A vary over a wavelength due to the variation in η and in either P or B^2/μ_0 respectively, and hence one part of the wave overtakes another leading to a steepening of the profile and eventually to breaking.

MHD shock waves depend on a number of parameters which one is free to choose in a variety of ways. A convenient choice is as follows. First choose the frame such that the shock is at rest and the inflow velocity is normal to it, i.e. such that $\mathbf{v}_1$ is along $\mathbf{n}$. Let the incoming magnetic field $\mathbf{B}_1$ be at an angle θ to $\mathbf{n}$. Let β be the plasma beta for the upstream plasma, specifically

$$\beta = \frac{c_{1s}^2}{v_{1A}^2} \tag{8.72}$$

Let r be the density ratio

$$r = \eta_2/\eta_1, \tag{8.73}$$

and let M_A be the Alfvénic Mach number

$$M_A = \frac{v_1}{v_{1A}}, \tag{8.74}$$

where v_1 denotes v_{1n} with the subscript n now redundant.

Using (8.63) to (8.68) we may derive an equation for the Mach number:

$$\left(\frac{aM_A^2}{r} - \beta\right)\left(\frac{M_A^2}{r} - \cos^2\theta\right)^2 - \frac{M_A^2}{r}\sin^2\theta\left\{\frac{M_A^2}{r}\left(\frac{a}{r} - \frac{1-r}{2}\right) - a\cos^2\theta\right\} = 0$$

(8.75)

with

$$a = \frac{(\Gamma + 1) - (\Gamma - 1)r}{2}.$$

(8.76)

Note that for $r = 1$, and hence $a = 1$, (8.75) reduces to the dispersion equation (8.71) (for $v_A^2 \ll c^2$) for small-amplitude MHD waves; in this case one identifies M_A as ω/kv_A. For $r \neq 1$ we may identify solutions of (8.75) as Alfvén, slow or fast mode solutions by analogy with the small-amplitude limit.

In general (8.75) is a cubic equation for M_A^2 with the sign of the term $\alpha(M_A^2)^3$ being determined by the sign of a, and with the term independent of M_A^2 being strictly positive. For $r = 1$ and $a = 1$ all three solutions for M_A^2 are positive. As r is allowed to vary the three solutions remain positive only for $a > 0$ and hence for

$$r < \frac{\Gamma + 1}{\Gamma - 1}$$

(8.77)

For a monatomic gas with $\Gamma = 5/3$, $r = \eta_2/\eta_1$ in (8.77) implies that the maximum density compression in a shock is by a factor of four.

Equation (8.75) simplifies in the cases of parallel ($\sin\theta = 0$) and perpendicular ($\sin\theta = 1$) propagation. For parallel propagation the three solutions of (8.75) are a double solution at

$$M_A^2 = r$$

(8.78a)

and another solution at

$$M_A^2 = r\beta/a.$$

(8.78b)

We may rewrite (8.78b) as $v_1^2/c_{1s}^2 = r/a$, which is independent of the magnetic field. Thus (8.78b) corresponds to a sound-mode shock, as in an unmagnetized fluid. The magnetic field is unaffected by a parallel sound-mode shock. In interpreting (8.78a) let us first recall that we have oriented the coordinate axes such that $\mathbf{v}_1$ is normal to the shock, and the assumption $\sin\theta = 0$ implies that $\mathbf{B}_1$ is also normal to the shock. As a consequence, the solution of (8.78a) is referred to as a *switch-on shock*, cf. Figure 8.6. There are also switch-off shocks.

In the general case of arbitrary θ the shock is said to be oblique. The three solutions may be classified as Alfvén, fast and slow shocks. However the Alfvén solution has $r = 1$ and $M_A = 1$; one readily confirms that $r = 1$, $M_A = 1$ is indeed a solution of (8.75). This solution is not a shock because there is no compression. In the jump condition for energy (8.65) we have $[\eta] = 0$, $[v_n] = 0$, $[P] = 0$ and hence $[B_t^2] = 0$. Thus the magnitude of the magnetic field does not change, and then $v_n^2 = B_n^2/2\eta\mu_0$ and $v_n = v_{1A} = v_{2A}$ implies $v_t^2 = B_t^2/2\eta\mu_0$ and hence that the magnitude of the flow velocity also does not change. This implies that the solution

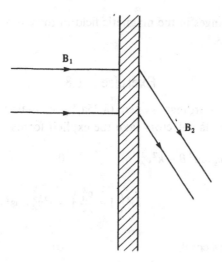

Fig. 8.6 In a switch-on shock the tangential component of magnetic field is zero downstream and non-zero upstream.

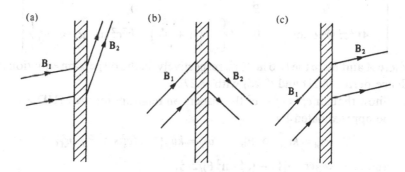

Fig. 8.7 The change in the direction of magnetic field is shown schematically for (a) a fast mode shock, (b) a rotational discontinuity, and (c) a slow mode shock.

is either trivial or corresponds to a reversal of the signs of the tangential components:

$$v_{2n} = v_{1n}, \qquad \mathbf{v}_{2t} = -\mathbf{v}_{1t}, \tag{8.79a}$$

$$B_{2n} = B_{1n}, \qquad \mathbf{B}_{2t} = -\mathbf{B}_{1t}. \tag{8.79b}$$

This type of solution is called a *rotational discontinuity*.

An important qualitative distinction between fast mode and slow mode shocks concerns the change in the magnitude of $\mathbf{B}_t$: in a fast mode shock it increases ($|\mathbf{B}_{2t}| > |\mathbf{B}_{1t}|$) and in a slow mode shock it decreases ($|\mathbf{B}_{2t}| < |\mathbf{B}_{1t}|$). The ratios of the relevant quantities are given by

$$\left| \frac{\mathbf{B}_{2t}}{\mathbf{B}_{1t}} \right| = \frac{r(M_A^2 - \cos^2 \theta)}{M_A^2 - r \cos^2 \theta}, \tag{8.80}$$

$$\left| \frac{\mathbf{v}_{2t}}{v_1} \right| = \frac{(r-1) \cos \theta \sin \theta}{M_A^2 - r \cos^2 \theta}. \tag{8.81}$$

The nature of the changes in the magnetic field in the cases of oblique shocks are illustrated in Figure 8.7.

Exercise set 8

8.1 Show that in the coordinate system (8.25a, b) the matrix Γ_{ij}, defined by (8.20), and the matrix of its cofactors take the explicit forms

$$
\begin{bmatrix}
\omega^2\left(1+\dfrac{v_A^2}{c^2}\right) - k^2c_s^2\sin^2\theta - k^2v_A^2 & 0 & -k^2c_s^2\sin\theta\cos\theta \\[2ex]
0 & \omega^2\left(1+\dfrac{v_A^2}{c^2}\right) - k^2v_A^2\cos^2\theta & 0 \\[2ex]
-k^2c_s^2\sin\theta\cos\theta & 0 & \omega^2 - k^2c_s^2\cos^2\theta
\end{bmatrix}
$$

and

$$
\begin{bmatrix}
A(\omega^2 - k^2c_s^2\cos^2\theta) & 0 & Ak^2c_s^2\sin\theta\cos\theta \\
0 & B & 0 \\
Ak^2c_s^2\sin\theta\cos\theta & 0 & A\left\{\omega^2\left(1+\dfrac{v_A^2}{c^2}\right) - k^2c_s^2\sin^2\theta - k^2v_A^2\right\}
\end{bmatrix}
$$

where A and B are the two factors, respectively, in the dispersion equation (8.21). Hence derive (8.21) and (8.26) with (8.27).

8.2 (a) Show that for $c_s^2 \ll v_A^2 \ll c^2$ the dispersion relations for the MHD waves may be approximated by

$$
\omega_A = kv_A|\cos\theta|, \qquad \omega_F = kv_A\{1 + (c_s^2\sin^2\theta)/2v_A^2\}
$$

and $\omega_s = kc_s|\cos\theta|\{1 - (c_s^2\sin^2\theta)/2v_A^2\}$.

(b) Find the directions of $\boldsymbol{\xi}$, $\mathbf{E}$ and $\mathbf{B}_1$ in the approximation made in part (a).

(c) Derive general expressions for the group velocity, cf. (8.25),

$$
v_{gM}(\mathbf{k}) = \frac{\partial\omega_M(\mathbf{k})}{\partial\mathbf{k}} = \boldsymbol{\kappa}\frac{\partial\omega_M(\mathbf{k})}{\partial k} + \frac{\mathbf{t}}{k}\frac{\partial\omega_M(\mathbf{k})}{\partial\theta}
$$

for each of the three MHD modes $M = A$, F & S, and show that the simplest approximation in the limit $c_s \ll v_A \ll c$ is

$$
\mathbf{v}_{gA} = \pm v_A\mathbf{b}, \qquad \mathbf{v}_{gF} = v_A\boldsymbol{\kappa}, \qquad \mathbf{v}_{gs} = \pm c_s\mathbf{b},
$$

where $\pm$ is the sign $\cos\theta/|\cos\theta|$.

8.3 Consider a fluid with two components (e.g. a thermal gas and a suprathermal or relativistic gas) with mass densities η_1 and η_2 and sound speeds c_{s1} and c_{s2}. Perpendicular to $\mathbf{B}$ the fluids move together with displacement $\boldsymbol{\xi}_\perp$, and parallel to $\mathbf{B}$ they move separately with displacements ξ_{1z} and ξ_{2z}.

(a) Derive a set of equations for the perturbations ξ_x, ξ_y, ξ_{1z}, ξ_{2z}, and set the coefficients equal to zero to find the dispersion equation.

(b) Show that for $v_A^2 \ll c^2$ the dispersion equation factorizes into the dispersion relation for Alfvén waves

$$\omega^2 = k^2 v_A^2 \cos^2 \theta, \qquad v_A^2 = \frac{B^2}{\mu_0(\eta_1 + \eta_2)},$$

and into a cubic equation for the phase speed squared $v_\phi^2 = \omega^2/k^2$:

$$(v_\phi^2 - v_A^2)(v_\phi^2 - c_{s1}^2 \cos^2 \theta)(v_\phi^2 - c_{s2}^2 \cos^2 \theta) - \frac{v_\phi^2}{1+r} \sin^2 \theta$$

$$\times \{c_{s1}^2(v_\phi^2 - c_{s2}^2 \cos^2 \theta) + rc_{s2}^2(v_\phi^2 - c_{s1}^2 \cos^2 \theta)\} = 0,$$

with $r = \eta_2/\eta_1$.

Remark: Solution of this cubic equation and applications to the effects of cosmic rays on MHD waves in the interstellar medium have been discussed by Parker (1965).

8.4 Consider an isothermal atmosphere in which P_0 and η_0 vary according to

$$P_0, \eta_0 \propto \exp\left[\frac{\mathbf{g}\cdot\mathbf{x}}{gH}\right],$$

where $\mathbf{g}$ is the acceleration due to gravity and $H = c_s^2/\Gamma g$ is the scale height.

(a) Show that the linearized fluid equations in the unmagnetized case may be written in the form (8.20) with

$$\Gamma_{ij}(\omega, \mathbf{k}) = \omega^2 \delta_{ij} + i(\Gamma - 1)g_i k_j + ik_i g_j - c_s^2 k_i k_j.$$

(b) Replace $\boldsymbol{\xi}$ by

$$\zeta = \xi \exp[\mathbf{g}\cdot\mathbf{x}/2gH]$$

and show that the wave equation for ζ leads to a dispersion equation

$$\omega^4 - \omega^2(\omega_1^2 + k^2 c_s^2) + \frac{|\mathbf{k} \times \mathbf{g}|^2}{g^2} c_s^2 \omega_2^2 = 0$$

with

$$\omega_1^2 = \frac{\Gamma^2 g^2}{4c_s^2} = \frac{c_s^2}{4H^2}, \qquad \omega_2^2 = \frac{(\Gamma - 1)g^2}{c_s^2} = \frac{\Gamma - 1}{\Gamma}\frac{g}{H}.$$

The frequency ω_2 is the Väisälä-Brunt frequency.

(c) Show that this equation has solutions (for $\omega_1^2 > \omega_2^2$) at $\omega > \omega_1$ and at $\omega < \omega_2$, but not for $\omega_1 < \omega < \omega_2$. Furthermore show that the solution for $\omega \gg \omega_1$ corresponds to sound waves, and the solution for $\omega < \omega_2$ has a dispersion relation $\omega^2 \approx \omega_2^2 |\mathbf{k} \times \mathbf{g}|^2/k^2 g^2$ for $k^2 c_s^2 \gg \omega^2, \omega_1^2, \omega_2^2$.

Remark: Waves corresponding to the lower frequency branch are called *internal gravity waves*. A fluid with a superadiabatic temperature gradient has $\omega_2^2 < 0$ and is "convectively" unstable; the internal gravity waves are then intrinsically growing, as in a reactive instability.

8.5 Consider surface waves in an incompressible (i.e. div $\boldsymbol{\xi} = 0$) unmagnetized fluid with a density gradient in the z-direction; there is a gravitational acceleration in the negative z-direction.

(a) Show that the linearized equations for a disturbance in the x–z plane with wavenumber k along the x-axis imply

$$\left(\omega^2\eta_0 + g\frac{\partial\eta_0}{\partial z}\right)\xi_z = \frac{\omega^2}{k^2}\frac{\partial}{\partial z}\left(\eta_0\frac{\partial\xi_z}{\partial z}\right).$$

(b) Show that if there is a density jump across the surface $z = 0$ between two homogeneous media, then on either side of $z = 0$, ξ_z varies as $\exp[-|kz|]$.

(c) Hence deduce the dispersion relation for surface waves

$$\omega^2 = kg\left(\frac{\eta_A - \eta_B}{\eta_A + \eta_B}\right),$$

where η_A and η_B are the densities of media A, below the surface, and B, above the surface, respectively.

Remarks: (i) The dispersion relation for water waves corresponds to $\eta_B/\eta_A \ll 1$ (density of air $\ll$ density of water), giving $\omega^2 = kg$. (ii) The instability implied for $\eta_A < \eta_B$ is called the Rayleigh-Taylor instability.

8.6 In an *interchange instability* two neighbouring flux tubes are interchanged, i.e. the matter in them interchanges position. Consider the stability of a planetary magnetosphere to interchanges assuming that the interchanged tubes have equal magnetic flux.

(i) Show that if initially the two tubes have pressure P and $P + \delta P$, and volumes V and $V + \delta V$, then the thermal energy W_T changes by

$$\delta W_T = \delta P\delta V + \frac{\Gamma P(\delta V)^2}{V}$$

for $\delta P/P \ll 1$ and $\delta V/V \ll 1$.

(ii) Assume a dipolar field such that $\delta V \propto r^4$, where r denotes radial distance, and show that one has

$$\delta W_T = P\delta V\frac{\delta r}{r}\left\{\frac{d\ln P}{d\ln r} + 4\Gamma\right\}$$

where the two flux tubes are at r and $r + \delta r$.

(iii) Hence show that $\Gamma = 5/3$ the magnetosphere is unstable to interchanges except when P falls off faster than r^{-n} with $n = 20/3$.

8.7 Apply the jump conditions (8.60) to (8.62) to shock waves in an unmagnetized fluid and hence deduce the conditions

$$r = \frac{\eta_2}{\eta_1} = \frac{(\Gamma+1)M^2}{2+(\Gamma-1)M^2},$$

$$\frac{P_2}{P_1} = \frac{2\Gamma M^2}{\Gamma+1} - \frac{\Gamma-1}{\Gamma+1},$$

$$\frac{T_2}{T_1} = \frac{\{2\Gamma M^2 - (\Gamma-1)\}\{(\Gamma-1)M^2 + 2\}}{(\Gamma+1)^2M^2}$$

with $M = v_{1n}/c_{1s}$.

9

MHD instabilities

9.1 Ideal MHD instabilities

Macroturbulence in plasma is generated by MHD instabilities. These instabilities can be classfied as 'ideal' and 'resistive': dissipative processes play no role in ideal MHD and play an essential role in resistive MHD. Ideal MHD instabilities can be further divided according to the source of free energy and to the geometric structure of the plasma. In this Section we summarize the ideal MHD instabilities which depend solely on the geometric structure of the plasma. This topic is of central importance for laboratory devices, such as tokamaks, which need to be designed to maximize the confinement time. It is also of interest in astrophysical applications, e.g. to the stability of magnetic loop structures in the solar atmosphere. Our primary interest in this Chapter is in the generation and dissipation of MHD turbulence; the ideal MHD instabilities discussed in this Section are not particularly effective in generating turbulence because the supply of free energy is limited to that made available from changing the geometric configuration of the plasma. The discussion of these instabilities here involves little more than an introduction to the terminology used to describe the various instabilities and brief summaries of their properties.

A plasma confined by a magnetic field is intrinsically unstable. This follows from the fact that the only stable solution of the Vlasov equation is a uniform Maxwellian distribution, and this is independent of the presence or absence of a magnetic field. Hence magnetic confinement of a plasma can only be maximized; it cannot be made indefinite. The maximization procedure is to identify the fastest growing instabilities and to modify the geometric structure to inhibit or eliminate these instabilities. In an unsheared magnetic field the fastest growing instabilities tend to be interchange (or flute), sausage (or pinch) and kink instabilities. Each of these can be inhibited or eliminated by shearing the magnetic field. However the shear itself can cause another type of instability: the resistive tearing mode (§9.5)

Interchange motions involve the matter in two neighbouring magnetic flux tubes interchanging position. An example of an interchange motion in the Earth's magnetosphere is illustrated in Figure 9.1. In a low-β plasma the dominant energy density is magnetic, and the energy density in the gas is insufficient to cause changes

(a)

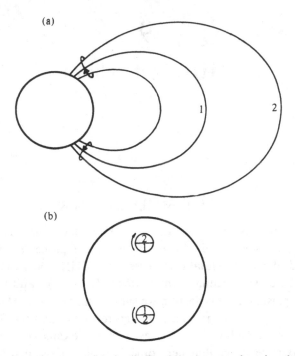

(b)

Fig. 9.1 An interchange motion in the Earth's magnetosphere involves two flux tubes 1 and 2 interchanging position. A possible direction for the initial motion is illustrated by the arrows in (a). The projection of the footpoints on the Earth's surface is shown in (b). A rotation of the footpoints through π radians in the directions indicated by the arrows in (b) effects the interchange.

which would increase the magnetic energy significantly. This requires that the two interchanged flux tubes must have equal magnetic flux threading them; only then does the total magnetic energy remain unchanged. The system is unstable to the interchange motion if the change in the energy δW, cf. (8.54), is negative, and here the only important contribution to δW is the thermal energy δW_T. It can be shown [Exercise 8.6] that δW_T is negative only if the change δP in the pressure and the change δV in the volume have opposite signs: we have $\delta W_T \approx \delta P \delta V$. The pressure P decreases outwards from the centre of the confined plasma, and an interchange motion which moves plasma outwards has $\delta P < 0$. Instability then occurs if the volume of the flux tube increases as it moves outwards. The volume may be evaluated by integrating the cross-sectional area A along the length l. The magnetic flux is proportional to the product AB, and hence we have

$$\delta V \propto \delta \int \frac{dl}{B} \tag{9.1}$$

Suppose the curvature of the field lines is such that the centre of curvature is inside the plasma ('convex' case); then the length of the field lines increases during an outward motion implying $\delta V > 0$ and hence that the configuration is unstable. If the centre of curvature is outside the plasma ('concave' case) then we have $\delta V < 0$, and the configuration is stable.

Mirror and pinch configurations with unsheared fields have a convex configuration and are unstable to interchange motions. These instabilities are inhibited by shear in the magnetic field because this limits the possibilities for making interchanges without altering the magnetic structure.

Another effect which can inhibit interchange motions is 'line tying'. In the discussion above it is implicit that the boundary conditions at either ends of the interchange flux tubes allow the interchange motion without twisting the magnetic field. This is the case in a mirror machine or in a planetary atmosphere where there is a non-conducting region which allows the interchange motion, cf. Figure 9.1b. However if the ends of the magnetic field lines are 'tied', e.g. to a rigid conductor, then the interchange motion must twist the magnetic field. This twist causes $\delta W_M > 0$ and instability requires $\delta W = \delta W_M + \delta W_T < 0$. Thus it is more difficult to satisfy the condition $\delta W < 0$ for instability when the lines are tied: line tying tends to inhibit instability.

Amongst the earliest ideal MHD instabilities to be studied in detail are those of a cylindrical plasma column. Let the z-axis be along the cylinder and let r and θ be polar coordinates in a plane $z = $ constant. In the simplest model there is no axial magnetic field ($B_z = 0$) and the radial pressure gradient is balanced by the $\mathbf{J} \times \mathbf{B}$ force due to the axial current ($J_z \neq 0$) and the poloidal magnetic field ($B_\theta \neq 0$). The stability can be investigated using either the energy principle or a normal mode analysis. In the latter one writes the MHD equations in cylindrical polar coordinates r, θ, z and seeks wave solutions which vary as $\exp[-i(\omega t - m\theta - kz)]$ with m an integer. The wave equation reduces to a second order differential equation in r, and the solutions are Bessel functions of order m and argument $\propto r$. The dispersion equation, cf. Exercise 9.1, is obtained by imposing the boundary conditions, and it involves Bessel functions and their derivatives evaluated at the edge of the column, i.e. at $r = a$ where a is the radius of the column. The $m = 0$ mode, called a 'pinch' or 'sausage' mode, and the $m = \pm 1$ modes, called 'kink' modes, are unstable. These are illustrated in Figure 9.2.

The 'pinch' and 'kink' models can be stabilized by an axial magnetic field. The 'pinch' instability may be attributed to a localized construction in the plasma column causing a localized increase in J_z and the value of B_θ at the (deformed) edge of the column; the increased B_θ provide a pressure imbalance which tends to push the plasma towards the axis thereby enhancing the constriction. An axial magnetic field is stabilizing because it provides a magnetic pressure which is increased inside the plasma and is decreased outside the plasma as a localized constriction develops; this additional pressure opposes that driving the instability. The 'kink' instability may be attributed to a localized twist in the column pushing the field lines together on the edge of the twist nearer the axis and pushing the field lines apart on the edge further from the axis (cf. Figure 9.2b); then the external magnetic pressure has a gradient which tends to push the column away from the

(a)   (b)

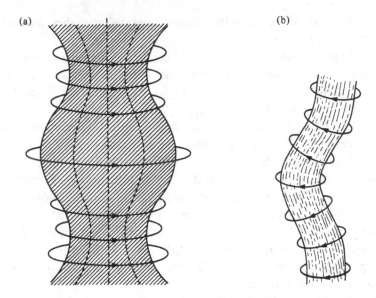

Fig. 9.2(a) The 'pinch' or 'sausage' and (b) the 'kink' modes for a cylindrical plasma column. The increases and decreases in the external magnetic pressure due to the field lines being pushed together and apart provide a feedback mechanism which tends to increase the deformations causing them.

axis, thereby increasing the amplitude of the 'kink'. An axial magnetic field provides a tension along the column, and this opposes any motion which tends to increase the length of the column. The length of the column increases in a 'kink' motion, and so again an axial magnetic field is stabilizing.

The criterion for stabilizing kink instabilities involves the q-value,

$$q(r) = \frac{kr B_z(r)}{B_\theta(r)} \tag{9.2}$$

The *Kruskal–Shafranov* criterion is that for stability q must exceed unity at the edge of the column, i.e. $q(a) > 1$.

The importance of the modes with $|m| > 1$ is more easily seen in toroidal geometry than in cylindrical geometry. A toroid may be regarded as a cylinder (of length L say) bent around and joined at its ends to form a doughnut-shaped object with major radius $R\,(= L/2\pi)$ and minor radius a. The axial coordinate z is then replaced by a toroidal angle $\phi\,(= 2\pi z/L)$. Thus a toroid is analogous to a cylindrical column of finite length L; the natural modes involve only discrete values of $k = nL/2\pi$ with n an integer. The natural modes of a toroidal system therefore vary like $\exp[-i(\omega t - m\theta - n\phi)]$ where m and n are the poloidal and toroidal mode numbers.

In a toroidal plasma device a given magnetic field line lies on a magnetic surface, which is a toroid in the idealized case of a toroidally symmetric system. A given magnetic surface is characterized by a q value, where q is the ratio of the number of windings around the major axis to the number of windings around the minor axis; the q-value is also called the 'safety factor' in this context. A surface on which q is equal to the ratio of two integers is called a *rational surface*. A toroidal wave mode has particular values of m and n; the surface with $q = n/m$ is called the *mode-rational surface* for this mode. In a given toroidal structure, q takes on a range of continuous values as a function of minor radius $0 \leqslant r \leqslant a$. It follows that there is an infinite (countable) number of rational surfaces. There are instabilities associated with the mode-rational surfaces, and these can be eliminated by a sufficiently strong shear, i.e. by $q(r)$ varying sufficiently rapidly with r. The condition for stabilization is called the Suydam criterion; which is

$$\left(\frac{q'}{q}\right)^2 + \frac{8\mu_0 P'}{r B_\phi^2}(1 - q^2) > 0, \tag{9.3}$$

where a prime denotes differentiation with respect to r. A large $q' = dq(r)/dr$ implies that the mode-rational surfaces are densely packed. This inhibits the growth of any one mode (given m, n) due to the interference with the modes of neighbouring surfaces. The derivation of (9.3) and its generalization (the Mercier criterion) has been discussed in detail by Bateman (1978).

These instabilities are of particular relevance to laboratory plasma machines. The design of a machine is chosen to minimize or eliminate the fastest growing or most undesirable instabilities.

The emphases in astrophysical applications of MHD theory are noticeably different from those in laboratory applications. For example, in the solar atmosphere there are structures such as filaments, prominences and coronal loops which persist for long times (from hours to months) and so are stable in some meaningful MHD sense. In solar MHD theory and in other astrophysical applications of MHD theory, the most important class of problems involves modelling specific structures, such as solar prominences, the solar wind, accretion discs, and so on. 'Instabilities' in such systems are readily observable only if they lead to massive disruptions of the system. Solar prominences sometimes disrupt, lifting off from the solar surface and ejecting a large mass of plasma from the solar atmosphere. Disruptions are thought to result from a secular change in some parameter causing the system to pass a critical value of that parameter beyond which no equilibrium configuration exists. Such a situation has been called 'non-equilibrium': it is inappropriate to call the resulting disruption an instability because instability implies a growing mode in an otherwise stable or, more correctly, metastable) configuration. One author (Pringle 1981), discussing accretion discs, remarked that 'instabilities' really means 'self-inconsistencies' in the initial assumptions of the astrophysical model.

Thus one emphasis in astrophysical applications of MHD theory is in modelling specific structures and in discussing their stability. Another emphasis is on processes which release free energy. Important sources of free energy are mass motions and non-potential magnetic fields. Two examples are discussed below, the Kelvin–Helmholtz instability (§9.2) which converts kinetic energy in mass flow into wave energy, and tearing instabilities (§9.5) which release magnetic energy.

9.2 The Kelvin–Helmholtz instability

Turbulence can be generated at an interface between two media flowing relative to each other. The classical example is the generation of water waves by wind blowing over the surface of the water. The particular instability involved is called the *Kelvin–Helmholtz instability*, and this name is also the generic name for the class of such instabilities. Another example is a jet of fluid injected into a stationary fluid; the instability disrupts the jet and converts the kinetic energy of its directed motion into turbulent energy. Quite spectacular jets are observed in some astrophysical sources, notably some quasars, active galactic nuclei, radio galaxies and a stellar object SS433.

The simplest example of the Kelvin–Helmholtz instability is that of two incompressible unmagnetized media separated by an interface with one medium flowing (with velocity **v**) relative to the other. The situation envisaged is illustrated in Figure 9.3: medium A is at rest, medium B is moving and the z-axis is normal to the interface. The equation of fluid motion is

$$\eta \frac{d^2 \xi}{dt^2} = -\operatorname{grad} P. \tag{9.4}$$

We assume that ξ varies as $\exp[-i(\omega t - kx)]$, i.e. that there is a ripple with wavenumber k along the x-axis. Then in medium A one replaces d/dt by $-i\omega$ and in medium B by $-i(\omega - kv\cos\theta)$, where **v** is at an angle θ to the x-axis. Linearizing (9.4), the perturbation in the pressure $P_1 \propto \exp[-i(\omega t - kx)]$ has no gradient in the y-direction and hence we have $\xi_y = 0$. Then (9.4) gives

$$-\eta_A \omega^2 \xi_x = -ikP_{1A}, \tag{9.5a}$$

$$-\eta_B (\omega - kv\cos\theta)^2 \xi_x = -ikP_{1B}, \tag{9.5b}$$

and

$$-\eta_A \omega^2 \xi_z = -\frac{\partial P_{1A}}{\partial z}, \tag{9.6a}$$

$$-\eta_B (\omega - kv\cos\theta)^2 \xi_z = -\frac{\partial P_{1B}}{\partial z}. \tag{9.6b}$$

The equation of state $\operatorname{div} \xi = 0$ implies

$$ik\xi_x + \frac{\partial \xi_z}{\partial z} = 0. \tag{9.7}$$

One then finds that ξ and P_1 vary with z as $\exp[-|kz|]$, as one expects for a surface wave, cf. §8.3. Continuity of ξ_z and P_1 across the interface then implies a

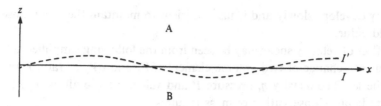

Fig. 9.3 An interface I between two media A and B develops a ripple with wavenumber k along the x axis due to relative flow of B to A with velocity
$$\mathbf{v} = v \cos \theta \mathbf{e}_x + v \sin \theta \mathbf{e}_y.$$

dispersion equation

$$\frac{1}{\eta_A \omega^2} + \frac{1}{\eta_B (\omega - kv \cos \theta)^2} = 0. \tag{9.8}$$

The solutions of (9.8) for ω are

$$\omega = \frac{kv \cos \theta}{1 + r} (1 \pm i r^{1/2}) \tag{9.9}$$

with $r = \eta_A/\eta_B$. The growing solution corresponds to the Kelvin–Helmholtz instability.

The generalization of (9.9) to include the effect of gravity is

$$\omega = \frac{kv \cos \theta}{1 + r} \pm \left[gk \frac{1 - r}{1 + r} - \frac{rk^2 v^2 \cos^2 \theta}{(1 + r)^2} \right]^{1/2}. \tag{9.10}$$

In the case of wind over water one has $r \ll 1$. Then for a given v there is a minimum wavenumber $g/rv^2 \cos^2 \theta$ below which instability does not occur, i.e. wavelengths longer than $2\pi rv^2 \cos^2 \theta/g$ are not excited.

The basic model for a jet is that of a cylinder of fluid flowing through a stationary fluid. Complicating factors included in discussions of astrophysical jets include compressibility, relativistic effects, a magnetic field and velocity shear. A dispersion relation for unsheared flow was written down by Ferrari, Trussoni & Zaninetti (1981) for the case where the magnetic field is parallel to the flow velocity. Possible instabilities include those of a plasma column without flow, e.g. the pinch and kink modes. The properties and growth rates of these modes are modified by the flow. However, intrinsically new effects appear only for supersonic flow: then there arises an additional class of 'reflection modes' whose existence depends on reflection from the boundaries. This additional branch leads to growing modes for flow speeds $v \gtrsim 2.5 c_s$; the maximum growth occurs for $ka \approx 1$ where a is the jet radius.

The development of such instabilities in astrophysical jets need not disrupt them. The instability is driven by a shear layer at the surface of the jet, and the development of the instability broadens the shear layer and reduces the velocity shear. The system can adjust by developing a smooth velocity gradient $\partial v/\partial r$ across its radial profile. When the velocity shear becomes sufficiently small the instability suppresses itself. Thus it is possible for the jet to approach a marginally stable state in which the instability develops slowly and is just sufficient to maintain the shear close to the threshold value.

The effect of velocity shear may be seen from the following simplified treatment. Consider interchange of two volume elements δV_1 and δV_2 at radii r and $r + \delta r$ across the jet. The density η, pressure P and velocity v are all assumed to have radial gradients. Conservation of mass requires

$$\delta V_1 - \delta V_2 = \frac{1}{\eta} \frac{\partial \eta}{\partial r} \delta r \, \delta V, \qquad (9.11)$$

where we set $\delta V_1 = \delta V_2 = \delta V$ to first order in small quantities. The change in the thermal energy is then

$$\delta W_T = \frac{\partial P}{\partial r} \delta r (\delta V_1 - \delta V_2) = \frac{1}{\eta} \frac{\partial \eta}{\partial r} \frac{\partial P}{\partial r} (\delta r)^2 \, \delta V. \qquad (9.12)$$

Free energy is made available when flows at velocities v and $v + \delta v$ are smoothed into an averaged flow at $\frac{1}{2}(v + v + \delta v)$; specifically an energy per unit mass of

$$\tfrac{1}{2}\{v^2 + (v + \delta v)^2 - \tfrac{1}{2}(2v + \delta v)^2\} = \tfrac{1}{4}(\delta v)^2$$

becomes available. Hence we have a change

$$\delta W_K = -\tfrac{1}{4}\eta \left(\frac{\partial v}{\partial r}\right)^2 (\delta r)^2 \, \delta V \qquad (9.13)$$

in the kinetic energy. The instability occurs only if $\delta W_K + \delta W_T$ is negative and this requires

$$\left|\frac{\partial v}{\partial r}\right| > \frac{2}{\eta} \left(\frac{\partial \eta}{\partial r} \frac{\partial P}{\partial r}\right)^{1/2}. \qquad (9.14)$$

That is, the Kelvin–Helmholtz instability requires a velocity shear which exceeds a minimum value, as given by (9.14). If the instability develops too fast, the resulting turbulent mixing can reduce the shear to below this threshold, thereby suppressing the instability.

9.3 Collisions and transport coefficients

So far we have been concerned with ideal MHD in which the electrical conductivity σ is assumed infinite. Implicitly we have also assumed that the thermal conductivity is zero (adiabatic equation of state) or infinite (isothermal equation of state) and that the viscosity is zero. Dissipation in MHD theory is due to finite (and/or non-zero) values of these transport coefficients. Of particular importance is a finite value for σ, as this leads to a decoupling between the plasma and the magnetic field, allowing magnetic 'merging', 'reconnection' or 'tearing'. Actually there are also non-collisional effects included in a 'generalized Ohm's law' which have the same consequences, and 'collisions' due to particle-wave interactions in a microturbulent plasma can lead to 'anomalous' transport coefficients. When these collisional or collision-like processes are taken into account one refers to 'resistive MHD theory'. There is a broad class of resistive MHD instabilities which differ qualitatively from ideal MHD instabilities in that the topology of the magnetic field can change (§9.5).

The MHD equations, and their generalization to include finite transport coefficients, may be derived from kinetic theory by including the effects of collisions in the kinetic equation (2.19). Let superscript (α) denote quantities for species of particle α. The kinetic equation for species α is

$$\left(\frac{\partial}{\partial t} + \mathbf{v} \cdot \frac{\partial}{\partial \mathbf{x}} + \mathbf{F}^{(\alpha)} \cdot \frac{\partial}{\partial \mathbf{p}}\right) f^{(\alpha)} = \sum_{\beta} C^{(\alpha,\beta)} \tag{9.15}$$

where $C^{(\alpha,\beta)}$ is a contribution due to collisions between species α and β. One then takes moments of (9.15) and the equations so obtained are the fluid equations.

Let us denote the average of a quantity M over the distribution function for species α by

$$M^{(\alpha)} = \langle M \rangle = \frac{1}{n^{(\alpha)}} \int d^3 \mathbf{p}\, M f^{(\alpha)}(\mathbf{p}). \tag{9.16}$$

Relevant averaged quantities are the mean velocity $\mathbf{v}^{(\alpha)}$, (isotropic) pressure $P^{(\alpha)}$, stress tensor $\Pi_{ij}^{(\alpha)}$, heat flux $\mathbf{q}^{(\alpha)}$ and source terms $\mathbf{R}^{(\alpha)}$ and $Q^{(\alpha)}$ for momentum and energy:

$$\mathbf{v}^{(\alpha)} = \langle \mathbf{v} \rangle, \quad P^{(\alpha)} = n^{(\alpha)} m^{(\alpha)} \langle (\mathbf{v} - \mathbf{v}^{(\alpha)})^2/3 \rangle,$$

$$\Pi_{ij}^{(\alpha)} = n^{(\alpha)} m^{(\alpha)} \langle (\mathbf{v} - \mathbf{v}^{(\alpha)})_i (\mathbf{v} - \mathbf{v}^{(\alpha)})_j - (\mathbf{v} - \mathbf{v}^{(\alpha)})^2 \delta_{ij}/3 \rangle,$$

$$\mathbf{q}^{(\alpha)} = \tfrac{1}{2} n^{(\alpha)} m^{(\alpha)} \langle (\mathbf{v} - \mathbf{v}^{(\alpha)})^2 \mathbf{v} \rangle,$$

$$\mathbf{R}^{(\alpha)} = \sum_{\beta} n^{(\alpha)} m^{(\alpha)} \langle (\mathbf{v} - \mathbf{v}^{(\alpha)}) C^{(\alpha,\beta)} \rangle, \tag{9.17}$$

$$Q^{(\alpha)} = \sum_{\beta} \tfrac{1}{2} n^{(\alpha)} m^{(\alpha)} \langle (\mathbf{v} - \mathbf{v}^{(\alpha)})^2 C^{(\alpha,\beta)} \rangle.$$

The moments of (9.15) corresponding to $M = 1$, v_i and $v_i v_j$ lead to the fluid equations

$$\frac{\partial n^{(\alpha)}}{\partial t} + \mathrm{div}\,(n^{(\alpha)} \mathbf{v}^{(\alpha)}) = 0, \tag{9.18}$$

$$n^{(\alpha)} m^{(\alpha)} \left(\frac{\partial}{\partial t} + \mathbf{v}^{(\alpha)} \cdot \mathrm{grad}\right) v_i^{(\alpha)} = -\frac{\partial}{\partial x_j} (P^{(\alpha)} \delta_{ij} + \Pi_{ij}^{(\alpha)})$$
$$+ q^{(\alpha)} n^{(\alpha)} (E_i + [\mathbf{v}^{(\alpha)} \times \mathbf{B}]_i) + R_i^{(\alpha)} \tag{9.19}$$

and

$$\tfrac{3}{2} n^{(\alpha)} \left(\frac{\partial}{\partial t} + \mathbf{v}^{(\alpha)} \cdot \mathrm{grad}\right) T^{(\alpha)} = -\mathrm{div}\,\mathbf{q}^{(\alpha)} - \Pi_{ij}^{(\alpha)} \frac{\partial V_i^{(\alpha)}}{\partial x_j} + Q^{(\alpha)} \tag{9.20}$$

respectively, where $T^{(\alpha)} = P^{(\alpha)}/n^{(\alpha)}$ is the temperature for species α.

First consider a charge-neutral electron-ion plasma. Assuming singly charged ions, the mass density, fluid velocity and current density are then given by, respectively,

$$\eta = m^{(i)}n^{(i)} + m^{(e)}n^{(e)}, \quad \mathbf{v} = \frac{1}{\eta}(n^{(i)}m^{(i)}\mathbf{v}^{(i)} + n^{(e)}m^{(e)}\mathbf{v}^{(e)}), \qquad (9.21a, b)$$

$$\mathbf{J} = en^{(e)}(\mathbf{v}^{(i)} - \mathbf{v}^{(e)}). \qquad (9.21c)$$

The simplest model we may use for the effects of collisions is one which leads to a frictional drag between the electrons and the ions. This corresponds to

$$\mathbf{R}^{(e)} = - \nu_0 m^{(e)}n^{(e)}(\mathbf{v}^{(e)} - \mathbf{v}^{(i)}), \qquad (9.22a)$$
$$\mathbf{R}^{(i)} = - \nu_0 m^{(i)}n^{(i)}(\mathbf{v}^{(i)} - \mathbf{v}^{(e)}). \qquad (9.22b)$$

An approximate expression for the electron-ion collision-frequency is

$$\nu_0 = \frac{\omega_P}{4\pi n_e \lambda_{De}^3} \ln \Lambda. \qquad (9.23)$$

The Coulomb logarithm $\ln \Lambda$ may be approximated by, cf. Appendix C,

$$\ln \Lambda \approx \begin{cases} 16.0 - \frac{1}{2}\ln n_e + \frac{3}{2}\ln T_e & T_e \lesssim 7 \times 10^4\,K \\ 21.6 - \frac{1}{2}\ln n_e + \ln T_e & T_e \gtrsim 7 \times 10^4\,K \end{cases} \qquad (9.24)$$

where n_e is per cubic metre and T_e is in kelvins.

Neglecting corrections of order m_e/m_i, terms involving squares of v and neglecting the stress tensor Π_{ij}, (9.19) summed over the electrons and ions gives the linearized equation of fluid motion

$$\eta \frac{\partial \mathbf{v}}{\partial t} = - \operatorname{grad} P - \mathbf{J} \times \mathbf{B}. \qquad (9.25)$$

Differentiating (9.21c) and using (9.19) with (9.22a, b), one obtains the *generalized Ohm's Law*

$$\frac{m_e}{n_e e^2} \frac{\partial \mathbf{J}}{\partial t} = \frac{1}{n_e e} \operatorname{grad} P^{(e)} + \mathbf{E} + \mathbf{v} \times \mathbf{B} - \frac{1}{n_e e} \mathbf{J} \times \mathbf{B} - \frac{\mathbf{J}}{\sigma}, \qquad (9.26)$$

where we now write $n^{(e)}$ as n_e and introduce an electrical conductivity

$$\sigma = \frac{e^2 n_e}{m_e \nu_0} = \varepsilon_0 \frac{\omega_p^2}{\nu_0}. \qquad (9.27)$$

Comparing with the form (8.8) for Ohm's law shows that (9.26) contains three additional terms: the $\partial \mathbf{J}/\partial t$ term, the grad $P^{(e)}$ term and the $\mathbf{J} \times \mathbf{B}$ term, called the inertial, pressure-gradient and Hall terms respectively. The pressure-gradient term may be rewritten in terms of a temperature gradient, and then (9.26) includes the thermo-electric effect written down as the final term in (4.33a).

The general procedure for calculating the transport coefficients involves assuming a form for the collision term $C^{(\alpha, \beta)}$ and using it to determine the deviation in each distribution $f^{(\alpha)}$ from a uniform Maxwellian due to the flows and gradients implied by $\mathbf{J} \neq 0$, grad $T^{(\alpha)} \neq 0$ etc. The coefficients introduced in this expansion can be re-expressed in terms of the averages introduced in (9.17). In general this procedure is cumbersome. The simplest case, leading to the electrical conductivity (9.27), is outlined in Exercise 9.4.

The transport coefficients for a collisional plasma have been derived in this way by Braginskii (1965) using the Landau form of the collision term $C^{(\alpha,\beta)}$. The results are summarized in Appendix C.

9.4 Magnetic merging and reconnection

In ideal MHD the magnetic field lines and the plasma are constrained to move with each other: the magnetic field lines are 'frozen-in' to the plasma. Consider the curl of $\mathbf{E} + \mathbf{v} \times \mathbf{B} = \mathbf{J}/\sigma$, i.e.

$$\frac{\partial \mathbf{B}}{\partial t} = \text{curl}\,(\mathbf{v} \times \mathbf{B}) + \frac{1}{\mu_0 \sigma} \nabla^2 \mathbf{B}. \tag{9.28}$$

For $\sigma = \infty$, (9.28) reduces to $\partial \mathbf{B}/\partial t = \text{curl}\,(\mathbf{v} \times \mathbf{B})$, which implies that the field lines move with the fluid velocity $\mathbf{v}$. One way of seeing this is to write $\mathbf{B}$ in terms of Euler potentials α and β, i.e.

$$\mathbf{B} = (\text{grad}\,\alpha) \times (\text{grad}\,\beta), \tag{9.29}$$

so that a field line is defined by the intersection of two surfaces $\alpha = \text{constant}$ and $\beta = \text{constant}$. It is then clear that the field lines move with velocity $\mathbf{v}$ if the surfaces α and β move with velocity $\mathbf{v}$. Suppose the surfaces do move with velocity $\mathbf{v}$. Then on differentiating $\alpha = \text{constant}$ with respect to time one has

$$\frac{d\alpha}{dt} = \frac{\partial \alpha}{\partial t} + \mathbf{v} \cdot \text{grad}\,\alpha = 0, \tag{9.30}$$

and similarly for β. Operating with $\partial/\partial t$ on (9.29) and using (9.30) then gives $\partial \mathbf{B}/\partial t = \text{curl}\,(\mathbf{v} \times \mathbf{B})$, cf. Exercise 9.6. This confirms that (9.28) for $\sigma = \infty$ implies that the magnetic field lines are 'frozen-in' to the motion of the plasma.

Now consider (9.28) for $\mathbf{v} = 0$ and $\sigma \neq \infty$. It then describes diffusion of magnetic field lines. We always have div $\mathbf{B} = 0$, and the curl of (9.28) implies

$$\frac{\partial \mathbf{J}}{\partial t} = \frac{1}{\mu_0 \sigma} \nabla^2 \mathbf{J}, \tag{9.31}$$

where the displacement current is neglected. Hence the current decays away, due to ohmic dissipation, and the magnetic field relaxes towards a potential field structure.

For semi-quantitative purpose it is useful to define two characteristic times: the diffusion time

$$\tau_R = \mu_0 \sigma L^2 \tag{9.32}$$

and the Alfvén time

$$\tau_A = L/v_A, \tag{9.33}$$

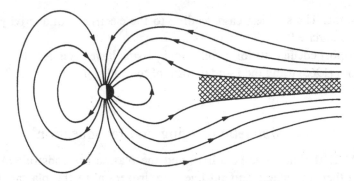

Fig. 9.4 The shaded region is a neutral sheet (called the plasma sheet) which separates regions of oppositely directed magnetic field lines in the Earth's magnetotail.

where L is a characteristic length. The dimensionless ratio

$$S = \tau_R/\tau_A \tag{9.34}$$

is called the magnetic Reynolds number. If L is taken to be characteristic of the dimensions of the system then for all plasmas of interest one has $S \gg 1$ and magnetic diffusion is very slow compared with the transfer of stresses through Alfvén waves. Put another way, the time τ_R is much longer than other timescales of interest. Consequently resistive effects can be important only in localized regions where the current profile has a large gradient, implying a small L.

Magnetic diffusion, and the associated current dissipation, occur in localized regions called *current sheets* or current layers. On a global scale these structures are essentially two-dimensional. As we shall see, on a finer scale they tend to break up into *current lines* which are essentially one-dimensional. The simplest example is a current sheet between two oppositely directed magnetic fields, as illustrated in Figure 9.4. This situation is relevant to the Earth's magnetotail and to the solar wind. In both cases outflowing plasma draws out magnetic field lines with opposite polarity from the northern and southern hemispheres of the central object (the Earth or the Sun). Between the regions of opposite polarity the magnetic field passes through zero, and pressure balance requires the plasma in this neutral region to be denser and/or hotter than in the surrounding regions where the magnetic field contributes significantly to the pressure.

An idealized model for a current sheet involves a magnetic field $B_z(x)$ which passes through zero at $x = 0$. The current density is along the y-axis with

$$J_y(x) = -\frac{1}{\mu_0}\frac{\partial B_z(x)}{\partial x}. \tag{9.35}$$

If B_z is approximately constant on either side of a layer in the y–z plane centred on $x = 0$, then the current is effectively confined to this layer, and any dissipation can occur only in this layer. This basic model is unaltered by the inclusion of a constant field along the y-axis. With $B_y \neq 0$, the magnetic field changes direction

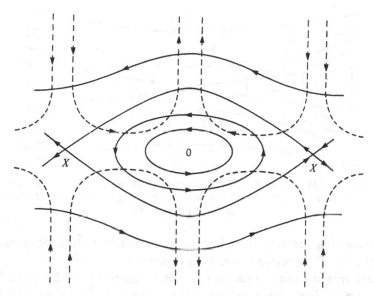

Fig. 9.5 A magnetic island is a region of closed field lines each encircling the 0-type neutral point with a separatrix which passes through two X-type neutral points. Flow lines are indicated by the dotted curves.

from an angle $\psi = \arctan (B_y/B_z(\infty))$ to the y-axis at $x = \infty$ through $\psi = 0$ at $x = 0$ to $-\psi$ at $x = -\infty$. This corresponds to a shear in the magnetic field.

Shear in cylindrical or toroidal geometry is closely analogous to shear in Cartesian geometry. In cylindrical geometry shear corresponds to q, as defined by (9.2), being a function of the radial coordinate r. Suppose we have $q = q_0$ at $r = r_0$. We may define an unsheared field with $q = q_0$ at all r, denoted $\mathbf{B}_{q0}$ say. Then $\mathbf{B} - \mathbf{B}_{q0}$ has an azimuthal component which passes through zero at $r = r_0$. Similarly in toroidal coordinates, a shear corresponds to a variation in q with the minor-radial coordinate r. Near a rational surface $q = n/m$ we may define an un-sheared field $\mathbf{B}_{n/m}$ with this q-value, and then $\mathbf{B} - \mathbf{B}_{n/m}$ has a zero in its poloidal component at the rational surface. In both these cases the geometry is locally equivalent to that described above using Cartesian coordinates. The x-direction is re-interpreted at the r-direction and the y-direction as the direction along the axis or around the toroid, respectively.

In practice magnetic field lines are created and destroyed at 0-type and X-type neutral points (or lines) which occur in pairs. It is possible in principle for field lines to appear and disappear at a neutral plane, as illustrated in Figure 9.4, but this is not favoured. The uniform (in z) current in a current sheet tends to break up into current lines. Each current line is at the centre of an 0-type neutral point, and the X-type neutral points are between the current lines, as illustrated in Figure 9.5. The magnetic field is then structured as a set of *magnetic islands*, each with a separatrix which passes through the X-type points at either end. Growth of such magnetic islands is the basic feature of a *tearing mode*. The magnetic topo-

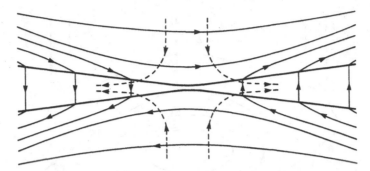

Fig. 9.6 Petchek's (1964) model for steady-state reconnection. The surfaces of discontinuity are interpreted as slow mode shocks. The dashed lines indicate the direction of plasma flow.

logy is changed in a tearing mode from (in Cartesian geometry) anti-parallel field lines to the set of magnetic islands, as in Figure 9.5.

Magnetic merging and dissipation occur in the neutral sheet. Plasma drifts into the neutral plane from above and below it carrying oppositely directed field lines. These field lines become distorted around the magnetic islands and drift into the X-type point where they reconnect. The closed field lines inside a magnetic island drift into the 0-type neutral point where they disappear. The plasma flows as indicated in Figure 9.5, i.e. in towards the X-type neutral points, then towards the 0-type neutral point and then vertically up or down away from the 0-type neutral point. This flow pattern does not allow the plasma to escape from the vicinity of the neutral plane to infinity. As a consequence the development of a tearing instability does not necessarily release most of the stored magnetic energy.

In order for magnetic energy in a relatively large volume of plasma to be dissipated, the entire volume must be processed by passing through one or more localized regions where magnetic reconnection occurs. One regards the situation as being in a steady state, with the plasma inflow transporting oppositely directed field lines, and the plasma outflow carrying away the energy released in the form of heat and kinetic energy. Although steady reconnection has much in common with a tearing instability, there are important differences. The most notable is that in any model for steady reconnection of the plasma outflow must transport the reprocessed plasma essentially to infinity, whereas the flow in a tearing mode remains localized.

A variety of models for steady-state reconnection have been proposed in the context of solar flares and reconnection in the Earth's magnetotail. A critical review of these models has been given by Vasyliunas (1975). The relative importance of steady reconnection and tearing is unclear; qualitatively a reconnection model is like a single magnetic island (the dissipation region) with a flow pattern which involves a shock discontinuity between the inflow and the outflow, as illustrated in Figure 9.6. There is some observational evidence for outflow at about the Alfvén speed along the separatrix of such a reconnection region in the Earth's magnetotail (e.g. Lin, Anderson, McCoy & Russell 1977, Sarris & Axford 1979). For solar flares it seems that release of magnetic free energy is more likely to occur through multiple tearing instabilities (e.g. Spicer 1981) than in terms of steady-state reconnection as envisaged in the models like that illustrated in Figure 9.6. Models for the energy release remain unsatisfactory in how they relate the global structure of the volume containing magnetic free energy to the localized regions where reconnection occurs.

9.5 Tearing instabilities

Tearing instabilities, in which tearing modes grow, are examples of resistive MHD instabilities. There are resistive versions of ideal MHD instabilities, such as the kink and interchange instabilities, but the tearing modes are the most important in practice. In toroidal plasma devices, specifically in tokamaks, sawtooth oscillations, Mirnov oscillations and disruptions are all attributed at least partly to the onset or saturation of a tearing instability. Tearing instabilities are also thought to be important in solar flares and other astrophysical phenomena.

A detailed mathematical treatment of tearing instabilities is impracticable here. However, it is practicable to derive a reasonably general set of equations for the tearing instability, and to use them to deduce the relevant properties of tearing modes through semi-quantitative arguments. We follow this course.

The situation envisaged is illustrated in Figure 9.5. The unperturbed magnetic field is assumed to be of the form

$$B_x = 0, \quad B_y = \text{constant}, \quad B_z = B_0 f(x) \tag{9.36}$$

with $f(x) \to \pm 1$ at $x \to \pm \infty$ and with $f(x) \approx x/L$ for $|x| \ll L$. The length L characterizes the variation in the magnetic field across the neutral layer. We can Fourier transform in the y- and z-directions. Now B_x in a magnetic island is an even function of x, and hence $\partial B_x/\partial x$ is zero at $x = 0$. This, together with div $\mathbf{B} = 0$, requires that $\mathbf{k}$ satisfy $\mathbf{k} \cdot \mathbf{B} = 0$ at $x = 0$. Here this reduces to $k_y = 0$, $k_z \neq 0$. In toroidal geometry the pitch of the tearing mode must be opposite to that of the magnetic field at the rational surface (i.e. for $q = m/n$ the tearing mode varies as $\exp[-i(m\theta - n\phi)]$), this being the appropriate form of $\mathbf{k} \cdot \mathbf{B} \propto (mq - n) = 0$ at the rational surface $q = n/m$. The plasma is assumed incompressible, implying div $\mathbf{v} = 0$.

The component of $\mathbf{B}$ in the x–z plane may be described in terms of Euler potentials; we set $\alpha = \psi(x, z)$ and $\beta = y$ so that (9.29) implies

$$\mathbf{B} = (\text{grad } \psi) \times \mathbf{e}_y + B_y \mathbf{e}_y, \tag{9.37}$$

where $\mathbf{e}_y$ is a unit vector along the y-axis. It is assumed that ψ varies with z as $\exp[ikz]$. Then the current density is given by

$$\mathbf{J} = \frac{1}{\mu_0} \text{curl } \mathbf{B} = \frac{1}{\mu_0} \nabla^2 \psi \, \mathbf{e}_y, \tag{9.38}$$

with $\nabla^2 = \partial^2/\partial x^2 - k^2$ here. The equation of fluid motion (9.41) together with div $\mathbf{v} = 0$ implies

$$\frac{\partial \psi}{\partial t} + \mathbf{v} \cdot \text{grad } \psi = \frac{1}{\mu_0 \sigma} \nabla^2 \psi, \tag{9.39}$$

with grad $= (\partial/\partial x, 0, ik)$ here.

The fluid velocity $\mathbf{v}$ is in the x–z plane and it may be written in terms of a stream function $\phi(x, z)$, which is also assumed to vary as $\exp[ikz]$:

$$\mathbf{v} = (\text{grad } \phi) \times \mathbf{e}_y. \tag{9.40}$$

The vorticity is curl $\mathbf{v} = \nabla^2 \phi \mathbf{e}_y$. The curl of the equation of fluid motion (8.7), with $\eta = $ constant and $\mathbf{f} = 0$, then implies

$$\eta \left(\frac{\partial}{\partial t} \nabla^2 \phi + \mathbf{v} \cdot \text{grad } \nabla^2 \phi \right) = \frac{\mathbf{e}_y}{\mu_0} \cdot [(\text{grad } \psi) \times (\text{grad } \nabla^2 \psi)]. \tag{9.41}$$

We are interested in wave motions of small amplitude, and so we linearize (9.39), with (9.40), and (9.41) in the oscillating parts of the fields ψ and ϕ. To satisfy (9.36) we write

$$\psi = \psi_0 + \delta \psi, \quad \frac{\partial \psi_0}{\partial x} = B_0 f(x), \tag{9.42}$$

and similarly we write $\phi = \phi_0 + \delta \phi$ with $\phi_0 = 0$ for zero unperturbed flow. The waves of interest turn out to be purely growing, and we anticipate this by assuming temporal variations as $\exp[\omega_i t]$. The linearized forms of (9.39) and (9.41) then reduce to

$$\omega_i \delta \psi + ik B_0 \delta \phi f(x) = \frac{1}{\mu_0 \sigma} \left(\frac{\partial^2}{\partial x^2} - k^2 \right) \delta \psi \tag{9.43}$$

and

$$\omega_i \eta \left(\frac{\partial^2}{\partial x^2} - k^2 \right) \delta \phi = \frac{ik B_0}{\mu_0} \left[\delta \psi \frac{\partial^2 f(x)}{\partial x^2} - f(x) \left(\frac{\partial^2}{\partial x^2} - k^2 \right) \delta \psi \right] \tag{9.44}$$

On eliminating $\delta \phi$ between these coupled equations, one obtains a fourth order differential equation for $\delta \psi$. At this stage it is appropriate to resort to semi-quantitative arguments to determine the form of the solution.

The system is separated into a resistive layer about $x = 0$ where resistivity is important, and ideal MHD regions above and below this layer. In the ideal MHD regions, (9.43) and (9.44) may be rewritten as

$$\delta \phi = \frac{i \omega_i}{k B_0} \frac{\delta \psi}{f(x)} \tag{9.45}$$

and

$$\left[\frac{\partial^2}{\partial x^2} - k^2 - \frac{1}{f(x)} \frac{\partial^2 f(x)}{\partial x^2} \right] \delta \psi = -\frac{(\omega_i \tau_A)^2}{f(x)} \delta \psi \approx 0, \tag{9.46}$$

with $\tau_A = 1/k v_A$, $v_A^2 = B_0^2/\mu_0 \eta$. We anticipate $\omega_i \tau_A \ll 1$ in setting the right hand side to zero. Equation (9.46) is singular at $x = 0$ and $\delta \psi$ has a discontinuous derivative there. Physically this corresponds to the magnetic field component in the z-direction appearing to have a discontinuity across the current layer when viewed on a scale over which the thickness of the current layer is negligible. Let us write

$$\Delta' = (dln\delta\psi/dx)_+ - (dln\delta\psi/dx)_- \tag{9.47}$$

for the difference across the layer (above $= +$, below $= -$). The quantity Δ' characterizes the properties of the resistive layer when it is viewed as a surface separating the two ideal MHD regions. There is a discontinuous change in the derivative of the x-component of the field (dB_x/dx changes by $\Delta'B_x$). The actual current distribution in the resistive layer is approximated by the surface current $J_y \propto d^2\delta\psi/dx^2 = \Delta'\delta(x)$, cf. (9.38), when the layer is approximated by a surface.

Inside the resistive layer the gradients in $\delta\psi$ and $\delta\phi$ are large and (9.43) and (9.44) may be approximated by

$$\omega_i\delta\psi + ikB_0\delta\phi\frac{x}{L} \approx \frac{1}{\mu_0\sigma}\frac{\partial^2\delta\psi}{\partial x^2} \tag{9.48}$$

and

$$\omega_i\frac{\partial^2\delta\phi}{\partial x^2} \approx -\frac{ikB_0}{\mu_0}\frac{x}{L}\frac{\partial^2\delta\psi}{\partial x^2}, \tag{9.49}$$

where we set $f(x) = x/L$.

Suppose the resistive layer has a width W (along the x-axis). For the purpose of semi-quantitative argument we estimate (9.48) and (9.49) at the edge of the layer by setting

$$x = W, \quad \frac{\partial^2\delta\psi}{\partial x^2} = \frac{\Delta'}{W}\delta\psi, \quad \frac{\partial^2\delta\phi}{\partial x^2} = \frac{\partial\phi}{W^2}. \tag{9.50}$$

The growth of the magnetic perturbations involves both the convective and diffusive terms, and hence the three terms in (9.48) should be comparable in magnitude:

$$\omega_i\delta\psi \approx ikB_0\delta\phi\frac{W}{L} \approx \frac{\Delta'\delta\psi}{\mu_0\sigma W}, \tag{9.51}$$

with $\delta\phi$ determined by (9.49):

$$\eta\frac{\omega_i}{W^2}\delta\phi = -\frac{ikB_0}{\mu_0}\frac{W}{L}\frac{\Delta'}{W}\delta\psi. \tag{9.52}$$

These equations then determine W and ω_i:

$$\left(\frac{W}{L}\right)^5 \approx \frac{\Delta'L}{S^2} \tag{9.53}$$

and

$$\omega_i \approx \frac{(\Delta'L)^{4/5}}{\tau_R}S^{2/5} \tag{9.54}$$

with $\tau_R = \mu_0\sigma L^2$ the resistive timescale and $S = \tau_L/\tau_A$ the magnetic Reynolds number. The fact that S is large (roughly of order 10^6 in a tokamak plasma and of order 10^{10} in the solar corona) implies that the width W is much smaller than the characteristic length L, and the growth time ω_i^{-1} is much shorter than the resistive timescale τ_R.

It might be remarked that although the classic paper on resistive MHD instabilities was published over two decades ago (Furth, Killeen & Rosenbluth 1963), the topic is still under development. Recently there has been emphasis on nonlinear development of tearing instabilities. One idea is that the development of one tearing mode can trigger another due to coupling between magnetic islands (e.g. Biskamp 1982).

A double tearing mode was identified by Pritchett, Lee & Drake (1980), and their work involved the identification of a 'fast' or 'non-constant-ψ' regime for the tearing mode investigated above, now called the 'slow' or 'constant-ψ' tearing mode. The fast regime is relevant where the growth time ω_i^{-1}, as given by (9.54), would be shorter than the time $\tau_R(W/L)^2$ for magnetic field lines to diffuse through the width W of the resistive layer. Then $\delta\psi$ varies more rapidly across the layer than implied by (9.50); specifically $\partial^2\delta\psi/\partial x^2 \approx \Delta'\delta\psi/W$ should be replaced by $\partial^2\delta\psi/\partial x^2 \approx \delta\psi/W^2$. Then (9.53) is replaced by $(W/L)^6 \approx 1/S^2$ and (9.54) is replaced by

$$\omega_i \approx \frac{S^{2/3}}{\tau_R} \qquad (9.55)$$

This growth rate corresponds to a growth time equal to the diffusion time $\tau_R(W/L)^2$ across the layer.

The 'resistive' effect in the foregoing discussion is a finite electrical conductivity σ. This allows the magnetic field lines to diffuse through the plasma. 'Resistive' instabilities, including the tearing mode, can be due to other effects, e.g. due to other terms in the generalized Ohm's Law (9.26) and other transport coefficients such as viscocity. Viscosity may be regarded as causing the stream lines of the plasma to diffuse, allowing a relative diffusion between the fluid and the magnetic field. Also, the transport coefficients may be anomalous, due to scattering of electrons by microturbulent fields rather than through collisions. For example, Coppi (1983) has discussed a tearing mode in a collisionless plasma due to either anomalous electrical conductivity or anomalous viscosity.

Some of the applications of tearing modes to the interpretation of phenomena in tokamaks have already been mentioned. The main astrophysical application is to solar flares and to other forms of magnetic reconnection in the solar corona (e.g. Priest 1984). Spicer (1981) has emphasized the importance of multiple tearing modes, i.e. of the coupling of magnetic islands to produce secondary islands; he also argued that the 'fast' tearing mode is likely to be relevant to the solar case. Very many localized sites for magnetic reconnection are required to release the stored magnetic energy in any realistic modes for a solar flare. Coupling between tearing modes to produce secondary magnetic islands does seem to be required, and this needs to occur rapidly throughout the entire region. That is tearing must be triggered at very many localized sites almost simultaneously, e.g. on a timescale comparable with that for an Alfvén wave to propagate across the region.

Exercise set 9

9.1 The dispersion relation for a cylindrical mode $\propto \exp[-i(\omega t - m\theta - kz)]$ in a plasma column of radius a with pressure P_0, density η_0, internal axial magnetic field B_p, and external axial and poloidal fields B_{z0} and $B_\theta = B_{\theta 0} a/r$ is

$$\mu_0 \eta_0 \omega^2 = k^2 B_p^2 - \left(kB_{z0} + \frac{m}{a}B_{\theta 0}\right)^2 \frac{\alpha}{k} \frac{I_m'(\alpha a)}{I_m(\alpha a)} \frac{K_m(ka)}{K_m'(ka)} - \frac{B_{\theta 0}^2}{a^2} \frac{\alpha a I_m'(\alpha a)}{I_m(\alpha a)}$$

where I_m and K_m are modified Bessel functions, with

$$\alpha^2 = \frac{(k^2 - \omega^2/c_s^2)(k^2 - \omega^2/v_A^2)}{(k^2 - \omega^2/c_s^2 - \omega^2/v_A^2)}$$

and $v_A^2 = B_p^2/\mu_0\eta_0$ and $c_s^2 = \Gamma P_0/\eta_0$.
  (a) Show that for $c_s^2 = 0$ the $m = 0$ mode is unstable for $B_p^2 < B_{\theta 0}^2/2$. (*Hint:* the maximum value of $I_0'(x)/xI_0(x)$ is $\frac{1}{2}$).
  (b) Show that for $\alpha \approx k$ and $ka \ll 1$ (implying $I_m'/I_m \approx m/ka \approx -K_m'/K_m$), the $m = 1$ mode is unstable for $|k| < B_{\theta 0}/aB_{z0}$.
  (c) Discuss the stability in the case of an unmagnetized column ($B_p = 0$) with zero external axial field ($B_{z0} = 0$).

9.2 Show that δW_F, as given by (8.55a), may be rewritten in the form·

$$\delta W_F = \frac{1}{2}\int_V d^3x \left\{ \frac{|\mathbf{Q} \times \mathbf{b}|^2}{\mu_0} + \mu_0 \left|\frac{\mathbf{Q}\cdot\mathbf{b}}{\mu_0} - \frac{\boldsymbol{\xi}\cdot\text{grad}\,P}{B}\right|^2 + \Gamma P(\text{div}\,\boldsymbol{\xi})^2 \right.$$

$$\left. - \mathbf{J}\cdot\mathbf{b}(\mathbf{b} \times \boldsymbol{\xi})\cdot\mathbf{Q} - 2(\boldsymbol{\xi}\cdot\text{grad}\,P)\boldsymbol{\xi}\cdot\boldsymbol{\rho} \right\}$$

where

$$\boldsymbol{\rho} = \frac{\mu_0}{B^2}\left\{\mathbf{b} \times \text{grad}\left(P + \frac{B^2}{2\mu_0}\right)\right\} \times \mathbf{b}$$

describes the curvature of the field lines, and where $\mathbf{b}$ is a unit vector along $\mathbf{B}$. *Remark*: The first three terms are stabilizing and are related to Alfvén, magnetoacoustic and sound waves respectively; the fourth term is destabilizing for the kink mode, and the final term is destabilizing for interchange modes.

9.3 A cylindrical jet of velocity $\mathbf{v}$ with density η_i, pressure P_i and axial magnetic field $\mathbf{B}$ is confined by an external medium with density η_e and pressure P_e. Both media are assumed incompressible.
  (a) Show that the linearized MHD equations, for variations of the form $\exp[-i(\omega t - m\theta - kz)]$ in cylindrical polar coordinates, imply

$$\left(\frac{d}{dr^2} + \frac{2}{r}\frac{d}{dr} - \frac{m^2}{r^2} - k^2\right)P_1 = 0$$

for the pressure fluctuations in either media and

$$\xi_r = \begin{cases} \dfrac{1}{\eta_i}\dfrac{\partial P_{1i}}{\partial r} \Big/ [(\omega - kv)^2 - k^2 v_A^2] \\[12pt] \dfrac{1}{\eta_e}\dfrac{\partial P_{1e}}{\partial r} \Big/ \omega^2 \end{cases}$$

for the radial component of the fluid displacement in the two media, (i = internal, e = external).

(b) The solutions for P_1 which vanish at $r = 0$ and $r = \infty$ are

$$P_{1i} \propto I_m(kr), \quad P_{1e} \propto K_m(kr).$$

Show that the continuity of P_1 and ξ_r at the edge ($r = a$) of the cylinder imply the dispersion relation

$$\frac{I'_m(ka)K_m(ka)}{I_m(ka)K'_m(ka)} = \frac{\eta_i}{\eta_e}\frac{(\omega - kv)^2 - k^2 v_A^2}{\omega^2}$$

(c) Assuming $ka \gg 1$ (cf. part (b) of Exercise 9.1) determine the conditions for growth of the Kelvin–Helmholtz instability, and find the growth rate.

9.4 Show that the temporal Fourier transform of the expression (9.26) for the generalized Ohm's Law, with $\mathbf{v} = 0$, i.e.

$$-\frac{i\omega n_e}{m_e e^2}\mathbf{J} = \frac{1}{n_e e}\,\mathrm{grad}\,P^{(e)} + \mathbf{E} - \frac{1}{n_e e}\mathbf{J} \times \mathbf{B} - \frac{m_e v_0}{n_e e^2}\mathbf{J}$$

implies

$$J_i = \sigma_{ij}E_j^{eff}$$

with

$$\mathbf{E}^{eff} = \mathbf{E} + \frac{1}{n_e e}\,\mathrm{grad}\,P^{(e)}$$

and, in a coordinate system with $\mathbf{B}$ along the z-axis,

$$\sigma_{ij} = \begin{bmatrix} a & b & 0 \\ -b & a & 0 \\ 0 & 0 & c \end{bmatrix}$$

$$a = \frac{i\varepsilon_0\omega_p^2(\omega + iv_0)}{(\omega + iv_0)^2 - \Omega_e^2}, \quad b = -\frac{i\Omega_e}{\omega + iv_0}a, \quad c = \frac{i\varepsilon_0\omega_p^2}{\omega + iv_0}.$$

9.5 Assume the collisional term $C^{(e,i)}$ for the electron distribution in (9.15) is

$$C^{(e,i)} = -v_0(f^{(e)} - f_M^{(e)})$$

where $f_M^{(e)}$ is a Maxwellian distribution.

(a) Show that the steady solution of (9.15) in the presence of an electric field $\mathbf{E}$ is

$$f^{(e)} - f_M^{(e)} = -\frac{e\mathbf{E}\cdot\mathbf{v}}{v_0 T_e}f_M^{(e)}$$

(b) Hence evaluate $\mathbf{R}^{(e)}$, as given by (9.17), and show that (9.21c), with $\mathbf{v}^{(i)} = 0$, implies $\mathbf{J} = \sigma\mathbf{E}$ with σ given by (9.27).

(c) Assume that there is a gradient in temperature T_e (with $P^{(e)} = n_e T_e$ = constant) and show that the result of part (a) is replaced by

$$f^{(e)} - f_M^{(e)} = \frac{\mathbf{v} \cdot \text{grad} \, T_e}{v_0 T_e} \left(-\frac{m_e v^2}{2T_e} + \frac{5}{2} \right) f_M^{(e)}$$

(d) Hence evaluate $\mathbf{q}^{(e)}$, as given by (9.17), and show that one has

$$\mathbf{q}^{(e)} = -\kappa^{(e)} \, \text{grad} \, T_e,$$

with the electron thermal conductivity given by

$$\kappa^{(e)} = \frac{5}{2} \frac{n_e T_e}{m_e v_0}$$

9.6 (a) Show that inserting

$$\frac{d\alpha}{dt} = \frac{\partial \alpha}{\partial t} + \mathbf{v} \cdot \text{grad} \, \alpha = 0$$

and

$$\frac{d\beta}{dt} = \frac{\partial \beta}{\partial t} + \mathbf{v} \cdot \text{grad} \, \beta = 0$$

in

$$\mathbf{B} = (\text{grad} \, \alpha) \times (\text{grad} \, \beta)$$

implies

$$\frac{\partial \mathbf{B}}{\partial t} = \text{curl} \, (\mathbf{v} \times \mathbf{B}).$$

(b) Assume that $\mathbf{B}$ is in the x–z plane and represent it by Euler potentials $\alpha = \psi$ and $\beta = y$. Show that

$$\frac{\partial \mathbf{B}}{\partial t} = \text{curl} \, (\mathbf{v} \times \mathbf{B}) + \frac{1}{\mu_0 \sigma} \nabla^2 \mathbf{B}$$

and div $\mathbf{v} = 0$ then imply

$$\frac{\partial \psi}{\partial t} + \mathbf{v} \cdot \text{grad} \, \psi = \frac{1}{\mu_0 \sigma} \nabla^2 \psi.$$

PART IV

Instabilities in magnetized collisionless plasmas

10

Dispersion in a magnetized plasma

10.1 Dielectric tensors

The inclusion of a magnetic field leads to a considerable increase in the richness and variety of the wave motions which can exist in a plasma. It also leads to a qualitative change in the orbits of the particles, which become spirals about the magnetic field lines. This affects the nature of particle-wave interactions. Not surprisingly, the generalization from unmagnetized to magnetized plasmas involves a marked increase in algebraic complexity of the relevant formulas. However the basic principles do not change.

In this Chapter the generalization (to the magnetized case) of the calculations of the dielectric tensor (§10.1) and of the quasilinear equations for wave-particle interactions (§10.5) are presented, and the properties of important classes of waves are discussed. The case of cold plasma wave modes is treated in a formal way in §10.2, the magnetoionic wave modes are discussed in §10.3, and low frequency wave modes are treated in §10.4. The waves discussed in detail in this chapter can nearly all be regarded either as magnetized versions of the waves in an unmagnetized plasma or as collisionless analogs of the MHD waves. There are other wave modes which are intrinsic to collisionless magnetized plasmas; some of these are discussed in Chapter 12.

For an unmagnetized plasma there are three equivalent methods for calculating the response tensors: the cold-plasma method (§2.1), generalized as discussed following (2.25), the Vlasov approach (§2.2) and the forward-scattering method (§5.5). In the magnetized case only the latter two methods are equivalent; the cold plasma method is less general in that it is incapable of taking account of non-zero particle gyroradii. The cold-plasma approach is much simpler than the other two from an algebraic viewpoint, and there are many applications in which the gyroradii can indeed be approximated by zero. As a consequence, cold plasma theory is used as widely as possible with 'warm-plasma' effects being taken into account in those limiting cases where the cold-plasma approximation breaks down.

In the cold-plasma approximation each species α of particle is described by its plasma parameters (q_α, m_α, n_α, $\omega_{p\alpha}$, $\Omega_\alpha = |q_\alpha|B_0/m_\alpha$) and a fluid velocity $\mathbf{v}^{(\alpha)}$. In the nonrelativistic case the equation of fluid motion is

$$m_\alpha \left[\frac{\partial}{\partial t} \mathbf{v}^{(\alpha)} + \mathbf{v}^{(\alpha)} \cdot \mathrm{grad}\, \mathbf{v}^{(\alpha)} \right] = q_\alpha [\mathbf{E} + \mathbf{v}^{(\alpha)} \times \mathbf{B}] \qquad (10.1)$$

where $\mathbf{B}$ includes both the background field $\mathbf{B}_0$ and the magnetic field associated with the test field $\mathbf{E}$. On linearizing (10.1) and Fourier transforming in time one finds

$$[-i\omega\delta_{ij} - \epsilon_\alpha \Omega_\alpha \varepsilon_{ijk} b_k] v_j^{(\alpha)} = \frac{q_\alpha}{m_\alpha} E_i \qquad (10.2)$$

where $\epsilon_\alpha = q_\alpha / |q_\alpha|$ is the sign of the charge, and where $\mathbf{b} = \mathbf{B}_0/B_0$ is a unit vector along $\mathbf{B}_0$. The permutation symbol is defined by

$$\varepsilon_{ijk} = \begin{cases} 1 & \text{for } ijk \text{ an even permutation of } xyz \\ -1 & \text{for } ijk \text{ an odd permutation of } xyz \\ 0 & \text{otherwise} \end{cases} \qquad (10.3)$$

Let us define $\tau_{ij}^{(\alpha)}(\omega)$ by

$$[-i\omega\delta_{ij} - \epsilon_\alpha \Omega_\alpha \varepsilon_{ijk} b_k] \tau_{jl}^{(\alpha)}(\omega) = -i\omega\delta_{il}. \qquad (10.4)$$

The solution of (10.2) is then

$$v_i^{(\alpha)} = \frac{q_\alpha}{m_\alpha} \frac{i}{\omega} \tau_{ij}^{(\alpha)}(\omega) E_j. \qquad (10.5)$$

The current density is

$$\mathbf{J} = \sum_\alpha q_\alpha n_\alpha \mathbf{v}^{(\alpha)}. \qquad (10.6)$$

Then using (1.26) and (1.27) one identifies the *cold plasma dielectric tensor* as

$$K_{ij}(\omega) = \delta_{ij} - \sum_\alpha \frac{\omega_{p\alpha}^2}{\omega^2} \tau_{ij}^{(\alpha)}(\omega). \qquad (10.7)$$

Explicit evaluation of $\tau_{ij}^{(\alpha)}(\omega)$ gives

$$\begin{aligned}
\tau_{ij}^{(\alpha)}(\omega) &= \frac{\omega^2}{\omega^2 - \Omega_\alpha^2} \left[\delta_{ij} - \frac{\Omega_\alpha^2}{\omega^2} b_i b_j + i\epsilon_\alpha \frac{\Omega_\alpha}{\omega} \varepsilon_{ijk} b_k \right] \\
&= \begin{bmatrix} \dfrac{\omega^2}{\omega^2 - \Omega_\alpha^2} & \dfrac{i\epsilon_\alpha \omega \Omega_\alpha}{\omega^2 - \Omega_\alpha^2} & 0 \\[2ex] -\dfrac{i\epsilon_\alpha \omega \Omega_\alpha}{\omega^2 - \Omega_\alpha^2} & \dfrac{\omega^2}{\omega^2 - \Omega_\alpha^2} & 0 \\[2ex] 0 & 0 & 1 \end{bmatrix},
\end{aligned} \qquad (10.8)$$

where the matrix form is for $\mathbf{b}$ along the z-axis. The unmagnetized case is reproduced by replacing $\tau_{ij}^{(\alpha)}(\omega)$ by δ_{ij}.

Before proceeding to the more general methods for calculating the dielectric tensor, let us write down the counterpart of the cold-plasma quadratic response tensor (6.76). Only the electronic contribution is retained, and for simplicity the superscript (e) on $\tau_{ij}(\omega)$ is omitted. The relevant result is

$$\alpha_{ijl}(k, k', k'') = -\frac{e^3 n_e}{2m_e^2} \left\{ \frac{k_r}{\omega'} \tau_{rj}(\omega')\tau_{il}(\omega'') + \frac{k_r}{\omega''} \tau_{rl}(\omega'')\tau_{ij}(\omega') + \frac{k_r'}{\omega} \tau_{ir}(\omega)\tau_{jl}(\omega'') \right.$$

$$\left. + \frac{k_r''}{\omega} \tau_{ir}(\omega)\tau_{lj}(\omega') - \frac{k_r''}{\omega'} \tau_{rj}(\omega')\tau_{il}(\omega) - \frac{k_r'}{\omega''} \tau_{rl}(\omega'')\tau_{ij}(\omega) \right\} \quad (10.9)$$

General forms for the response tensors can be evaluated explicitly after expanding in Bessel functions. It is convenient to introduce this expansion for the single particle current. The current (5.51), viz.

$$\mathbf{J}(\omega, \mathbf{k}) = q \int dt \, \mathbf{v}(t) \exp \left[i\omega t - i\mathbf{k} \cdot \mathbf{X}(t) \right] \quad (10.10)$$

is to be evaluated for a spiralling charge. The equation of motion

$$\frac{d\mathbf{p}}{dt} = q\mathbf{v} \times \mathbf{B} \quad (10.11)$$

may be solved by writing ($\epsilon = q/|q|$)

$$\mathbf{p} = \mathbf{p}(\phi) = (p_\perp \cos \phi, -\epsilon p_\perp \sin \phi, p_\parallel) \quad (10.12)$$

with $p_\perp, p_\parallel, v_\perp = p_\perp/m\gamma, v_\parallel = p_\parallel/m\gamma$ and $\gamma = (1 - v_\perp^2/c^2 - v_\parallel^2/c^2)^{-1/2}$ all constants of the motion. Then (10.11) reduces to

$$\frac{d\phi}{dt} = \epsilon \Omega, \qquad \Omega = \frac{|q|B}{m\gamma}, \quad (10.13)$$

where Ω is the gyrofrequency of the particle. We assume $\phi = \phi_0$ at $t = t_0$ and write the solution for the orbit as

$$\mathbf{X}(t) = \mathbf{x}_0 + (R \sin (\phi_0 + \Omega t), \epsilon R \cos (\phi_0 + \Omega t), v_\parallel t) \quad (10.14)$$

where $\mathbf{x}_0$ describes the initial position of the centre of gyration and where

$$R = \frac{v_\perp}{\Omega} = \frac{p_\perp}{|q|B} \quad (10.15)$$

is the radius of gyration. On substituting (10.14) into (10.10), and writing

$$\mathbf{k} = (k_\perp \cos \psi, k_\perp \sin \psi, k_\parallel), \quad (10.16)$$

one uses the generating function for Bessel functions

$$e^{iz \sin \phi} = \sum_{s=-\infty}^{\infty} e^{is\phi} J_s(z) \quad (10.17)$$

to write

$$\exp \left[i\omega t - i\mathbf{k} \cdot \mathbf{X}(t) \right] = \sum_{s=-\infty}^{\infty} J_s(k_\perp R) \exp \left[i(\omega - s\Omega - k_\parallel v_\parallel)t - i\mathbf{k} \cdot \mathbf{x}_0 - is(\phi_0 + \epsilon \psi) \right] \quad (10.18)$$

Then the integral in (10.10) gives a δ-function and one has

$$\mathbf{J}(\omega, \mathbf{k}) = q e^{-i\mathbf{k} \cdot \mathbf{x}_0} \sum_{s=-\infty}^{\infty} e^{-is(\phi_0 + \epsilon \psi)} \mathbf{V}(\mathbf{k}, \mathbf{p}; s) 2\pi\delta(\omega - s\Omega - k_\parallel v_\parallel) \quad (10.19)$$

with

$$\mathbf{V}(\mathbf{k}, \mathbf{p}; s) = (\tfrac{1}{2}v_{\perp}\{e^{i\epsilon\psi}J_{s-1}(k_{\perp}R) + e^{-i\epsilon\psi}J_{s+1}(k_{\perp}R)\},$$

$$-\frac{i\epsilon v_{\perp}}{2}\{e^{i\epsilon\psi}J_{s-1}(k_{\perp}R) - e^{-i\epsilon\psi}J_{s+1}(k_{\perp}R)\}, v_{\parallel}J_s(k_{\perp}R)). \qquad (10.20)$$

The form of the dielectric tensor which results when one uses the Vlasov approach and then expands in Bessel functions is [cf. Exercise 10.1]

$$K_{ij}(\omega, \mathbf{k}) = \delta_{ij} + \sum \frac{q^2}{\varepsilon_0 \omega^2} \int d^3\mathbf{p} \left[\frac{v_{\parallel}}{v_{\perp}}\left(v_{\perp}\frac{\partial}{\partial p_{\parallel}} - v_{\parallel}\frac{\partial}{\partial p_{\perp}} \right) f(\mathbf{p}) b_i b_j \right.$$

$$\left. + \sum_{s=-\infty}^{\infty} \frac{V_i(\mathbf{k}, \mathbf{p}; s)V_j^*(\mathbf{k}, \mathbf{p}; s)}{\omega - s\Omega - k_{\parallel}v_{\parallel}} \left\{ \frac{\omega - k_{\parallel}v_{\parallel}}{v_{\perp}}\frac{\partial}{\partial p_{\perp}} + k_{\parallel}\frac{\partial}{\partial p_{\parallel}} \right\} f(\mathbf{p}) \right], \qquad (10.21)$$

with $\mathbf{b}$ a unit vector along the background magnetic field (the z-axis here). The distribution function is assumed time-independent, and as a consequence it cannot depend on ϕ; that is, $f(\mathbf{p})$ depends only on $p_{\perp}$ and $p_{\parallel}$. The sum in (10.21) implies a sum over (unlabelled) species of particle, as in (2.24). The dissipative part of (10.21) is the anti-hermitian (A) part obtained by imposing the causal condition

$$K_{ij}^A(\omega, \mathbf{k}) = \sum -\frac{i\pi q^2}{\varepsilon_0 \omega^2} \int d^3\mathbf{p} \sum_{s=-\infty}^{\infty} V_i(\mathbf{k}, \mathbf{p}; s)V_j^*(\mathbf{k}, \mathbf{p}; s)\delta(\omega - s\Omega - k_{\parallel}v_{\parallel})$$

$$\times \left\{ \frac{s\Omega}{v_{\perp}}\frac{\partial}{\partial p_{\perp}} + k_{\parallel}\frac{\partial}{\partial p_{\parallel}} \right\} f(\mathbf{p}) \qquad (10.21')$$

The forward-scattering method (§5.5) leads to the following form in the time-asymptotic limit:

$$K_{ij}(\omega, \mathbf{k}) = \delta_{ij} - \sum \frac{q^2}{\varepsilon_0 m \omega^2} \int \frac{d^3\mathbf{p}}{\gamma} f(\mathbf{p}) \sum_{s=-\infty}^{\infty} \left[J_s^2(k_{\perp}R)\tau_{ij}(\omega_s) \right.$$

$$+ \frac{J_s(k_{\perp}R)}{\omega_s}\left\{ \tau_{im}(\omega_s)k_m V_j^*(\mathbf{k}, \mathbf{p}; s) + V_i(\mathbf{k}, \mathbf{p}; s)k_m\tau_{mj}(\omega_s) \right\}$$

$$\left. + \frac{1}{\omega_s^2}\left(k_l k_m \tau_{lm}(\omega_s) - \frac{\omega^2}{c^2} \right) V_i(\mathbf{k}, \mathbf{p}; s)V_j^*(\mathbf{k}, \mathbf{p}; s) \right] \qquad (10.22)$$

with $\omega_s = \omega - s\Omega - k_{\parallel}v_{\parallel}$. The form (10.22) is the appropriate generalization of (2.25). One may also obtain (10.22) from (10.21) by a partial integration, which however is particularly tedious.

In the *small gyroradius limit* $k_{\perp}R = 0$, $J_s(k_{\perp}R)$ is zero except for $J_0(0) = 1$. Then for $\psi = 0$ (10.20) simplifies to

$$\mathbf{V}(\mathbf{k}, \mathbf{p}; s) = \begin{cases} (0, i\epsilon v_{\perp}k_{\perp}R, v_{\parallel}) & s = 0 \\ \tfrac{1}{2}v_{\perp}(1, -is\epsilon, 0) & s = \pm 1 \\ 0 & |s| > 1. \end{cases} \qquad (10.23)$$

(The correction of order $k_{\perp}R$ in the y-component for $s = 0$ is important in the derivation of the form (10.74) below.) Formally the small gyroradius limit corresponds to $v_{\perp} = 0$, and then only the contribution from $s = 0$ remains in (10.23).

The resulting approximate form may be derived by generalizing the cold-plasma method, cf. Exercise 10.2.

The counterpart of the approximate form (6.54) for the nonlinear response tensor is

$$\alpha_{ijl}(k, k_1, k_2) \approx \frac{e\varepsilon_0\omega_2}{m_e} \tau_{ij}^{(e)}(\omega)k_{2l}\chi^{L(e)}(\omega_2, \mathbf{k}_2),\tag{10.24}$$

which applies for $\omega \approx \omega_1 \gg \omega_2$ and $\omega_2/k_2 \ll V_e \ll \omega/k, \omega_1/k_1$. An expression for $\chi^{L(e)}$ for a thermal magnetized plasma follows from (12.4) below; the real part of $\chi^{L(e)}$ is identified as the electronic contribution on the right hand side of (12.4).

10.2 The cold plasma modes

In this Section the formal properties of waves in a cold plasma are summarized. These properties are required for later reference.

The cold plasma dielectric tensor (10.7) may be written in the form (Stix 1962)

$$K_{ij}(\omega) = \begin{bmatrix} S & -iD & 0 \\ iD & S & 0 \\ 0 & 0 & P \end{bmatrix}\tag{10.25}$$

with

$$S = \tfrac{1}{2}(R_+ + R_-), \quad D = \tfrac{1}{2}(R_+ - R_-),$$

$$R_\pm = 1 - \sum_\alpha \frac{\omega_{P\alpha}^2}{\omega^2} \frac{\omega}{\omega \pm \epsilon_\alpha\Omega_\alpha}, \quad P = 1 - \sum_\alpha \frac{\omega_{P\alpha}^2}{\omega^2}\tag{10.26}$$

The tensors Λ_{ij} and λ_{ij}, defined by (1.29) and (2.55) respectively, are

$$\Lambda_{ij} = \begin{bmatrix} S - N^2\cos^2\theta & -iD & N^2\sin\theta\cos\theta \\ iD & S - N^2 & 0 \\ N^2\sin\theta\cos\theta & 0 & P - N^2\sin^2\theta \end{bmatrix}\tag{10.27}$$

and

$$\lambda_{ij} = \begin{bmatrix} (S - N^2)(P - N^2\sin^2\theta) & iD(P - N^2\sin^2\theta) & -(S - N^2)N^2\sin\theta\cos\theta \\ -iD(P - N^2\sin^2\theta) & -N^2A + PS & iDN^2\sin\theta\cos\theta \\ -(S - N^2)N^2\sin\theta\cos\theta & -iDN^2\sin\theta\cos\theta & N^4\cos^2\theta - N^2S(1 + \cos^2\theta) + S^2 - D^2 \end{bmatrix}\tag{10.28}$$

The quantity A appears as the coefficient of N^4 in

$$\Lambda = \det[\Lambda_{ij}] = AN^4 - BN^2 + C,\tag{10.29}$$

with

$$A = P\cos^2\theta + S\sin^2\theta$$
$$B = (S^2 - D^2)\sin^2\theta + PS(1 + \cos^2\theta)\tag{10.30}$$
$$C = P(S^2 - D^2).$$

The solutions of the dispersion equation $\Lambda = 0$ are

$$N^2 = N_\pm^2(\omega, \theta) = \frac{B \pm F}{2A}\tag{10.31a}$$

with

$$F^2 = (PS - S^2 + D^2)^2 \sin^4 \theta + 4P^2 D^2 \cos^2 \theta. \tag{10.31b}$$

For the present the two modes are labelled $M = \pm$. Their polarization vectors may be written in the form

$$\mathbf{e}_M = \frac{K_M \boldsymbol{\kappa} + T_M \mathbf{t} + i\mathbf{a}}{(K_M^2 + T_M^2 + 1)^{1/2}}, \tag{10.32}$$

with $\boldsymbol{\kappa}$, $\mathbf{t}$ and $\mathbf{a}$ given by (8.25) for $\psi = 0$ in (10.16):

$$\boldsymbol{\kappa} = (\sin \theta, 0, \cos \theta), \quad \mathbf{t} = (\cos \theta, 0, -\sin \theta), \quad \mathbf{a} = (0, 1, 0). \tag{10.33}$$

The middle column of (10.28) gives

$$T_M = \frac{DP \cos \theta}{AN_M^2 - PS}, \quad K_M = \frac{(P - N_M^2)D \sin \theta}{AN_M^2 - PS}. \tag{10.34}$$

It is sometimes convenient to invert the relations (10.34) and (10.29), in the sense that one solves

$$T^2 - \frac{(PS - S^2 + D^2) \sin^2 \theta}{PD \cos \theta} T - 1 = 0 \tag{10.35}$$

for $T = T_\pm$. This gives

$$T_\pm = \frac{(PS - S^2 + D^2) \sin^2 \theta \pm F}{2PD \cos \theta} = \frac{-2PD \cos \theta}{(PS - S^2 + D^2) \sin^2 \theta \mp F}. \tag{10.36}$$

One then finds N_M^2 and K_M by substituting the relevant solution $T = T_M$ in

$$N^2 = \frac{P}{A} S + \frac{D \cos \theta}{T} = \frac{S^2 - D^2}{S - DT \cos \theta} \tag{10.37}$$

and

$$K = \frac{\sin \theta}{A} \{(P - S)T \cos \theta - D\} = \frac{\sin \theta}{P} \frac{(PS - S^2 + D^2)T \cos \theta - PD}{S - DT \cos \theta} \tag{10.38}$$

respectively.

The representation (10.32) for the polarization vector corresponds to a longitudinal part, described by K_M, and a transverse part, described by T_M. In general the transverse part is elliptical, and T_M is the axial ratio of the polarization ellipse, cf. Figure 10.1. The special values $T_M = \pm 1$ corresponds respectively to right and left circular polarizations, $T_M = 0$ to linear polarization along $\mathbf{a}$ and $T_M = \infty$ to linear polarization along $\mathbf{t}$. The 'orthogonality' of the modes requires that their transverse parts be orthogonal:

$$T_+ T_- = 1. \tag{10.39}$$

The polarization vectors themselves are not orthogonal in general. That is

$$\mathbf{e}_+ \cdot \mathbf{e}_-^* = \frac{K_+ K_-}{(K_+^2 + T_+^2 + 1)^{1/2}(K_-^2 + T_-^2 + 1)^{1/2}}$$

is not zero in general.

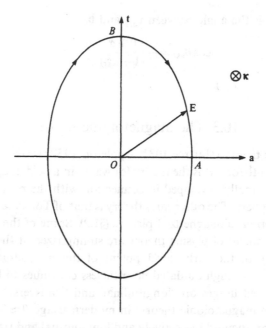

Fig. 10.1 In an elliptically polarized wave the tip of the electric vector traces out an ellipse in the plane orthogonal to κ. In the figure κ is into the page and the directions $\mathbf{t}$ and $\mathbf{a}$ are shown. The axial ratio T_M is positive if the sense of rotation is clockwise (as indicated) and the magnitude of T_M is equal to the axial ratio OB/OA.

The energy flux in waves is equal to the group velocity

$$\mathbf{v}_{gM} = \frac{c}{\dfrac{\partial}{\partial\omega}(\omega N_M)}\left(\kappa - \frac{1}{N_M}\frac{\partial N_M}{\partial\theta}\mathbf{t}\right)$$

(10.40)

times the energy density in the waves. If the medium is not spatially dispersive, i.e. if K_{ij} does not depend on $\mathbf{k}$, then the energy flux is also given by the Poynting vector, implying

$$\mathbf{v}_g \propto \mathrm{Re}\,[\mathbf{E}^* \times \mathbf{B}].$$

The components of this relation along $\mathbf{k}$ and $\mathbf{t}$ give, on comparison with (10.40) [cf. Exercise 10.3],

$$R_M = \frac{1}{(1-|\kappa\cdot\mathbf{e}_M|^2)2N_M\dfrac{\partial}{\partial\omega}(\omega N_M)},$$

(10.41)

where R_M is defined in (2.66), and

$$\frac{1}{N_M}\frac{\partial N_M}{\partial\theta} = \frac{K_M T_M}{1+T_M^2}.$$

(10.42)

The ray angle θ_{rM} is the angle between $\mathbf{v}_{gM}$ and $\mathbf{b}$:

$$\cos\theta_{rM} = \frac{\cos\theta + \alpha_M \sin\theta}{(1+\alpha_M^2)^{1/2}},\tag{10.43}$$

with $\alpha_M = K_M T_M/(1+T_M^2)$.

10.3 The magnetoionic modes

The magnetoionic theory (Hartree 1931, Appleton 1932) is one of the two oldest branches of plasma theory. It is the theory for waves in a cold, magnetized electron gas, and it was originally developed in connection with the propagation of radio waves in the ionosphere. The other early theory is that of Tonks & Langmuir (1929) for waves in a thermal unmagnetized plasma (§1.2). Some of the highlights in the subsequent development of plasma theory are summarized at the end of §1.2. An unfortunate legacy of the early development of the magnetoionic theory is a terminology which although outdated nevertheless continues to be used. Notable examples of outdated usages are 'longitudinal' and 'transverse' to refer to $\theta \approx 0$ and $\theta \approx \pi/2$ in the magnetoionic theory: in modern usage $\theta \approx 0$ and $\theta \approx \pi/2$ are 'parallel' and 'perpendicular' respectively and longitudinal and transverse are used only to refer to directions relative to $\mathbf{k}$, e.g. 'longitudinal polarization' means $\mathbf{E}$ along $\mathbf{k}$. Even the name 'magnetoionic' is outdated: 'ions', in the modern sense, play no role in the theory. As far as practicable the outdated terminology of the magnetoionic theory is avoided here.

There are only two plasma parameters in the magnetoionic theory: ω_p and Ω_e. These are often written in terms of magnetoionic parameters

$$X = \frac{\omega_p^2}{\omega^2}, \quad Y = \frac{\Omega_e}{\omega}.\tag{10.44a, b}$$

The cold plasma theory of §10.2 becomes the magnetoionic theory on writing

$$R_\pm = 1 - \frac{X}{1 \mp Y}, \quad P = 1 - X,\tag{10.45a, b}$$

$$S = \frac{1-X-Y^2}{1-Y^2}, \quad D = \frac{-XY}{1-Y^2}, \quad S^2 - D^2 = R_+ R_- = \frac{(1+X)^2 - Y^2}{1-Y^2}.$$

$$\tag{10.45c, d, e}$$

The two solutions $N_\pm^2(\omega,\theta)$ plotted as a function of ω/ω_p separate into four branches of real modes ($N^2 > 0$) and three branches of evanescent modes (which are of no further interest here). These are illustrated in Figure 10.2. The two magnetoinic modes are generally labelled as 'extraordinary' and 'ordinary' modes. Here the higher and lower frequency branches of the extraordinary mode are called the x-mode and the z-mode respectively, and the higher and lower frequency branches of the ordinary mode are called the o-mode and the whistler mode respectively, (Another common terminology for these modes is $R-X, L-X, L-0$

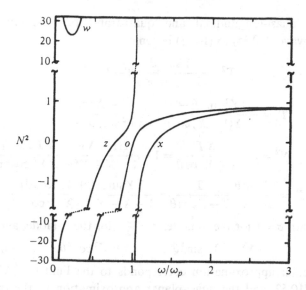

Fig. 10.2 The refractive index curves for the magnetoionic modes are illustrated schematically as a function of ω/ω_p for $\Omega_e/\omega_p < 1$ and θ not close to either 0 or $\pi/2$. The four branches are labelled (o = o-mode, x = x-mode, z = z-mode, w = whistler mode).

and $R - 0$, respectively, with R and L referring to right and left respectively; this notation is outdated in that right and left polarizations in modern usage are defined as screw senses relatively to **k**, whereas R and L in the magnetoionic theory are defined as screw senses relative to **b**.)

The various branches of the curves end at cutoffs (zeros of N^2) or resonances (infinities of N^2) which separate real solutions ($N^2 > 0$) from evanescent solutions ($N^2 < 0$). The cutoffs are at $C/A = 0$, i.e. $(1 - X)\{(1 - X)^2 - Y^2\} = 0$, which implies $\omega = \omega_p$, $\omega = \omega_x$ or $\omega = \omega_z$:

$$\omega_x = \tfrac{1}{2}\Omega_e + \tfrac{1}{2}\{\Omega_e^2 + 4\omega_p^2\}^{1/2}, \quad \omega_z = \omega_x - \Omega_e. \tag{10.46a, b}$$

The resonances are at $A/C = 0$, i.e. $1 - X - Y^2 + XY^2 \cos^2\theta = 0$, implying $\omega = \omega_{\pm}(\theta)$ with

$$\omega_{\pm}^2(\theta) = \tfrac{1}{2}(\omega_p^2 + \Omega_e^2) \pm \tfrac{1}{2}\{(\omega_p^2 + \Omega_e^2)^2 - 4\omega_p^2\Omega_e^2 \cos^2\theta\}^{1/2}. \tag{10.47}$$

The *o-mode* and *x-mode* exist at frequencies $\omega > \omega_p$ and $\omega > \omega_x$, respectively. Both have $N^2 < 1$. The polarization ellipse of the *o-mode* has a handedness opposite to that of a spiralling electron and that for the *x-mode* is in the same sense as a spiralling electron. In the limit $\Omega_e/\omega_p \to 0$ the *o-mode* and *x-mode* reduce to two oppositely circularly polarized waves with the same refractive index $N^2 = 1 - X = 1 - \omega_p^2/\omega^2$ as in an unmagnetized plasma.

In deriving approximate expressions for the refractive indices for the two modes it is convenient to start from the assumption that the polarization may be regarded as either circular or linear (planar). We refer to these as the *quasi-circular* and *quasi-planar* approximations respectively. (In the older magnetoionic literature these are referred to as 'quasi-longitudinal' and 'quasi-transverse' respectively.) For the magnetoionic waves (10.35) to (10.38) become

$$T^2 + \frac{Y \sin^2 \theta}{(1-X)\cos\theta} T - 1 = 0, \tag{10.48}$$

$$T_\sigma = -\frac{\frac{1}{2}Y^2 \sin^2\theta - \sigma\Delta}{Y(1-X)\cos\theta} = \frac{Y(1-X)\cos\theta}{\frac{1}{2}Y^2\sin^2\theta - \sigma\Delta}, \tag{10.49}$$

$$N^2 = 1 - \frac{XT}{T - Y\cos\theta} = 1 - \frac{X(1-X)(1 + YT\cos\theta)}{1 - X - Y^2 + XY^2\cos^2\theta}, \tag{10.50}$$

$$K = \frac{XY\sin\theta}{1-X} \frac{T}{T - Y\cos\theta} = \frac{XY\sin\theta(1 + YT\cos\theta)}{1 - X - Y^2 + XY^2\cos^2\theta}, \tag{10.51}$$

respectively, with $\sigma = 1$ for the o-mode, $\sigma = -1$ for the x-mode, and with

$$\Delta^2 = \frac{1}{4}Y^4 \sin^4\theta + (1-X)^2 Y^2 \cos^2\theta. \tag{10.52}$$

The quasi-circular approximation corresponds to the limit $(1-X)^2 Y^2 \cos^2\theta \gg \frac{1}{4}Y^4\sin^4\theta$ in (10.52) and the quasi-planar approximation to the opposite limit. The value of θ, $(\theta_0$ say) at which the two terms on the right hand side of (10.52) are equal i.e. the solution of $\frac{1}{2}Y^2\sin^2\theta_0 = (1-x)Y\cos\theta_0$, is

$$\theta_0 = \arcsin\left[\frac{\sqrt{2}}{Y}\{[(1-X)^4 + Y^2(1-X)^2]^{1/2} - (1-X)^2\}^{1/2}\right]$$

$$\approx \arccos\tfrac{1}{2}Y \quad \text{for} \quad X \ll 1, Y \ll 1. \tag{10.53}$$

The quasi-circular approximation corresponds to $|T| = 1$, and the signs are such that one has

$$T_\sigma = -\sigma\frac{\cos\theta}{|\cos\theta|}\frac{1-X}{|1-X|}. \tag{10.54}$$

Then (10.50) and (10.51) imply

$$N_\sigma^2 = 1 - \frac{X}{1 + \sigma Y|\cos\theta|}, \quad K_\sigma = \frac{XY\sin\theta}{1-X}\frac{1}{1 + \sigma Y|\cos\theta|} \tag{10.55}$$

for the o-mode ($\sigma = 1$) and x-mode ($\sigma = -1$) (these modes exist only for $X < 1$). The opposite quasi-planar approximation corresponds to

$$T_o = \infty, \quad N_o^2 = 1 - X, \qquad\qquad K_o = XY\sin\theta/(1-X), \tag{10.56a}$$

$$T_x = 0, \quad N_x^2 = 1 - \frac{X(1-X)}{1 - X - Y^2 + XY^2\cos^2\theta}, \quad K_x = \frac{XY\sin\theta}{1 - X - Y^2 + XY^2\cos^2\theta} \tag{10.56b}$$

The *z-mode* exists in the frequency range $\omega_z \leqslant \omega \leqslant \omega_+(\theta)$. In the range $\omega_p < \omega \leqslant \omega_+(\theta)$ it has $N_z^2 > 1$ and its polarization becomes predominantly longitudinal for $N_z^2 \gg 1$ as the frequency approaches the resonant frequency. For $\Omega_e \ll \omega_p$, the inclusion of thermal corrections leads to an approximate dispersion relation $\omega = \omega_z(k,\theta)$, with

$$\omega_z^2(k,\theta) \approx \omega_p^2 + 3k^2 V_e^2 + \Omega_e^2 \sin^2\theta \tag{10.57}$$

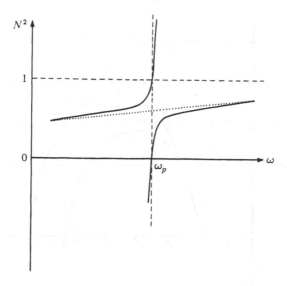

Fig. 10.3 The refractive index curves for the *o*-mode and the *z*-mode are illustrated schematically for $\omega \approx \omega_p$ and $\theta \approx 0$. As θ approaches zero the solid curve becomes distorted and approaches the refractive index curve (dotted curve) for the combined *o*-mode – *z*-mode for $\theta = 0$.

near the resonance. Thus the resonant *z*-mode may be regarded as a magnetized form of the Langmuir mode. The wave properties change rapidly with frequency and angle for $\omega_z < \omega \lesssim \omega_p$. The point $\omega = \omega_p$, $\theta = 0$ or $\pi/2$ is a singular point in the magnetoionic theory. For $\theta \neq 0$ one has $N_z^2 = 1$, $T_z = 0$ and $N_o^2 = 0$, $T_o = \infty$ at $\omega = \omega_p$, and for $\theta = 0$ or π one has $N_z^2 = N_o^2 = Y/(1 + Y)$ and $T_z = T_o = -\cos\theta/|\cos\theta|$. These properties are illustrated schematically in Figure 10.3.

The peculiar properties of the *z*-mode for $\omega \approx \omega_p$, $\theta \approx 0$ is related to the origin of the name '*z*-mode'. In ionospheric sounding experiments, a signal directed into the ionosphere splits into *o*-mode and *x*-mode components which reflect at different heights (Figure 10.4). At a receiver on the ground the returning signals lead to an '*o*-trace' and a '*x*-trace' which are separated by their time of arrival and have opposite polarizations. Sometimes a third '*z*-trace' is observed later than the other two and with the same polarization as the *o*-trace. The accepted interpretation is that if the *o*-component approaches the plasma level ($\omega = \omega_p$) at $\theta \approx 0$ then it can couple into the *z*-mode; the *z*-component reflects at a higher layer (near $\omega \approx \omega_z$) couples back into the *o*-mode at the plasma level on its return path, and then propagates to the ground to be observed as the *z*-trace.

Apart from the nearly-resonant case (10.57) there is no widely used approximate form for the dispersion relation of the *z*-mode.

The fourth branch corresponds to the *whistler mode*. It exists in the range $\omega \leqslant \omega_-(\theta)$. For most purpose the quasi-circular approximation is adequate, and

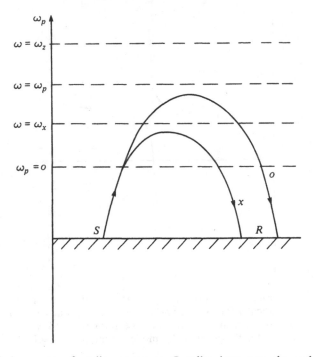

Fig. 10.4 A source of radio waves at S splits into o-mode and x-mode components on entering the ionosphere, and they are refracted along different ray paths as ω_p increases with increasing height. The layers $\omega = \omega_x, \omega = \omega_p$ and $\omega = \omega_z$ corresponding to the cutoffs of the x-mode, o-mode and z-mode, respectively, are indicated schematically. The z-trace occurs when the initial direction is such that as the o-component approaches the layer $\omega = \omega_p$ its angle of propagation approaches $\theta = 0$.

in this approxmation one has

$$N_W^2 = 1 - \frac{X}{1 - Y|\cos\theta|}. \tag{10.58}$$

The refractive index has a minimum,

$$(N_W^2)_{\min} = 1 + \frac{\omega_p^2}{\omega^2}, \quad \text{at} \quad \omega = \frac{\Omega_e|\cos\theta|}{2}, \tag{10.59}$$

and hence it is always greater than unity. For $X, Y|\cos\theta| \gg 1$, corresponding to frequencies well below this minimum, the wave properties simplify further to

$$N_W^2 = \frac{\omega_p^2}{\omega\Omega_e|\cos\theta|}, \quad \mathbf{e}_W = \frac{(1, i|\cos\theta|, 0)}{(1+\cos^2\theta)^{1/2}}, \quad R_W = \frac{\omega\Omega_e}{2\omega_p^2}\frac{1+\cos^2\theta}{|\cos\theta|}. \tag{10.60}$$

At even lower frequencies the motion of the ions cannot be neglected. In particular, for $\Omega_e \ll \omega_p$ the resonant frequency $\omega_-(\theta)$ may be approximated by

$$\omega_-(\theta) = \begin{cases} \Omega_e|\cos\theta| & |\cos\theta| > \omega_{LH}/\Omega_e \\ \omega_{LH} & |\cos\theta| \lesssim \omega_{LH}/\Omega_e, \end{cases} \tag{10.61}$$

where ω_{LH} is the *lower hybrid frequency*. The hybrid frequencies are defined formally as the zeros of S. In the magnetoionic theory $S = 0$ implies $1 - X - Y^2 = 0$, which gives the *upper hybrid frequency* $\omega_{UH} = (\omega_p^2 + \Omega_e^2)^{1/2}$. For $\Omega_i^2 \ll \omega_{pi}^2$, when one includes the motion of (a single species of) ions, $S = 0$ has another solution at $\omega = \omega_{LH}$ with

$$\omega_{LH} \approx \left[\frac{1}{\omega_{pi}^2} + \frac{1}{\Omega_e \Omega_i} \right]^{-1/2} \tag{10.62}$$

$$\approx \min \left[\omega_{pi}, (\Omega_e \Omega_i)^{1/2} \right].$$

The name of the whistler mode arises from 'whistling atmospherics'. These are generated when a lightening flash produces a pulse of broad band radio noise. The noise at frequencies $\omega \lesssim \omega_-(\theta)$ propagates in the whistler mode along the Earth's magnetic field lines to the opposite hemisphere of the Earth, where it may be observed. It may also reflect and bounce between hemispheres several times. The 'whistling' is due to the dispersion causing the higher frequencies to arrive before the lower frequencies, thereby producing a falling tone. The guiding of the whistlers along the field lines is due in part to the ray angle being restricted to $\theta_{rW} \lesssim 20°$ (cf. Exercise 10.4) and due in part to field-aligned density irregularities. The guiding of waves along field-aligned density irregularities is referred to as *ducting*.

10.4 Low frequency modes

At sufficiently low frequencies where the motion of the ions dominates that of the electrons, there are three wave modes in a collisionless magnetized plasma. These are analogous to the three MHD modes (§8.2). The MHD theory applies only to a collision-dominated plasma, and the MHD treatment of the waves applies only at frequencies below the ion collision frequency. The collisionless theory for the analogous waves applies above the electron collision frequency. (In the range between the electron and ion collision frequencies the wave properties are derived assuming collisionless ions and collision-dominated electrons.)

Two of the three low frequency modes are present in cold plasma theory; these are the Alfvén and magnetoacoustic modes. The third mode is a magnetized version of the ion sound mode (§1.2). We start here by treating the cold plasma modes. We then treat all three modes together in an approximate way after introducing the relevant approximation to the dielectric tensor for a thermal electron-ion plasma.

In the low frequency limit in cold plasma theory one has $P = 1 - \omega_p^2/\omega^2$ large and negative. Therefore let us expand in powers of $1/P$. To lowest order in $1/P$, (10.30) with (10.32) gives

$$T_\pm = \frac{S\sin^2\theta \pm [S^2\sin^4\theta + 4D^2\cos^2\theta]^{1/2}}{2D\cos\theta} = \frac{-2D\cos\theta}{S\sin^2\theta \mp [S^2\sin^4\theta + D^2\cos^2\theta]^{1/2}}$$

$$\tag{10.63}$$

and (10.37) and (10.38) simplify to

$$N^2 = \frac{1}{\cos^2\theta}\left(S + \frac{D\cos\theta}{T}\right) = \frac{S^2 - D^2}{S - DT\cos\theta},$$ (10.64)

$$K = T\tan\theta = \sin\theta\left(\frac{ST\cos\theta - D}{S - DT\cos\theta}\right)$$ (10.65)

respectively. Approximate expressions for S and D may be obtained from (10.26) by assuming $\omega^2 \ll \Omega_e^2$, $\omega_p^2/\Omega_e^2 \ll \omega_{pi}^2/\Omega_i^2$. From the definition (8.2) of the Alfvén speed one has

$$\sum_i \frac{\omega_{pi}^2}{\Omega_i^2} = \frac{c^2}{v_A^2}$$ (10.66)

and, from charge neutrality, one has

$$\frac{\omega_p^2}{\Omega_e} = \sum_i \frac{\omega_{pi}^2}{\Omega_i}.$$ (10.67)

These lead to the identities

$$S = 1 + \frac{c^2}{v_A^2} + \sum_i \frac{\omega_{pi}^2\omega^2}{\Omega_i^2(\Omega_i^2 - \omega^2)}$$ (10.68)

and

$$D = -\sum_i \frac{\omega_{pi}^2\omega}{(\Omega_i^2 - \omega^2)\Omega_i}.$$ (10.69)

Then in the limit $\omega \ll \Omega_i$, D becomes small. The $+$ mode becomes the Alfvén (A) mode and the $-$ mode becomes the magnetoacoustic (M) mode. For $4D^2\cos^2\theta \ll S^2\sin^4\theta$ one finds

$$T_A \approx \frac{S\sin^2\theta}{D\cos\theta}, \quad N_A^2 \approx \frac{S}{\cos^2\theta} + \frac{D^2\cos^2\theta}{S\sin^2\theta},$$ (10.70)

and

$$T_M \approx -\frac{D\cos\theta}{S\sin^2\theta}, \quad N_M^2 \approx S - \frac{D^2}{S\sin^2\theta}.$$ (10.71)

Comparison with the analogous formulas of MHD theory, cf. (8.22) to (8.24) for $v_A^2 \gg c_s^2$, confirms the identification of these modes as analogs of the MHD modes.

The results (10.70) and (10.71) apply for $4D^2\cos^2\theta \ll S^2\sin^4\theta$. This approximation breaks down for sufficiently small angles θ, and it also breaks down over a much wider range of angles for $\omega \approx \Omega_i$ where both S^2 and D^2 become large and approximately equal. In the opposite limit $4D^2\cos^2\theta \gg S^2\sin^4\theta$ the modes become oppositely circularly polarized with

$$T_A = \frac{D\cos\theta}{|D\cos\theta|}, \quad N_A^2 = R_-,$$ (10.72)

$$T_M = -\frac{D\cos\theta}{|D\cos\theta|}, \quad N_M^2 = R_+.$$ (10.73)

As the cyclotron frequency of any ionic species is approached R_- approaches infinity and the parallel Alfvén mode (10.72) has a resonance. The resonant waves are called 'ion-cyclotron waves', or specifically 'parallel ion-cyclotron waves' to distinguish them from 'perpendicular ion-cyclotron waves', also known as 'the ion Bernstein waves'. The dispersion curves for the two low frequency modes in a plasma with two or more ionic species have some complicating features; these are discussed in §12.3.

The refractive index of the magnetoacoustic mode is not greatly affected as the frequency is increased from $\omega < \Omega_i$ to $\omega > \Omega_i$. The dispersion curve for this mode extends to higher frequencies where it joins continuously onto the whistler mode. The typical frequency separating the magnetoacoustic and whistler regions is the lower hybrid frequency. This frequency corresponds to a zero of S, and for $S = 0$ the relevant solution $T_M = -D\cos\theta/|D\cos\theta|$ of (10.63) implies $N_M^2 = |D/\cos\theta|$ from (10.64). The expression (10.69) for D may be rewritten for $\omega^2 \gg \Omega_i^2$ using (10.67) so that one has $|D/\cos\theta| = \omega_p^2/\omega\Omega_e|\cos\theta|$. This confirms that the magnetoacoustic mode and the whistler mode, cf. (10.60), coincide for $\omega = \omega_{LH}$.

Now let us include thermal motions. An appropriate approximate form for the dielectric tensor is

$$
K_{ij} = \frac{c^2}{v_A^2}
\begin{bmatrix}
1 & \dfrac{i\omega}{\Omega_i} & 0 \\[2ex]
-\dfrac{i\omega}{\Omega_i} & 1 & -\dfrac{i\Omega_i}{\omega}\tan\theta \\[2ex]
0 & \dfrac{i\Omega_i}{\omega}\tan\theta & -\dfrac{\Omega_i^2}{\omega^2}\left(1 - \dfrac{c^2}{N^2 v_s^2 \cos^2\theta}\right)
\end{bmatrix}
\tag{10.74}
$$

where we assume $v_A^2 \ll c^2$ and that there is only one ionic species. The approximation (10.74) may be derived from the general expression for $K_{ij}(\omega, \mathbf{k})$ for a thermal plasma, cf. Exercise 10.5. The leading terms (for $i, j = x, y$) are just the cold plasma terms for $\omega \ll \Omega_i$. The thermal correction appears explicitly only in the final entry which may be written

$$
K_{zz} = -\frac{\omega_{pi}^2}{\omega^2} + \frac{1}{k^2 \lambda_{De}^2 \cos^2\theta},
\tag{10.75}
$$

which is the magnetized version of the approximate form $K^{(L)} = 1 - \omega_{pi}^2/\omega^2 + 1/k^2\lambda_{De}^2$ used to derive the dispersion relation $\omega = kv_s$ for ion sound waves in an unmagnetized plasma. The remaining terms K_{yz} in (10.74) may be interpreted as follows. The motion of the electrons is assumed so rapid that they adjust instantaneously to the electric field. Then $\mathrm{grad}(\delta n_e T_e) = \mathbf{J} \times \mathbf{B}$, from pressure balance, leads to $J_y = ik\sin\theta\, \delta n_e T_e/B$. The fluctuation δn_e in the electron density is determined by the longitudinal response of the electrons as described by (10.75), i.e. by $\delta n_e = -\varepsilon_0 E_z/e\lambda_{De}^2 k\cos\theta$. This gives a component $J_y \propto E_z$ which leads to the K_{yz} term in (10.54). The Onsager relations imply $K_{xy} = -K_{yx}$, $K_{xz} = K_{zx}$, $K_{yz} = -K_{zy}$ when $\mathbf{k}$ and $\mathbf{B}$ are in the x–z plane, and hence the K_{zy} component is implied by the K_{yz} component. A direct calculation of the K_{zy} component may be made in terms of the guiding-centre approximation, e.g. using the drift kinetic equation (§12.5). Such interpretations are useful for understanding the physical significance of (10.74); however its formal derivation here is as an approximation to a more general expression (specifically to (11.27) below).

The dispersion equation for a dielectric tensor of the form (10.74) is

$$(N^2\cos^2\theta - K_{xx})\{(N^2 - K_{yy})K_{zz} - K_{yz}^2\} + [N^4 K_{xx}\sin^2\theta - N^2 K_{xx}K_{yy}\sin^2\theta$$
$$+ 2N^2 K_{xy}K_{yz}\sin\theta\cos\theta + K_{xy}^2 K_{zz}] - N^2 K_{xy}^2 \sin^2\theta = 0 \qquad (10.76)$$

where the terms are arranged according to increasing powers of ω^2/Ω_i^2. To lowest order in ω^2/Ω_i^2 the solutions of (10.76) are

$$N_A^2 = \frac{K_{xx}}{\cos^2\theta} = \frac{c^2}{v_A^2\cos^2\theta}, \qquad (10.77)$$

and the solutions of $(N^2 - K_{yy})K_{zz} - K_{yz}^2 = 0$ may be written

$$N_\pm^2 = \frac{c^2}{v_\pm^2(\theta)} \qquad (10.78a)$$

with

$$v_\pm^2(\theta) = \tfrac{1}{2}(v_A^2 + v_s^2) \pm \tfrac{1}{2}[(v_A^2 + v_s^2)^2 - 4v_A^2 v_s^2 \cos^2\theta]^{1/2}. \qquad (10.78b)$$

The analogy with the MHD modes, cf. (8.22) to (8.24), is obvious.

The polarization vectors for the three modes are, to first order in $\omega/\Omega_i \ll 1$ and in $v_s^2/v_A^2 \ll 1$,

$$\mathbf{e}_A = \left(1, -\frac{i\omega}{\Omega_i}\cot^2\theta, \frac{\omega^2}{\Omega_i^2}\frac{v_s^2}{v_A^2}\frac{1}{\sin\theta\cos\theta}\right), \qquad (10.79a)$$

$$\mathbf{e}_M = \left(\frac{\omega}{\Omega_i}\cosec^2\theta, i, \frac{\omega}{\Omega_i}\frac{v_s^2}{v_A^2}\sin\theta\cos\theta\right), \qquad (10.79b)$$

$$\mathbf{e}_s = \frac{\left(\sin\theta, -i\frac{\Omega_i}{\omega}\frac{v_s^2}{v_A^2}\sin\theta\cos^2\theta, \cos\theta\right)}{\left(1 + \frac{\Omega_i^2}{\omega^2}\frac{v_s^4}{v_A^4}\sin^2\theta\cos^4\theta\right)^{1/2}}, \qquad (10.79c)$$

where M and s are the $+$ and $-$ modes in (10.78) for $v_A^2 \gg v_s^2$. The other wave properties are

$$R_A \approx \frac{v_A^2}{2c^2} \approx R_M, \quad R_s \approx \frac{v_A^2}{2c^2}\frac{\omega^2}{\Omega_i^2\cos^2\theta}\left(1 + \frac{\Omega_i^2}{\omega^2}\frac{v_s^4}{v_A^4}\sin^2\theta\cos^4\theta\right) \quad (10.80a,b)$$

$$\gamma_A = \omega\left(\frac{\pi}{2}Z_i\frac{m_e}{m_i}\right)^{1/2}\frac{v_s}{v_A}\frac{\omega^2}{\Omega_i^2}(\tan^2\theta + \cot^2\theta) \qquad (10.81a)$$

$$\gamma_M = \omega\left(\frac{\pi}{2}Z_i\frac{m_e}{m_i}\right)^{1/2}\frac{v_s}{v_A}\frac{\sin^2\theta}{|\cos\theta|}, \qquad (10.81b)$$

$$\gamma_s = \omega\left(\frac{\pi}{2}Z_i\frac{m_e}{m_i}\right)^{1/2}, \qquad (10.81c)$$

where a single ionic species with charge $Z_i e$ is assumed. These damping rates are due to Landau damping (at $s = 0$) by thermal electrons.

10.5 Quasilinear equations for **B** $\neq 0$

Two steps are involved in generalizing the semiclassical treatment of quasilinear theory (§6.3) to include the effects of a background magnetic field. First the relevant probability for emission needs to be derived, and second, the quantization of the particle states needs to be treated appropriately.

The *probability for gyromagnetic emission* may be derived by substituting the current (10.19) in the emission formula and then identifying the probability, as in the identification of (6.14) in the unmagnetized case. The result is

$$w_M(\mathbf{k}, \mathbf{p}) = \sum_{s=-\infty}^{\infty} w_M(\mathbf{k}, \mathbf{p}; s), \tag{10.82}$$

$$w_M(\mathbf{k}, \mathbf{p}; s) = \frac{2\pi q^2 R_M(\mathbf{k})}{\varepsilon_0 \hbar |\omega_M(\mathbf{k})|} |\mathbf{e}_M^*(\mathbf{k}) \cdot \mathbf{V}(\mathbf{k}, \mathbf{p}; s)|^2 \delta(\omega_M(\mathbf{k}) - s\Omega - k_\parallel v_\parallel) \tag{10.83}$$

The probability $w_M(\mathbf{k}, \mathbf{p}; s)$ is referred to as that for emission at the sth harmonic. The resonance condition

$$\omega - s\Omega - k_\parallel v_\parallel = 0 \tag{10.84}$$

is sometimes called the Doppler condition. Emission at $s > 0$ is said to be by the normal Doppler effect, and that at $s < 0$ to be by the anomalous Doppler effect.

The quantization of the particle motion is readily understood in the non-relativistic case. Then the motion perpendicular to the magnetic field lines is circular motion, in the classical limit, and circular motion is simple harmonic motion. The energy of a simple harmonic oscillator with frequency $\Omega_0 = |q|B/m$ is quantized as $(n + \frac{1}{2})\hbar\Omega_0$ with $n = 0, 1, 2, \ldots$. For an electron there is also a spin quantum number which contributes $\pm\frac{1}{2}\hbar\Omega_0$ to the energy. The classical limit corresponds to large n, and here we ignore the terms $\frac{1}{2}$ in comparison with n. (Formally the energy eigenvalues for an electron are $n\hbar\Omega_0$ with the ground state ($n = 0$) being non-degenerate and all other states being doubly degenerate due to the spin.) The component of momentum $p_\parallel$ along the magnetic field is a continuous quantum number. The relativistic quantum case is treated using Dirac's equation; the energy eigenvalues are

$$\varepsilon = (m^2 c^4 + p_\parallel^2 c^2 + 2neB\hbar c^2)^{1/2}. \tag{10.85}$$

For particles other than electrons the form (10.85) applies with e replaced by $|q|$.

Now consider an electron $(\varepsilon, p_\parallel, n)$ which emits a wave quantum $(\omega, \mathbf{k})$ so that its final state is $(\varepsilon', p_\parallel', n')$. The final state must satisfy

$$\varepsilon' = (m^2 c^4 + p_\parallel'^2 c^2 + 2n'eB\hbar c^2)^{1/2} \tag{10.85'}$$

and conservation of energy and parallel momentum require

$$\varepsilon' = \varepsilon - \hbar\omega, \quad p_\parallel' = p_\parallel - \hbar k_\parallel, \quad n' = n - s. \tag{10.86}$$

In (10.86) we also note that n' can differ from n only by an integer which is written as s. For $|\hbar k_\parallel| \ll |p_\parallel|$ and $|s| \ll n$, (10.86) in (10.85′) with (10.85) leads to

$$\varepsilon - \hbar\omega = (1 - \hat{D}_s + \tfrac{1}{2}\hat{D}_s^2 + \cdots)\varepsilon \tag{10.87}$$

with

$$\hat{D}_s = \hbar k_\parallel \frac{\partial}{\partial p_\parallel} + s\frac{\partial}{\partial n}. \tag{10.88}$$

In the classical limit $2neB\hbar c^2$ in (10.85) is replaced by $p_\perp^2 c^2$, implying $n = p_\perp^2/2eB\hbar$. With this identification (10.88) becomes

$$\hat{D}_s = \hbar\left(\frac{s\Omega}{v_\perp}\frac{\partial}{\partial p_\perp} + k_\parallel\frac{\partial}{\partial p_\parallel}\right). \tag{10.88′}$$

To lowest order in $\hbar(10.87)$ implies the Doppler condition (10.84). The term arising from the second derivative in (10.87) describes the quantum recoil (Exercise 10.6).

The derivation of the quasilinear equations now proceeds as in the derivations of (6.39) and (6.42). The form of the kinetic equation for the waves is unchanged from (6.39). One finds

$$\frac{dN_M(\mathbf{k})}{dt} = \alpha_M(\mathbf{k}) - \gamma_M(\mathbf{k})N_M(\mathbf{k}) \tag{10.89}$$

with

$$\begin{bmatrix} \alpha_M(\mathbf{k}) \\ \gamma_M(\mathbf{k}) \end{bmatrix} = \sum_{s=-\infty}^{\infty} \int d^3\mathbf{p}\, w_M(\mathbf{k}, \mathbf{p}; s) \begin{bmatrix} f(\mathbf{p}) \\ -\hat{D}_s f(\mathbf{p}) \end{bmatrix} \tag{10.90}$$

The form for the equation for $f(\mathbf{p})$ which results is, in terms of the cylindrical components $p_\perp$ and $p_\parallel$,

$$\frac{\partial f}{\partial t} = \frac{1}{p_\perp}\frac{\partial}{\partial p_\perp}\left[p_\perp\left\{A_\perp f + \left(D_{\perp\perp}\frac{\partial}{\partial p_\perp} + D_{\perp\parallel}\frac{\partial}{\partial p_\parallel}\right)f\right\}\right]$$
$$+ \frac{\partial}{\partial p_\parallel}\left\{A_\parallel f + \left(D_{\parallel\perp}\frac{\partial}{\partial p_\perp} + D_{\parallel\parallel}\frac{\partial}{\partial p_\parallel}\right)f\right\}. \tag{10.91}$$

An alternative form is in terms of spherical polar coordinates p, α with $p_\perp = p\sin\alpha, p_\parallel = p\cos\alpha$:

$$\frac{\partial f}{\partial t} = \frac{1}{p^2}\frac{\partial}{\partial p}\left[p^2\left\{A_p f + \left(D_{pp}\frac{\partial}{\partial p} + D_{p\alpha}\frac{\partial}{\partial \alpha}\right)f\right\}\right]$$
$$+ \frac{1}{\sin\alpha}\frac{\partial}{\partial \alpha}\left[\sin\alpha\left\{A_\alpha f + \left(D_{\alpha p}\frac{\partial}{\partial p} + D_{\alpha\alpha}\frac{\partial}{\partial \alpha}\right)f\right\}\right]. \tag{10.92}$$

Let λ, μ run over $p_\perp, p_\parallel$ or p, α; then one has (cf. Exercise 6.3)

$$\begin{bmatrix} A_\lambda \\ D_{\lambda\mu} \end{bmatrix} = \sum_{s=-\infty}^{\infty} \int \frac{d^3\mathbf{k}}{(2\pi)^3} w_M(\mathbf{k}, \mathbf{p}; s) \begin{bmatrix} \Delta\lambda \\ \Delta\lambda\Delta\mu N_M(\mathbf{k}) \end{bmatrix} \tag{10.93}$$

with

$$\Delta\lambda = \hat{D}_s\lambda. \tag{10.94}$$

The components of momentum perpendicular to the field line are not conserved in the usual sense. However, there is an additional effect which arises from a con-servation-type relation for the perpendicular components. The canonical momentum is $\mathbf{P} = \gamma m v + q\mathbf{A}$ where $\mathbf{A}$ is the vector potential. If we choose $\mathbf{A} = (-By, 0, 0)$ then the Lagrangian for the system does not depend on x(or z) and hence P_x(and $P_z = p_\parallel$) is a constant of the motion. This constant specifies the y-coordinate of the centre of gyration of the particle as being determined by $P_x = -qBy$. On emitting a wave quantum the centre of gyration moves from $y = -P_x/qB$ to $y = -(P_x - \hbar k_\perp)/qB$, where $\mathbf{k}$ is in the x–z plane. Hence we have $\Delta y = \hbar k_\perp/qB$. In a coordinate-free form, the centre of gyration is displaced by

$$\Delta\mathbf{x} = -\frac{\hbar\mathbf{k}\times\mathbf{B}}{qB^2}. \tag{10.95}$$

The effect of the displacement of the gyrocentre may be included in the quasilinear equations by the replacement $\hat{D}_s$, as given by (10.88′) by

$$\hat{D}_s = \hbar\left(\frac{s\Omega}{v_\perp}\frac{\partial}{\partial p_\perp} + k_\parallel\frac{\partial}{\partial p_\parallel}\right) - \frac{\hbar\mathbf{k}\times\mathbf{B}}{qB^2}\cdot\mathrm{grad}_c, \tag{10.96}$$

where the subscript c is to emphasize that the gradient operates only on the position vector of the particle gyrocentres. The additional term (10.96) needs to be included when treating certain types or drift wave (cf. §12.7).

Exercise set 10

10.1 (a) Show that in the presence of a background magnetic field the solution of the linearized Vlasov equation (§2.2) may be written in the form

$$f^{(1)}(\mathbf{p},\omega,\mathbf{k}) = \frac{|q|}{\omega\Omega}\int^\phi d\phi' e^{P(\phi) - P(\phi')}g_{rj}(\omega,\mathbf{k};\mathbf{v}')E_j(\omega,\mathbf{k})\frac{\partial f^{(0)}}{\partial p_r'}(p_\perp, p_\parallel)$$

with $\mathbf{p}'$ given by replacing ϕ by ϕ' in (10.12), with

$$P(\phi) = \frac{i\epsilon}{\Omega}\int^\phi d\phi'(\omega - \mathbf{k}\cdot\mathbf{v}'),$$

and with $g_{rj}(\omega,\mathbf{k};\mathbf{v})$ defined by (5.20).

(b) Hence derive the dielectric tensor in the form

$$K_{ij}(\omega,\mathbf{k}) = \delta_{ij} + \sum\frac{iq}{\varepsilon_0\omega}\int d^3\mathbf{p}\frac{|q|}{\Omega\omega}\int^\phi_{\epsilon\infty} d\phi' e^{P(\phi) - P(\phi')}v_i g_{rj}(\omega,\mathbf{k};\mathbf{v}')\frac{\partial f^{(0)}}{\partial p_r'}(p_\perp, p_\parallel),$$

and show that the contribution from the limit $\phi' = \epsilon\infty$ gives zero in view of the causal condition.

(c) Show that on expanding in Bessel functions the result in part (b) reduces to the expression (10.21).

(d) Using the sum rules for the Bessel functions

$$\sum_{s=-\infty}^\infty J_s^2(z) = 1, \quad \sum_{s=-\infty}^\infty J_s'^2(z) = \sum_{s=-\infty}^\infty \frac{s^2}{z^2}J_s^2(z) = \tfrac{1}{2},$$

show that one has

$$\sum_{s=-\infty}^{\infty} V_i(\mathbf{k}\cdot\mathbf{p},s)V_j^*(\mathbf{k},\mathbf{p},s) = v_\perp^2(\delta_{ij} - b_i b_j) + v_\parallel^2 b_i b_j.$$

Hence derive the alternative expression to (10.21)

$$K_{ij}(\omega,\mathbf{k}) = \delta_{ij} + \sum \frac{q^2}{\varepsilon_0\omega^2}\int d^3\mathbf{p}\left[\left\{\tfrac{1}{2}(\delta_{ij} - b_i b_j)v_\perp \frac{\partial}{\partial p_\perp} + b_i b_j v_\parallel \frac{\partial}{\partial p_\parallel}\right\}f(\mathbf{p})\right.$$

$$\left. + \sum_{s=-\infty}^{\infty} \frac{V_i(\mathbf{k},\mathbf{p},s)V_j^*(\mathbf{k},\mathbf{p},s)}{\omega - s\Omega - k_\parallel v_\parallel}\left\{\frac{s\Omega}{v_\perp}\frac{\partial}{\partial p_\perp} + k_\parallel \frac{\partial}{\partial p_\parallel}\right\}f(\mathbf{p})\right]$$

10.2 (a) Generalize the cold plasma equation of motion (10.1) (*i*) by allowing $v^{(\alpha)}$ to have a zeroth order value $v_\parallel^{(\alpha)}\mathbf{b}$, and (*ii*) by including relativistic effects. Show that the resulting generalization of (10.7) is

$$K_{ij}(\omega,\mathbf{k}) = \delta_{ij} - \sum_\alpha \frac{q^2 n}{\varepsilon_0\gamma m\omega^2}\left[\tau_{ij} + \frac{v_\parallel k_m}{\omega - k_\parallel v_\parallel}(\tau_{im}b_j + b_i\tau_{mj})\right.$$

$$\left. + \frac{v_\parallel^2 b_i b_j}{(\omega - k_\parallel v_\parallel)^2}\left(k_l k_m \tau_{lm} - \frac{\omega^2}{c^2}\right)\right],$$

where the argument of τ_{ij} is $\omega - k_\parallel v_\parallel$ and with $\gamma = (1 - v_\parallel^2/c^2)^{-1/2}$.

(b) Show that the expression (10.22) reproduces the result of part (a) if one assumes $f(\mathbf{p}) \propto \delta(p_\perp)\,\delta(p_\parallel - p_{\parallel 0})$ and then sets $p_{\parallel 0} = \gamma m v_\parallel$.

10.3 (a) Show that the Poynting vector $\mathbf{E} \times \mathbf{B}/\mu_0$ for waves in the Mode M, with amplitude defined as in (2.56), may be reduced to the form

$$\int \frac{d^3\mathbf{k}}{(2\pi)^3}\frac{2|\mathbf{k}|c^2}{|\omega_M(\mathbf{k})|}\varepsilon_0|E_M(\mathbf{k})|^2 \operatorname{Re}\left[\boldsymbol{\kappa} - \boldsymbol{\kappa}\cdot\mathbf{e}_M(\mathbf{k})\mathbf{e}_M^*(\mathbf{k})\right]$$

(Note that on using (2.59), $\varepsilon_0|E_M(\mathbf{k})|^2$ may be rewritten as $R_M(\mathbf{k})W_M(\mathbf{k})$.)

(b) Show that if the Poynting vector is equal to the energy flux

$$\int \frac{d^3\mathbf{k}}{(2\pi)^3}\mathbf{v}_{gM}(\mathbf{k})W_M(\mathbf{k})$$

then, with the form (10.32) for $\mathbf{e}_M(\mathbf{k})$ and the form (10.40) for $\mathbf{v}_{gM}(\mathbf{k})$, the identities (10.41) and (10.42) follow.

10.4 (a) Show that for $T_W = \cos\theta/|\cos\theta|$ and $X, Y|\cos\theta| \gg 1$ the wave properties for the whistler mode imply the group velocity

$$\mathbf{v}_{gW} = \frac{2c}{N_W}\{\boldsymbol{\kappa} - (\tfrac{1}{2}\tan\theta)\mathbf{t}\}.$$

(b) Show that the ray angle θ_{rW} (between $\mathbf{v}_{gW}$ and $\mathbf{b}$) cannot exceed the value $\theta_{rW} = \arccos(2\sqrt{2}/3)$ which occurs for $\theta = \arccos(1/\sqrt{3})$.

10.5 Derive the approximate form (10.74) for the dielectric tensor as follows using (10.21) and assuming (non-relativistic) Maxwellian distributions for the electrons and ions. (i) Make the small gyroradius approximation (10.23). (ii) For the ions assume $\omega \gg \sqrt{2}/k_\parallel/V_i$ and show that the resulting expression is the same as for cold ions when the term involving $k_\perp R$ for $s = 0$ in (10.23) is neglected. (iii) For electrons assume $\omega \ll \sqrt{2}|k_\parallel| V_e$ and show that the contribution from $s = 0$ gives

$$K_{yz}^{(e)} = -K_{zy}^{(e)} = -i\frac{k_\perp}{k_\parallel}\frac{\omega_p^2}{\Omega_e}, \quad K_{zz}^{(e)} = \frac{\omega_p^2}{k_\parallel^2 V_e^2}.$$

(One may either expand $1/(\omega - k_\parallel v_\parallel) = (-1/k_\parallel v_\parallel)(1 + \omega/k_\parallel v_\parallel + \cdots)$ or evaluate the $v_\parallel$-integral in terms of the plasma dispersion function and then expand in $\omega/\sqrt{2}k_\parallel V_e$.)

10.6 (a) Show that when the first quantum recoil is included in (10.87) the resonance condition (10.84) is replaced by

$$\omega - s\Omega - k_\parallel v_\parallel + \frac{\hbar}{2\varepsilon}(k_\parallel^2 c^2 - \omega^2) = 0.$$

(Terms of order $\hbar$ correspond to the first quantum recoil and higher order terms are neglected.)

(b) Include the correction of order $\hbar$ in the δ-function in the probability (10.83). Then use the identity

$$\int dx\, F(x)\delta(x - x_0 - \Delta x) = \int dx\, F(x + \Delta x)\delta(x - x_0)$$

to show that the quantum recoil modifies the emission coefficient in (10.90) to

$$\alpha_M(\mathbf{k}) = \sum_{s=-\infty}^{\infty} \int d^3\mathbf{p}\, w_M(\mathbf{k}, \mathbf{p}, s)\{1 + \tfrac{1}{2}\hbar\hat{D}_s\} f(\mathbf{p}),$$

where $w_M(\mathbf{k}, \mathbf{p}, s)$ is the unmodified probability (10.83).

11

Electron cyclotron maser emission

11.1 Classification of gyromagnetic processes

The generic name given to the interaction between a wave and a particle spiralling in a magnetic field is 'gyromagnetic'. The gyromagnetic resonance condition separates into contributions from harmonics $s = 0, \pm 1, \pm 2$ etc. For a gyromagnetic interaction at the sth harmonic the Doppler condition (10.84), viz.

$$\omega - s\Omega - k_\parallel v_\parallel = 0, \tag{11.1}$$

is satisfied or nearly satisfied. Gyromagnetic interactions may be classified in several ways, including the classification according to harmonic number. We are primarily interested in those gyromagnetic interactions which involve cyclotron instabilities producing escaping radiation. Before discussing these cyclotron instabilities let us describe the various classes of gyromagnetic interaction, starting with the classification according to harmonic number s.

The particular case $s = 0$ is essentially a one-dimensional version of the Cerenkov condition in an unmagnetized plasma. As in the unmagnetized case, streaming motions along the magnetic field lines can lead to instabilities, but only for waves with refractive index greater than unity. Actually one requires $k_\parallel^2 c^2 / \omega^2 = N^2 \cos^2\theta > 1$. We do not discuss such instabilities here. The interaction at $s = 0$ can also be important in the Landau damping of some waves, as is the case in (10.81).

For $s \neq 0$, usually two of the terms in the Doppler condition are much greater than the third. The case $\omega \approx k_\parallel v_\parallel \gg s\Omega$ corresponds to a weakly magnetized plasma, with the resonance being similar to the Cerenkov resonance $\omega = \mathbf{k} \cdot \mathbf{v}$. The case $|k_\parallel v_\parallel| \approx |s\Omega| \gg \omega$ can be satisfied for waves with large refractive index. It is important for ions and relativistic electrons interacting with Alfvén waves and magneto-acoustic waves (for $v_A \gg c_s$) and for electrons interacting with whistlers. These interactions can lead to large changes in the momenta of the particles with little change in their energy; as a consequence they are called 'resonant scattering' processes. Resonant scattering is discussed in detail in §13.1 below.

The third possibility in (11.1) is $\omega \approx s\Omega \gg |k_\parallel v_\parallel|$. This corresponds to emission near the cyclotron frequency or its harmonics. This is of interest for 'resonant waves' near the cyclotron frequencies and its harmonics, as discussed in §12.2 & §12.4. The case $\omega \approx s\Omega \gg |k_\parallel v_\parallel|$ is also relevant to cyclotron emission or absorption of o-mode and x-mode waves. These waves have refractive index $N < 1$, and then for a nonrelativistic particle one has $k_\parallel v_\parallel = \omega N(v/c)\cos\alpha\cos\theta$ much less than ω in magnitude. As any escaping radiation must be in the o-mode or the x-mode the condition $\omega \approx s\Omega \gg |k_\parallel v_\parallel|$ is the appropriate one for electron cyclotron instabilities which produce escaping radiation.

Gyromagnetic emission is classified as 'cyclotron', 'synchrotron' and 'gyrosynchrotron' for nonrelativistic, ultrarelativistic and mildly relativistic particles respectively. The formal distinction between the cyclotron, synchrotron and gyrosynchrotron cases is in the approximations made to the Bessel functions. In synchrotron emission the important harmonics are very large, s is regarded as a continuous variable, and the Airy integral approximation is made to the Bessel functions. In analytic treatments of gyrosynchrotron radiation s is again regarded as a continuous variable and the Carlini approximation is made to the Bessel functions. These cases are not discussed here.

In the cyclotron approximation the Bessel functions are approximated by the leading term in their power series expansion, i.e. one retains only the term $k = 0$ in

$$J_s(k_\perp R) = \sum_{k=0}^{\infty} \frac{1}{(s+k)!}\left(\frac{k_\perp R}{2}\right)^{s+2k}. \tag{11.2}$$

The cyclotron approximation corresponds to a multipole expansion. In the dipole approximation only emission at $s = 1$ is allowed. Emission at the sth harmonic corresponds to the 2^s-electric multipole, i.e. $s = 2$ and 3 correspond to electric quadrupole and electric octupole, respectively. This multipole expansion converges rapidly, with each term being smaller than the preceding term by a factor of order $(k_\perp R)^2 = N^2(v_\perp/c)^2\sin^2\theta < (v/c)^2$ for escaping radiation. Hence in discussing cyclotron emission of escaping radiation we are concerned primarily with the case $s = 1$, with $s = 2$ or 3 included only as corrections or when $s = 1$ is forbidden. Let us now consider the possible type of electron cyclotron instabilities leading to escaping radiation.

Cyclotron instabilities may be classified as 'reactive' or 'kinetic', as for the instabilities discussed in chapters 3 and 4 for an unmagnetized plasma. Both classes of cyclotron instabilities may be further classified in terms of the instability mechanism. Reactive instabilities are due to bunching (§5.2), and there are two distinct types of bunching: *axial bunching*, that is bunching along the z-axis (parallel to the background magnetic field), and *azimuthal bunching*, that is bunching in the azimuthal angle ϕ associated with the gyration about the magnetic field lines. Azimuthal bunching is an intrinsically relativistic effect. Kinetic instabilities may be interpreted in terms of maser emission (§5.4). In the absence of a magnetic field the 'inverted population' corresponds to $\mathbf{k} \cdot \partial f/\partial \mathbf{p} > 0$. In the presence of a magnetic field the expression (10.90) with (10.88') suggests that maser emission could result either from $k_\parallel \partial f/\partial p_\parallel > 0$ or from $(s\Omega/v_\perp)\partial f/\partial p_\perp > 0$. Let us refer to these as 'parallel-driven' and 'perpendicular-driven' kinetic instabilities. There is usually a correspondence such that a parallel-driven instability is the kinetic version of an axial-bunching reactive instability, and that a perpendicular-driven instability is the kinetic version of an azimuthal-bunching reactive instability. The perpendicular-driven instability, like azimuthal bunching instability, involves an intrinsically relativistic effect.

Although this fourfold classification was evident in the early (circa 1960) literature on cyclotron instabilities, it has been obscured by the development of the subject in three largely independent contexts. The three contexts are: the theory of the gyrotron, the theory of cyclotron instabilities and cyclotron waves in plasmas, and the theory of cyclotron maser emission in space plasmas. The theory of the gyrotron was initiated following the identification of azimuthal bunching as an alternative to axial bunching by Gaponov (1959). Subsequent developments in the theory of the gyrotron were predominantly in the Russian literature. In a gyrotron one has $\Omega_e \gg \omega_p$ and the dispersive properties of the electrons are relatively unimportant, i.e. the refractive index differs little from unity.

The theory of cyclotron instabilities in plasmas due to anisotropic velocity-space distribution was initiated in the early 1960's (e.g. Harris 1959, 1961, Sagdeev & Shafranov 1961); it was discussed in some detail in Stix' book (Stix 1962). The main emphasis in this context has been the growth of waves with refractive index > 1 due to a temperature anisotropy or to a loss-cone anisotropy. The non-relativistic approximation is made, specifically the Doppler condition (11.1) is approximated by $\omega - s\Omega_0 - k_\parallel v_\parallel = 0$ with $\Omega_0 = |q|B/m$.

The first discussion of electron cyclotron maser emission was by Twiss (1958) who was interested in the possibility of negative absorption in radioastronomical sources. Twiss assumed $p_\parallel = 0$, so that the Doppler condition becomes $\omega = \Omega_e/\gamma$ for electrons ($\Omega_e = eB/m_e$). Waves at frequency ω resonate with electrons with $\gamma = \Omega_e/\omega$, i.e. with $v_\perp/c = (1 - \omega^2/\Omega_e^2)^{1/2}$, and negative absorption occurs if $\partial f/\partial p_\perp$ is positive for $p_\perp$ corresponding to this value of $v_\perp$. A major problem with Twiss' mechanism in practice is that the frequency Ω_e/γ is below the cutoff frequency ω_x for the x-mode, and hence any resulting radiation cannot escape directly in the x-mode. Relatively little attention was paid in the astrophysical literature to this mechanism until Wu & Lee (1979) noted that the full Doppler condition (11.1) implies $\omega = \Omega_e/\gamma + k_\parallel v_\parallel$, and this does allow emission at $\omega > \omega_x$ for $\Omega_e \gg \omega_p$ and $s = 1$. Following Wu & Lee's work, electron cyclotron maser emission has become the accepted mechanism for certain very bright radio emissions from the planets, the Sun and some flare stars (§11.5).

In discussing cyclotron instabilities we follow the same order of discussion as in Chapters 2 to 5, discussing reactive instabilities, then kinetic instabilities and finally presenting a model which includes both and allows one to treat wave trapping.

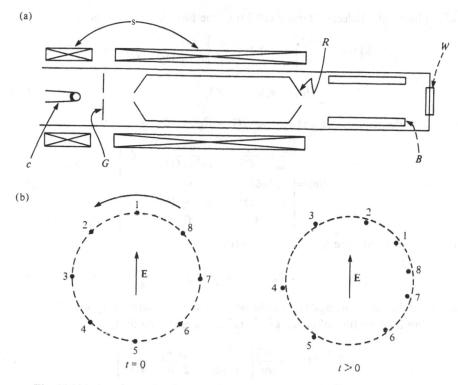

Fig. 11.1(a) A schematic diagram for a gyrotron indicating c-cathode, G-electron gun, s-solenoids, R-resonant cavity, B-beam collecting anode, W-output window. The magnetic field B increases away from G causing $v_\perp^2 \propto B$ to increase to $\gg v_\parallel^2$
(b) The positions of eight electrons around their orbit at $t = 0$ and $t = n2\pi/\omega$ are shown to illustrate azimuthal bunching (for $\omega > \Omega_e/\gamma$) due to the wave field. Electrons 2 to 4 gain energy and 6 to 8 lose energy leading to the phase slippage.

11.2 The gyrotron instability

The instability involved in laboratory gyrotrons is a reactive instability associated with azimuthal bunching. A schematic diagram of a gyrotron is indicated in Figure 11.1. The electrons entering the radiation cavity have $p_\perp \gg |p_\parallel|$ and are nearly mono-energetic. In practice the wave properties are affected by the cavity, that is one should treat the emission as being into cavity modes. However, this is a complicating detail which is not essential to the radiation mechanism. (This point was recognized by Gaponov (1959); the axial bunching instability requires $N^2 > 1$ and for it to occur the properties of the cavity or cavity-like modes play an essential role.)

To see how the *reactive cyclotron* instabilities arise, let us consider a distribution of electrons

$$f(\mathbf{p}) = \frac{n_e}{2\pi p_{\perp 0}} \delta(p_\perp - p_{\perp 0})\delta(p_\parallel) \tag{11.3}$$

and evaluate the dielectric tensor (10.22) in the limit $k_\perp R = 0$. One finds

$$K_{ij}(\omega, \mathbf{k}) = \delta_{ij} - \frac{\omega_p^2}{\omega^2}\left[\tau_{ij}(\omega) + \sum_{\pm}\frac{1}{\omega^2}\left\{\frac{(\omega \mp \Omega)^2 k_\perp^2}{(\omega \mp \Omega)^2 - \Omega^2} + k_\parallel^2 - \frac{\omega^2}{c^2}\right\}\right.$$

$$\left.\times \frac{v_{\perp 0}^2}{4}(\mathbf{e}_x + i\mathbf{e}_y)_i(\mathbf{e}_x - i\mathbf{e}_y)_j\right] \tag{11.4}$$

with $\omega_p^2 = e^2 n_e/\varepsilon_0 m_e \gamma_0$, $\Omega = \Omega_e/\gamma_0$, $\gamma_0 = (1 - v_{\perp 0}^2/c^2)^{-1/2}$, and with

$$\tau_{ij}(\omega) = \begin{bmatrix} \dfrac{\omega^2}{\omega^2 - \Omega^2} & -\dfrac{i\omega\Omega}{\omega^2 - \Omega^2} & 0 \\[3mm] \dfrac{i\omega\Omega}{\omega^2 - \Omega^2} & \dfrac{\omega^2}{\omega^2 - \Omega^2} & 0 \\[3mm] 0 & 0 & 1 \end{bmatrix} \tag{11.5}$$

Let us project onto the polarization vector

$$\mathbf{e}_R = \frac{1}{\sqrt{2}}(\mathbf{e}_x + i\mathbf{e}_y) \tag{11.6}$$

whose handedness corresponds to the sense in which electrons gyrate. Then, after neglecting a correction of order $k_\perp^2 v_{\perp 0}^2/\Omega^2$ for $\omega \approx \Omega$, one finds

$$K^R(\omega, \mathbf{k}) = 1 - \frac{\omega_p^2}{\omega^2}\left[\frac{\omega}{\omega - \Omega} + \frac{v_{\perp 0}^2}{2}\frac{k_\parallel^2 \omega^2/c^2}{(\omega - \Omega)^2}\right]. \tag{11.7}$$

It is relevant to note that if one makes the nonrelativistic approximation in evaluating the dielectric tensor (10.21) from the Vlasov equations, and then partially integrates before setting $f(\mathbf{p})$ equal to the value (11.3), the resulting expression differs from (11.7) in that the final term ω^2/c^2 is absent. This term is essential in the treatment of the reactive instability involved in the gyrotron.

The dispersion equation for waves with the polarization (11.6) is

$$\frac{k_\parallel^2 c^2}{\omega^2} = K^R(\omega, \mathbf{k}) \tag{11.8}$$

Assuming $|\omega - \Omega| \ll \Omega$, (11.8) with (11.7) reduces to a quadratic equation for $\omega - \Omega$:

$$\left(\frac{\omega - \Omega}{\omega}\right)^2 - \frac{\omega_p^2}{\omega^2 - k_\parallel^2 c^2}\left(\frac{\omega - \Omega}{\omega}\right) + \frac{v_{\perp 0}^2}{2c^2} = 0. \tag{11.9}$$

The solutions are

$$\frac{\omega - \Omega}{\omega} = \frac{\omega_p^2}{2(\omega^2 - k_\parallel^2 c^2)}\left[1 \pm \left\{1 - \frac{2v_{\perp 0}^2}{c^2}\frac{(\omega^2 - k_\parallel^2 c^2)}{\omega_p^4}\right\}^{1/2}\right]. \tag{11.10}$$

Growing solutions occur for

$$k_\parallel^2 c^2 - \omega^2 > \frac{\omega_p^2}{\sqrt{2}}\frac{c}{v_{\perp 0}} \quad \text{or} \quad \omega^2 - k_\parallel^2 c^2 > \frac{\omega_p^2}{\sqrt{2}}\frac{c}{v_{\perp 0}}. \tag{11.11a,b}$$

The regime (11.1a) corresponds to axial bunching and the regime (11.11b) to azimuthal bunching, as discussed in §11.6. Axial bunching requires $k_\parallel^2 c^2 > \omega^2$, and such waves cannot escape directly from a plasma.

The instability in a gyrotron corresponds to (11.11b). Now (11.11b) applies only for $k_\perp = 0$, and for $k_\perp \neq 0$ (11.8) generalizes to an equation of the form (cf. Exercise 11.1)

$$\omega^2 - k^2 c^2 - \omega_p^2 \left(\frac{\omega}{\omega - \Omega} - \frac{v_{\perp 0}^2}{2c^2} \frac{\omega^2 - k_\parallel^2 c^2}{(\omega - \Omega)^2} \right) = 0. \qquad (11.12)$$

Writing $\Delta\omega = \omega - kc$ and $\Delta\omega_0 = kc - \Omega$, (11.12) reduces to a cubic equation for $\Delta\omega$ when one neglects $\Delta\omega$ in comparison with $2kc$:

$$(2kc\Delta\omega - \omega_p^2)(\Delta\omega + \Delta\omega_0)^2 - \omega_p^2 \Omega(\Delta\omega + \Delta\omega_0) - \frac{v_{\perp 0}^2}{2c^2} \omega_p^2 (k_\parallel^2 c^2 - \omega^2) = 0. \qquad (11.13)$$

We assume $\omega_p^2 \ll 2kc\Delta\omega$ and neglect the middle term in comparison with the other two. Then one finds the growth rate, with $k_\parallel c = \omega N \cos\theta$,

$$|\gamma_{\max}| \approx \sqrt{3} \left\{ \frac{v_{\perp 0}^2}{4c^2} \omega_p^2 \Omega (1 - N^2 \cos^2\theta) \right\}^{1/3}. \qquad (11.14)$$

In the application to the gyrotron we may assume $N^2 \approx 1$ and $1 - N^2 \cos^2\theta \approx \sin^2\theta$. The neglect of the middle term in (11.13) is then justified for $\omega_p/\Omega \ll (v_{\perp 0}^2/c^2) \times \sin^2\theta$ and $\omega_p^2 \ll 2kc\Delta\omega$ requires $(\omega_p/\Omega)^4 \ll 2(v_{\perp 0}^2/c^2)\sin^2\theta$. These conditions are satisfied in laboratory gyrotrons. However they probably cannot be justified in applications to space plasmas; the kinetic version of the instability (§11.4) is relevant in such applications.

The foregoing discussion of reactive instabilities is based on a projection of the dielectric tensor onto the polarization vector $\mathbf{e}_R$, cf. (11.6). The underlying assumption may be stated in the following more general form. We are interested in the dispersive properties for waves under conditions where we believe that a given polarization vector is a reasonable approximation to the actual polarization vector. Let this approximate polarization vector be $\mathbf{e}_M$. Then the wave equation projected onto $\mathbf{e}_M$ becomes

$$\frac{k^2 c^2}{\omega^2} (1 - |\boldsymbol{\kappa} \cdot \mathbf{e}_M|^2) = K^M(\omega, \mathbf{k}), \qquad (11.15)$$

with

$$K^M(\omega, \mathbf{k}) = e_{Mi}^* e_{Mj} K_{ij}(\omega, \mathbf{k}). \qquad (11.16)$$

We solve (11.15) for the dispersion relation. In principle, this may be the first step in an iterative procedure. The next step is the substitution of the resulting dispersion relation into $\lambda_{ij}(\omega, \mathbf{k})$ which is then used to find a corrected form for $\mathbf{e}_M$. A second iteration may be started with this new value of $\mathbf{e}_M$.

We should comment on the choice of polarization vector (11.6): $\mathbf{e}_R$ describes the sense of gyration of electrons. In the case of parallel propagation one has $\boldsymbol{\kappa} \cdot \mathbf{e}_R = 0$ but for $k_\perp \neq 0$ the choice $\mathbf{e} = \mathbf{e}_R$ does not correspond to transverse waves. A better choice is then

$$e'_R = \frac{1}{\sqrt{2}}(t + ia) = \frac{1}{\sqrt{2}}(\cos\theta, i, -\sin\theta) \qquad (11.17)$$

corresponding to the sense of gyration of electrons projected onto the plane ortho-gonal to κ. The dispersion equation (11.12) is an approximation to the more general expression derived starting from (11.17) (cf. Exercise 11.1).

11.3 The parallel-driven cyclotron maser

We now turn to kinetic versions of cyclotron instabilities. As discussed in §11.1 these may be classified as parallel-driven and perpendicular-driven. Relativistic effects are unimportant for the parallel-driven instabilities, which we discuss first. Here we start with a qualitative discussion of the nonrelativistic approximation and of parallel-driven cyclotron instabilities for $s \gtrsim 1$; then we write down some general expressions for relevant dielectric tensors, and use them to discuss the parallel-driven cyclotron instability at $s = 1$ in the x-mode. In practice escaping radiation due to electron cyclotron maser emission is dominated by the x-mode emitted at $s = 1$ whenever this is allowed.

Let us refer to the 'nonrelativistic' approximation in the evaluation of the dielectric tensor (10.21) as that in which one sets $\gamma = 1$ everywhere. The resonance condition (11.1) then becomes

$$v_\parallel = \frac{\omega - s\Omega_0}{k_\parallel}, \qquad (11.18)$$

which is independent of $v_\perp$. The integrals over $dv_\parallel$ and $dv_\perp$ become independent of each other, and this allows one to evaluate $K_{ij}(\omega, \mathbf{k})$ in terms of known functions for some specific forms of the distribution function. In particular the dielectric tensor can be evaluated for a DGH distribution (Dory, Guest & Harris 1965)

$$f(\mathbf{p}) = \frac{n}{j!(2\pi)^{3/2}V_\perp^2 V_\parallel}\left(\frac{v_\perp^2}{2V_\perp^2}\right)^j \exp\left[-\frac{v_\perp^2}{2V_\perp^2} - \frac{(v_\parallel - U)^2}{2V_\parallel^2}\right], \qquad (11.19)$$

where $mV_\perp^2 = T_\perp$, $mV_\parallel^2 = T_\parallel$ and U are constants and $j \geq 0$ is an integer. For $j = 0$ (11.19) is called a bi-Maxwellian streaming distribution; $j \neq 0$ simulates a loss-cone anisotropy in which f is an increasing function of $v_\perp$ for small $v_\perp$.

This nonrelativistic approximation has been made in most treatments of instabilities due to velocity-space anisotropies in plasmas. It is found that instabilities at $s > 0$ can be driven by a temperature anisotropy $T_\perp > T_\parallel$ or a loss-cone anisotropy. The former has an excess of perpendicular momentum, compared with an isotropic distribution, and the latter has $\partial f / \partial v_\perp > 0$ at small $v_\perp$. These points might lead one to expect that the instabilities are driven by the $\partial/\partial p_\perp$ term and not the $\partial/\partial p_\parallel$ term in the absorption coefficient (10.87), viz.

$$\gamma_M(\mathbf{k}) = -\sum_{s=-\infty}^{\infty}\int d^3\mathbf{p}\, w_M(\mathbf{k}, \mathbf{p}, s)\hbar\left(\frac{s\Omega}{v_\perp}\frac{\partial}{\partial p_\perp} + k_\parallel\frac{\partial}{\partial p_\parallel}\right)f(\mathbf{p}). \qquad (11.20)$$

However this is not the case: instabilities for $s > 0$ and small $k_\perp R$ in the nonrelativistic approximation are driven by the $\partial/\partial p_\parallel$-term. That is, they are all 'parallel-driven'.

The proof that (for $s > 0$) instabilities in the nonrelativistic approximation are parallel-driven is as follows. First, the fact that the resonance condition (11.18) does not depend on $v_\perp$ allows one to separate the $v_\parallel$- and $v_\perp$-integrals. The $v_\parallel$-integral is performed over the δ-function, and the remaining integral involving $\partial f / \partial p_\perp$ is of the form

$$\gamma(\mathbf{k}) \propto - \int_0^\infty dv_\perp v_\perp A_s(v_\perp) \frac{s\Omega_0}{v_\perp} \frac{\partial f}{\partial v_\perp} \tag{11.21}$$

when only the contribution from the sth harmonic is retained. The factor $A_s(v_\perp)$ involves products of terms

$$v_\perp \frac{s}{k_\perp R} J_s(k_\perp R), \quad v_\perp J_s'(k_\perp R), \quad v_\parallel J_s(k_\perp R)$$

with $R = v_\perp/\Omega_0$. For small $k_\perp R$, $A_s(v_\perp)$ is an increasing function of $v_\perp$, and a partial integration of (11.21) leads to

$$\gamma(\mathbf{k}) \propto \int_0^\infty dv_\perp s\Omega_0 f \frac{\partial A_s(v_\perp)}{\partial v_\perp} > 0. \tag{11.21'}$$

Thus the $p_\perp$-derivative cannot lead to a negative contribution to the absorption coefficient in the nonrelativistic approximation, and any instability (for $s > 0$ and small $k_\perp R$) must be parallel-driven.

Consider the contributions from the $p_\perp$-and $p_\parallel$-dervatives for the DGH distribution (11.19). One has

$$\left(\frac{s\Omega_0}{v_\perp} \frac{\partial}{\partial v_\perp} + k_\parallel \frac{\partial}{\partial v_\parallel} \right) f = \left\{ s\Omega_0 \left(\frac{j}{v_\perp^2} - \frac{1}{V_\perp^2} \right) - \frac{k_\parallel (v_\parallel - U)}{V_\parallel^2} \right\} f$$

$$= \left\{ s\Omega_0 \left(\frac{j}{v_\perp^2} - \frac{1}{V_\perp^2} \right) - \frac{\omega - s\Omega_0 - k_\parallel U}{V_\parallel^2} \right\} f, \tag{11.22}$$

where in the final term we assume $\omega - s\Omega_0 - k_\parallel v_\parallel = 0$. For a Maxwellian distribution ($j = 0$, $V_\perp^2 = V_\parallel^2 = V^2$, $U = 0$) the right hand side reduces to $-\omega f/V^2$, and (11.21) then implies positive absorption $\gamma(\mathbf{k}) > 0$, as is necessarily the case for a Maxwellian distribution. For a bi-Maxwellian distribution ($j = 0$, $U = 0$, $V_\perp^2 \neq V_\parallel^2$) the right hand side is proportional to $-\omega/V_\perp^2 + (s\Omega_0/V_\parallel^2)(1 - V_\parallel^2/V_\perp^2)$ which can lead to negative absorption (for $s > 0$) for $V_\perp^2 > V_\parallel^2$. Any negative absorption is possible only on the low frequency side of the line, i.e. for $\omega < s\Omega_0$. The inclusion of $U \neq 0$ causes a shift in the line centre so that growth is possible only for $\omega < s\Omega_0 + k_\parallel U$; the inclusion of $j \neq 0$ has little effect in the case where instability is possible [cf. Exercise 11.2].

We may interpret these results as follows. First let us assume that f has a single maximum as a function of $v_\parallel$ and define the 'centre of the line' at the sth harmonic as the value of ω for which $\omega - s\Omega_0 - k_\parallel v_\parallel$ vanishes at this maximum in f. For the distribution (11.19) the centre of the line corresponds to $\omega - s\Omega_0 - k_\parallel U = 0$. The contribution from the $p_\parallel$-derivative in (11.20) is then stabilizing at frequencies above the centre of the line and destabilizing at frequencies below the centre of the line. The contribution from the $p_\perp$-derivative in (11.20) is stabilizing (for all $s > 0$). For a Maxwellian distribution this stabilizing contribution from the $p_\perp$-derivative overcomes the destabilizing contribution below the centre of the line from the $p_\parallel$-derivative. For an anisotropic distribution with $V_\perp^2 > V_\parallel^2$ or with a loss-cone anisotropy, the relative magnitude of the contribution from the $p_\perp$-derivative is reduced, and this can allow the destabilizing contribution from the $p_\parallel$-derivative to dominate below the centre of the line.

As already mentioned the dielectric tensor may be evaluated explicitly in the nonrelativistic approximation for the DGH distribution (11.19). The integral over $v_\parallel$ may be reduced by performing some sums, using sum rules such as

$$\sum_{s=-\infty}^{\infty} J_s^2(z) = 1, \quad \sum_{s=-\infty}^{\infty} \frac{s^2}{z^2} J_s^2(z) = \sum_{s=-\infty}^{\infty} J_s'^2(z) = \tfrac{1}{2} \tag{11.23}$$

and other sums which vanish. In this way the $v_\parallel$-integral may be reduced to the form which may be evaluated in terms of the plasma dispersion function (2.30):

$$\int_{-\infty}^{\infty} dv_\parallel \frac{\exp\left[-(v_\parallel - U)^2/2V_\parallel^2\right]}{\omega - s\Omega_0 - k_\parallel v_\parallel} = \frac{(2\pi)^{1/2} V_\parallel}{\omega - s\Omega_0 - k_\parallel U}\{\phi(y_s) - i\sqrt{\pi} y_s e^{-y_s^2}\} \tag{11.24}$$

with

$$y_s = \frac{\omega - s\Omega_0 - k_\parallel U}{\sqrt{2} k_\parallel V_\parallel}. \tag{11.25}$$

The integral over $v_\perp$ may be evaluated in terms of the standard integral

$$\int_0^\infty \frac{dv_\perp v_\perp}{V_\perp^2} J_s^2\left(\frac{k_\perp v_\perp}{\Omega_0}\right) \exp\left[-\frac{v_\perp^2}{2V_\perp^2}\right] = e^{-\lambda} I_s(\lambda) \tag{11.26}$$

with $\lambda = k_\perp^2 V_\perp^2/\Omega_0^2$ and where $I_s(\lambda)$ is a modified Bessel function. Other relevant results may be derived from (11.26), cf. Exercise 11.4.

The result for a bi-Maxwellian distribution, i.e. for $j = 0$ in (11.19), is

$$K_{ij}(\omega, \mathbf{k}) = \delta_{ij} - \sum \frac{\omega_p^2}{\omega^2}\left[- A a_{ij}(k) - \frac{\omega^2}{k_\parallel^2 V_\parallel^2} b_i b_j\right.$$

$$\left. + \sum_{s=-\infty}^{\infty} \{\phi(y_s) - i\sqrt{\pi} y_s e^{-y_s^2}\}\left\{\frac{\omega - k_\parallel U}{\omega - s\Omega_0 - k_\parallel U} + A\right\} N_{ij}(\omega, \mathbf{k}, s)\right], \tag{11.27}$$

where the unlabelled sum is over (unlabelled) species of particle, and with

$$a_{ij}(k) := \begin{bmatrix} 1 & 0 & -k_\perp/k \\ 0 & 1 & 0 \\ -k_\perp/k_\parallel & 0 & k_\perp^2/k^2 \end{bmatrix}. \tag{11.28}$$

The parameter A for each species is

$$A_\alpha = \frac{T_{\alpha\perp}}{T_{\alpha\parallel}} - 1 = \frac{V_{\alpha\perp}^2}{V_{\alpha\parallel}^2} - 1; \tag{11.29}$$

it is determined by the temperature anisotropy, being zero for an isotropic distribution. The final entry in (11.27) is

$$N_{ij}(\omega, \mathbf{k}, s) =$$

$$= e^{-\lambda}
\begin{bmatrix}
\dfrac{s^2}{\lambda} I_s & i\epsilon s(I_s' - I_s) & \dfrac{s\,k_\perp}{\lambda\,k_\parallel}\left(\dfrac{\omega - s\Omega_0}{\Omega_0}\right) I_s \\[2ex]
-i\epsilon s(I_s' - I_s) & \left(\dfrac{s^2}{\lambda} + 2\lambda\right) I_s - 2\lambda I_s' & -i\epsilon\dfrac{k_\perp}{k_\parallel}\left(\dfrac{\omega - s\Omega_0}{\Omega_0}\right)(I_s' - I_s) \\[2ex]
\dfrac{s\,k_\perp}{\lambda\,k_\parallel}\left(\dfrac{\omega - s\Omega_0}{\Omega_0}\right) I_s & i\epsilon\dfrac{k_\perp}{k_\parallel}\left(\dfrac{\omega - s\Omega_0}{\Omega_0}\right)(I_s' - I_s) & \dfrac{1}{\lambda}\dfrac{k_\perp^2}{k_\parallel^2}\left(\dfrac{\omega - s\Omega_0}{\Omega_0}\right)^2 I_s
\end{bmatrix}$$

$$(11.30)$$

where the argument λ of the modified Bessel functions is omitted.

The dielectric tensor can be evaluated for a DGH distribution for $j \neq 0$ [cf. Exercise 11.4]. The result of most interest is the anti-hermitian part of $K_{ij}(\omega, \mathbf{k})$ for $s \gtrsim 1$ in the small gyroradius limit $\lambda \ll 1$. Retaining the leading terms for each s, the nonrelativistic approximation to (10.21') for the distribution (11.19) gives

$$K_{ij}^A(\omega, \mathbf{k}) = \sum \frac{i\sqrt{\pi}}{V_\perp^2} \frac{(j+s)!}{s!\,j!} \left\{ \left(\frac{s\Omega_0}{\omega - s\Omega_0 - k_\parallel U} \right) \left(\frac{s}{j+s} \right) + \frac{V_\perp^2}{V_\parallel^2} \right\}$$

$$\times y_s e^{-y_s^2} \frac{1}{s!} \left(\frac{\lambda}{2} \right)^s v_i(\omega, \mathbf{k}, s) v_j^*(\omega, \mathbf{k}, s) \qquad (11.31)$$

with y_s given by (11.25) and with

$$\mathbf{v}(\omega, \mathbf{k}, s) = \left(\frac{s\Omega_0}{k_\perp}, \; -i\epsilon\frac{s\Omega_0}{k_\perp}, \; \frac{\omega - s\Omega_0}{k_\parallel} \right). \qquad (11.32)$$

Now let us consider the particular case of growth of x-mode waves at $s = 1$ due to electrons with a DGH distribution. The resonance condition is compatible with the condition $\omega > \omega_x$ that the emission be above the cutoff frequency for the x-mode only for $\Omega_e > \omega_p$. As already argued, growth is possible only below the centre of the line. For $U/c \ll 1$ and $\Omega_e^2/\omega_p^2 \ll 1$, the condition that the centre of the line be above $\omega_x \approx \Omega_e + \omega_p^2/\Omega_e$ reduces to

$$N_x \frac{U}{c} \cos\theta > \frac{\omega_p^2}{\Omega_e^2}. \qquad (11.33)$$

(We omit the subscripts e on U, $V_\perp$ and $V_\parallel$, which all refer to electrons here.) We also require that the absorption coefficient not be too small in magnitude; specifically we impose the condition $y_s^2 \lesssim 1$ so that the exponential factor in (11.31) is of order unity. This corresponds to

$$\left| 1 - \frac{\Omega_e}{\omega} - N_x \frac{U}{c} \cos\theta \right| \lesssim \sqrt{2} N_x \frac{\omega}{\Omega_e} \frac{V_\parallel}{c} |\cos\theta|. \qquad (11.34)$$

The sign of the absorption coefficient is determined by the sign of y_s times the factor in curly brackets in (11.31). Growth at $s = 1$ requires

$$\frac{\Omega_e}{\omega} + \frac{(j+1)V_\perp^2}{V_\parallel^2}\left(1 - \frac{\Omega_e}{\omega} - N_x\frac{U}{c}\cos\theta\right) < 0. \tag{11.35}$$

The condition (11.35) necessarily requires $1 - \Omega_e/\omega - N_x(U/c)\cos\theta < 0$, which is the condition for emission below the centre of the line.

The conditions (11.34) and (11.35), with $\omega \approx \Omega_e$, together require

$$\frac{(j+1)V_\perp^2}{V_\parallel^2}\sqrt{2}N_x\frac{V_\parallel}{c}|\cos\theta| \gtrsim 1. \tag{11.36}$$

This condition is not easily satisfied in practice. This is because it requires a very small spread in parallel velocities. To satisfy (11.36) one requires $V_\perp^2 > V_\parallel c$, where $V_\parallel$ is not the mean parallel velocity $\langle v_\parallel \rangle = U$, but the root mean square velocity $\langle (v_\parallel - \langle v_\parallel \rangle)^2 \rangle^{1/2} = V_\parallel$. Physically the driving term requires a narrow parallel velocity spread so that the gradient $\partial f/\partial p_\parallel$ is large enough to overcome the stabilizing contribution from the perpendicular derivative.

To within a factor of order unity the maximum growth rate for this parallel-driven instability is

$$|\gamma_x|_{max} \approx -\frac{n_1}{n_e}\frac{\omega_p^2(j+1)V_\perp^2}{\Omega_e V_\parallel^2} \tag{11.37}$$

where n_1 is the number density of the electrons in the DGH distribution. This growth rate is relatively large; however, it applies only under the quite restrictive conditions discussed above. It seems unlikely that the conditions required for (11.37) are satisfied in applications to space plasmas. The alternative perpendicular-driven instability seems to be more favourable for the interpretation of naturally occurring cyclotron maser emission.

There is a reactive couterpart to the instability (11.37). This may be treated by calculating the frequency shift $\Delta\omega_M(\mathbf{k})$ due to the distribution (11.19), and deriving a quadratic equation for it. Some details are outlined in Exercise 11.3.

11.4 The perpendicular-driven cyclotron maser

The step taken by Wu & Lee (1979) in initiating recent renewed interest in cyclotron maser emission was to make the 'semirelativistic' approximation

$$\Omega = \Omega_0/\gamma \approx \Omega_0(1 - v^2/2c^2). \tag{11.38}$$

The resonance condition (11.1) then becomes

$$\omega - s\Omega_0(1 - v^2/2c^2) - k_\parallel v_\parallel = 0. \tag{11.39}$$

With $v^2 = v_\parallel^2 + v_\perp^2$, (11.39) becomes a quadratic equation for $v_\parallel^2$. There are then two resonant values of $v_\parallel$ compared with the one resonant value $v_\parallel = (\omega - s\Omega_0)/k_\parallel$ in the nonrelativistic case $\gamma = 1$.

It is useful to interpret the resonance condition geometrically. For a wave with given ω and $k_\parallel$, and at a given harmonic s, the resonance condition (11.1) determines the values of $v_\parallel$ and γ for resonant particles. Suppose we construct a velocity space with axial coordinate $v_\parallel/c$ and cylindrical radial coordinate $v_\perp/c$. The physical region in our space is the interior of the unit sphere $v/c < 1$. The nonrelativistic approximation (11.18) to the resonance condition corresponds to a plane perpendicular to the axis at $v_\parallel = v_N$ with

$$\frac{v_N}{c} = \frac{\omega - s\Omega_0}{k_\parallel c}. \tag{11.40}$$

The semirelativistic approximation corresponds to a sphere centred at $v_\parallel/c = v_c/c$, $v_\perp/c = 0$ and of radius v_R/c with

$$\frac{v_c}{c} = \frac{k_\parallel c}{s\Omega_0}, \quad \frac{v_R}{c} = \left\{ \left(\frac{v_c}{c}\right)^2 - \frac{2(\omega - s\Omega_0)}{s\Omega_0} \right\}^{1/2}. \tag{11.41}$$

The general result for the exact resonance condition (11.1) is an ellipsoid of revolution. The intersection of this with a $v_\perp - v_\parallel$ plane is called a *resonance ellipse*. The ellipse is centred on the $v_\parallel$-axis, at $v_\parallel/c = v_c/c$ say, and its semi-major axis, v_R/c say, is perpendicular to the $v_\parallel$-axis. Let e be the eccentricity of the ellipse. One has

$$\frac{v_c}{c} = \frac{\omega k_\parallel c}{s^2\Omega_0^2 + k_\parallel^2 c^2},$$

$$\frac{v_R}{c} = \left(\frac{s^2\Omega_0^2 + k_\parallel^2 c^2 - \omega^2}{s^2\Omega_0^2 + k_\parallel^2 c^2}\right)^{1/2}, \quad e = \left(\frac{k_\parallel^2 c^2}{s^2\Omega_0^2 + k_\parallel^2 c^2}\right)^{1/2}. \tag{11.42}$$

For $\omega^2 > k_\parallel^2 c^2$ the resonance ellipse lies entirely within the unit sphere, for $\omega^2 = k_\parallel^2 c^2$ it touches the unit sphere on the $v_\parallel$-axis, and for $\omega^2 < k_\parallel^2 c^2$ it touches the unit sphere off the axis, and the outer segment of the ellipse is then nonphysical. Examples of resonance ellipses are illustrated in Figure 11.2.

The nonrelativistic approximation (11.40) is valid for nonrelativistic particles $(v^2/c^2 \ll 1)$ in the case $k_\parallel^2 c^2 \gg \omega^2$ when the resonance ellipse is highly elliptical. The segment of the ellipse near the origin is then well approximated by the straight line (11.40). A straight line is also a reasonable approximation in the semirelativistic case (i) provided one has $v_c \approx v_R \gg |v_c - v_R|$, and (ii) provided the resonant particles of interest have $v \approx |v_c - v_R|$. Comparison of (11.40) and (11.41) shows that one has $v_c^2 - v_R^2 = 2v_c v_N$, and $v_\parallel = v_c \pm v_R$ at the two points where the sphere crosses the axis; hence the condition stated in (i) implies $v_\parallel \approx v_N$ for the inner crossing point, as illustrated in Figure 11.3.

In early work on electron cyclotron maser emission, e.g. Twiss (1958) and Bekefi and his co-workers, cf. §9.3 of Bekefi's (1966) book, the term $k_\parallel v_\parallel$ was excluded from the resonance condition. The resonance ellipse then becomes a circle centred on the origin. In evaluating the emission or absorption coefficients one integrates around the resonance ellipse in velocity space; this integral simplifies greatly in the two limiting cases where the ellipse is approximated by the straight line $v_\parallel = v_N$ and by a circle centred on the origin. The case (11.41) of an off-centred circle introduces a qualitatively new effect: in principle it allows any distribution with

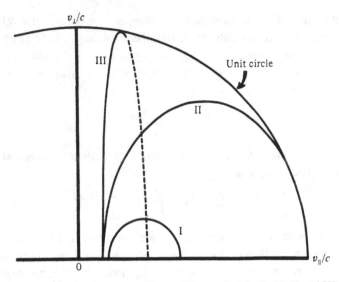

Fig. 11.2 Examples of resonance ellipses. The curves labelled I, II and III are for $\omega^2 > k_\parallel^2 c^2$, $\omega^2 = k_\parallel^2 c^2$ and $\omega^2 < k_\parallel^2 c^2$ respectively. The dashed part of the curve III is nonphysical.

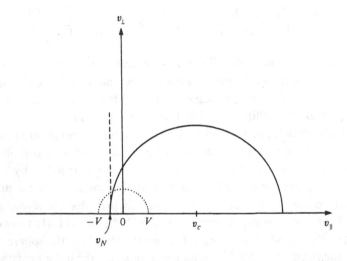

Fig. 11.3 For $|v_c| \approx v_R \ll c$ with $|v_c| - v_R \approx |v_N| \ll |v_c|$, the straight line $v_\parallel = v_N$ (dashed line) is a reasonable approximation to the resonant ellipse (solid curve) for an electron distribution which falls off rapidly outside the region $v = V$ (dotted curve) for $|v_N| \lesssim V \ll v_c$.

$\partial f/\partial p_\perp > 0$ at small $p_\perp$ to drive an instability. Such a perpendicular-driven instability applies in the cases discussed by Twiss and Bekefi, but in these cases one requires $\partial f/\partial p_\perp > 0$ at $p_\parallel = 0$ and $\partial f/\partial p > 0$ respectively. In the off-centred case (11.41) the driving term can be $\partial f/\partial p_\perp > 0$ for $p_\perp \approx 0$ at any $p_\parallel$. In particular a loss-cone distribution is favourable for driving the instability, cf. Figure 11.4.

In the semirelativistic approximation the integration around the resonance ellipse corresponds to a limited range of $v_\perp \lesssim v_R$. This is in contrast to the nonrelativistic case where the $v_\perp$-integral is unlimited. The proof that a $p_\perp$-derivative cannot drive an instability in the nonrelativistic case, cf. (11.21) and (11.21′), does not apply in the semirelativistic case. The form (11.21) is not strictly relevant in the semi-relativistic case but qualitatively one may use it with the upper limit $v_\perp = \infty$ to the integral replaced by $v_\perp = v_R$; then the partial integration leads to an additional destabilizing term, specifically $-A_s(v_\perp)s\Omega_0 f$ with $v_\perp = v_R$, in addition to the stabilizing term in (11.21′).

In evaluating the absorption coefficient (11.20) we may integrate around the resonance ellipse, approximated here by the off-centred circle (11.41). Writing

$$v_\parallel = v_c - v' \cos\phi', \quad v_\perp = v' \sin\phi', \qquad (11.43)$$

the δ-function in the approximation (11.39) becomes

$$\delta(\omega - s\Omega - k_\parallel v_\parallel) \approx \frac{c^2}{s\Omega_0 v_R}\delta(v' - v_R). \qquad (11.44)$$

The integral over momentum space may be replaced according to

$$\int d^3\mathbf{p} \to 2\pi m^3 \int dv_\perp v_\perp \int dv_\parallel = 2\pi m^3 \int dv' v'^2 \int_0^\pi d\phi', \qquad (11.45)$$

where we make the nonrelativistic approximation everywhere other than in the resonance condition. The absorption coefficient (11.20) with the probability (10.80) then reduces to

$$\gamma_M(\mathbf{k}) = -\sum_{s=-\infty}^{\infty} \frac{2\pi m^3 c^2 v_R}{s\Omega_0} \frac{2\pi q^2 R_M(\mathbf{k})}{\varepsilon_0 \omega} \int_0^\pi d\phi'$$
$$\times \left[|\mathbf{e}_M^* \cdot \mathbf{V}(\mathbf{k}, \mathbf{p}; s)|^2 \left\{ \frac{s\Omega}{v_\perp}\frac{\partial}{\partial p_\perp} + k_\parallel\frac{\partial}{\partial p_\parallel} \right\} f(\mathbf{p}) \right]_{v' = v_R}. \qquad (11.46)$$

A loss-cone distribution is particularly effective in deriving the instability; such a distribution is illustrated in Figure 11.4. The loss cone may be either one-sided or two-sided. A two-sided distribution would lead to two independent instabilities, one with $k_\parallel > 0$ and the other with $k_\parallel < 0$. The point about a loss-cone distribution is that $\partial f/\partial p_\perp$ is positive inside the loss cone. Hence if we choose a resonance ellipse which lies entirely within the loss cone the contribution from the $p_\perp$-derivative in (11.46) can be destabilizing at every point around the contour (the resonance ellipse) of integration. The maximum growth rate corresponds to the resonance ellipse which has the largest net positive contribution from the $p_\perp$-derivative term. This resonance ellipse is that which passes roughly through the edge of the loss cone, as illustrated in Figure 11.4.

For a loss-cone distribution the parameters of the resonance ellipse corresponding to maximum growth can be estimated as follows. The value of v_c should be such that $\frac{1}{2}mv_c^2$ is roughly the mean parallel energy $\frac{1}{2}mv_m^2$ say, of the electrons in the loss-cone distribution. The ratio v_R/v_c should be roughly equal to $\tan\alpha_0$

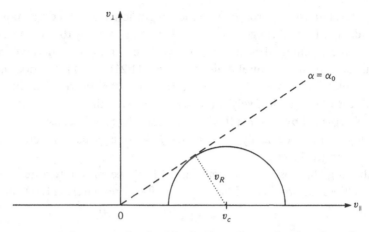

Fig. 11.4 For a loss-cone distribution $f(\mathbf{p})$ is an increasing function of $p_\perp$ for $\alpha \lesssim \alpha_0$, where $\alpha = \alpha_0$ defines the loss cone. For ω and $k_\parallel$ (at given s) such that the corresponding resonance ellipse lies entirely within the region $\alpha < \alpha_0$, the contribution from the term $(s\Omega/v_\perp)\,\partial f/\partial p_\perp$ in (11.20) is destabilizing everywhere around the ellipse.

where $\alpha = \alpha_0$ is the pitch angle defining the edge of the loss cone, cf. Figure 11.4. We may then estimate the maximum growth rate by introducing

$$n_1 \approx 2\pi m^3 v_c v_R^2 f_{\max},\tag{11.47}$$

where $f_{\max}$ is the value of f at $v_\parallel = v_c$, $v_\perp = v_R$. Then for the x-mode sufficiently above its cutoff such that one has $N_x \approx 1$, the growth rate at $s = 1$ has a maximum

$$|\gamma_x|_{\max} \approx \pi \frac{\omega_p^2}{\Omega_e} \frac{n_1}{n_e} \frac{c^2}{v_R v_c}.\tag{11.48}$$

This maximum growth rate is restricted to small ranges of v_c and v_R, which we estimate to be $\Delta v_c \approx v_R$ and $\Delta v_R \approx v_c \Delta\alpha_0$, where $\Delta\alpha_0$ is the range of pitch angles over which f falls off inside the loss cone (e.g. $|\partial f/\partial\alpha| \approx f/\Delta\alpha_0$ for $\alpha \lesssim \alpha_0$).

The kinematic restrictions, with $N_x \approx 1$, imply that growth is restricted to a small range of angles $\Delta\theta$ about a characteristic angle $\theta = \theta_m$, and to a small range $\Delta\omega$ of frequencies about a central frequency ω_m. We have

$$\cos\theta_m \approx \frac{v_R}{c}, \quad \Delta\theta \approx \frac{\Delta v_c}{c}\tag{11.49a, b}$$

$$\frac{\omega_m - \Omega_e}{\Omega_e} \approx \frac{1}{2}\left(\cos^2\theta_m - \frac{v_R^2}{c^2}\right),\tag{11.49c}$$

$$\frac{\Delta\omega}{\Omega_e} \approx \frac{v_R v_c}{c^2}\Delta\alpha_0.\tag{11.49d}$$

The frequency ω_m must be sufficiently above the cutoff frequency to have $N_x \approx 1$. A restriction on it can be inferred by plotting the curve $v_R = 0$, i.e. $\omega^2 = \Omega_0^2 + k_\parallel^2 c^2$, at which resonance ceases to be possible, with $k_\parallel$ related to ω through the x-mode

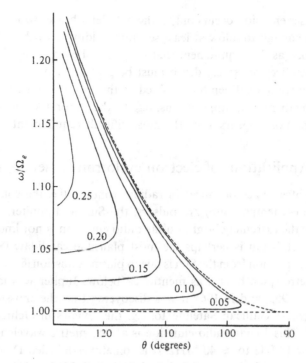

Fig. 11.5 The boundary curve $\omega^2 = \Omega_e^2 + k_\parallel^2 c^2$ for resonance at $s = 1$ is illustrated for the x-mode; the labels on the solid curve refer to values of ω_p/Ω_e, and the dashed curve is the limiting case $\omega_p/\Omega_e = 0$ (Hewitt *et al.* 1982).

dispersion relation. The relevant curve is illustrated in Figure 11.5. This curve has a 'nose' at $(\omega - \Omega_e)/\Omega_e \approx 2\omega_p^2/\Omega_e^2$, $\theta \approx \pi/2 - 2\omega_p/\Omega_e$ for $\omega_p/\Omega_e \ll 1$. For (11.49) to be compatible with this requirement on $v_R > 0$, one requires

$$\frac{\omega_p}{\Omega_e} \lesssim \frac{v_R}{2c}. \tag{11.50}$$

For ω_p/Ω_e larger than the limit allowed by (11.50) amplification of the x-mode at $s = 1$ is suppressed.

These and other estimates have been compared with the results of detailed numerical calculations by Hewitt, Melrose & Rönnmark (1982). Two other semi-quantitative results which are of interest are: (i) the growth rate for the o-mode is smaller than that for the x-mode at $s = 1$ by a factor of order $(v_m/c)^2$, and (ii) the growth rates at harmonics $s > 1$ are smaller than that at $s = 1$ by factors of order $(v_R/c)^{2s-2}$.

The growth rate (11.48) for the perpendicular-driven instability is comparable in magnitude to that (11.37) for the parallel-driven instability. The perpendicular-driven instability is the much more favourable in applications because it occurs under conditions which are much more easily satisfied. Essentially any nonrelativistic electron distribution with $\partial f/\partial p_\perp > 0$ at small $p_\perp$ can cause negative absorption of x-mode waves at $s = 1$ in a plasma with $\Omega_e/\omega_p \gg 1$.

Effective maser emission occurs only if the path length over which amplification occurs is large enough to allow at least several *e*-folding growths. This condition may be described as the requirement that the source be opticially thick to amplification, i.e. the effective optical depth must be greater than unity in magnitude, where the 'effective optical depth' is defined as the integral along the ray path of the spatial growth rate. In applications, e.g. to the auroral kilometric radiation (§11.5) this condition appears to be the most difficult requirement to satisfy.

11.5 Applications of electron cyclotron maser emission

The three brightest types of source in radioastronomy (i.e. the sources with the highest brightness temperature) are pulsars, the Sun and Jupiter. The emission mechanism for the extremely bright pulsar radio emission is not known, but free-electron maser emission is perhaps the most plausible candidate (§7.3). Most of the bright radio emission from the Sun is due to plasma emission (§6.6); an exception is solar decimetre spike bursts, as mentioned below. Jupiter was identified as a radio source in 1955, and in 1962 it was discovered that the emission correlated with the innermost Galilean Satellite Io, e.g. the reviews in Gehrels (1976) and Dessler (1983). This Jovian radio emission is at decametric wavelengths, actually from just below 1 MHz to $\approx$ 40 MHz. It is sometimes labelled DAM. (A higher-frequency emission at decimetric wavelengths, labelled DIM, is due to synchrotron emission and is not discussed further here.) A terrestrial counterpart of DAM was first seen by spacecraft in the late 1960's with radio receivers intended for the study of low-frequency extensions of solar radio bursts. This 'Earth noise' was initially considered a nuisance. A systematic study (Gurnett 1974) showed that it originates from the auroral zones, leading to the name 'auroral kilometric radiation', or AKR for short. The frequency range, from $\approx$ 100 kHz to $\approx$ 1 MHz, corresponds to kilometric wavelengths. There is also a counterpart of DAM and AKR from Saturn.

AKR correlates with a particular class of precipitating auroral electrons, called 'inverted-V' events and it is widely (but not universally) accepted that AKR is due to electron cyclotron maser emission due to inverted-V electrons. The pitch-angle distribution of these electrons has been studied both at low altitudes and later at heights $\gtrsim 1\,R_E$ (R_E = radius of the Earth) where the AKR originates. The source of free energy which drives the electron cyclotron maser emission is a one-sided loss cone distribution. This forms as downward propagating electrons are reflected. Those with sufficiently small pitch angles reach the denser layers of the atmosphere and are lost. The reflected upgoing electron distribution includes a 'hole' at small pitch angles, cf. Figure 11.6. The measured pitch angle distributions have been shown to lead to growth of *x*-mode radiation; that is, on inserting the measured distribution in a numerical code for calculating the absorption coefficient, growth is implied. Although this provides strong evidence that AKR is due to the perpendicular-driven cyclotron maser mechanism, the resulting growth rates are not necessarily very meaningful. The reason is that saturation of the instability,

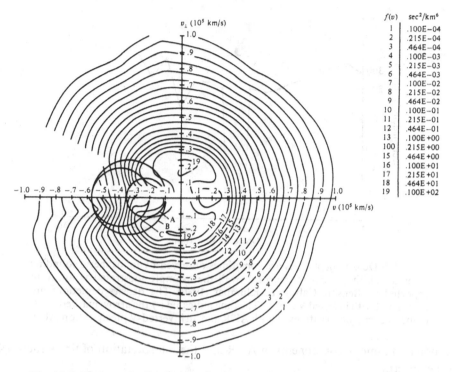

The following data table appears to the right of the figure:

$f(v)$	sec²/km⁶
1	.100E−04
2	.215E−04
3	.464E−04
4	.100E−03
5	.215E−03
6	.464E−03
7	.100E−02
8	.215E−02
9	.464E−02
10	.100E−01
11	.215E−01
12	.464E−01
13	.100E+00
100	.215E+00
15	.464E+00
16	.100E+01
17	.215E+01
18	.464E+01
19	.100E+02

Fig. 11.6 Pitch-angle distribution for inverted V electrons in an event reported by Croley, Mizera & Fennell (1978). The solid curves are resonance ellipses used in calculations performed by Omidi & Gurnett (1982); the source of free energy is the 'hole' in the distribution of upgoing electrons.

due to quasilinear relaxation, should occur on a timescale ($\lesssim 0.1$ s) much shorter than the time (≈ 4s) required to measure the pitch-angle distribution. The actual growth rate should be due to the unrelaxed distribution and should be much larger than that calculated.

As implied by (11.50), the condition $\omega_p/\Omega_e \ll 1$ is essential for growth at $s=1$ in the x-mode, and this condition is not normally satisfied in the terrestrial magnetosphere. However, it is satisfied in the source region for AKR. The reason for this is related to the acceleration of the inverted-V electrons. The flow of the solar wind past the Earth sets up a potential difference of several tens of kilovolts. When the sign of this difference is such that it drives electrons from the magnetosphere to the Earth along the auroral field lines, the magnetosphere can become depopulated. This can lead to localized extremely sharp depletions in ω_p, cf. Figure 11.7. The potential drop then develops in so-called double layers or electrostatic shocks, and drives the required current by accelerating the ambient electrons (thermal energies $\lesssim 10$ eV) to energies of several keV.

The study and interpretation of AKR and related phenomena are areas of active interest at the time of writing. Topics of continuing interest include harmonic,

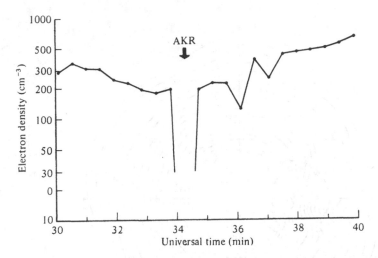

Fig. 11.7 Density depletions are sometimes observed by ionospheric sounding satellites as they cross the auroral oval; the example illustrated is an event reported by Benson, Calvert & Klumpar (1980). The density depletions are associated with inverted V events. The time indicated by the arrow marked AKR coincided with observations of both AKR and an inverted-V precipitation event.

o-mode and z-mode components in AKR and the interpretation of fine structures in the spectra.

The interpretation of DAM is thought to be analogous to that of AKR. The potential difference in this case is attributed to the relative motion of Io (which has a highly conducting ionosphere) through the Jovian magnetosphere. One particularly intriguing older result on DAM which has a favourable interpretation in terms of the cyclotron maser mechanism is the weird angular distribution of the emission. It was inferred from ground-based data (Dulk 1967) that DAM is emitted in a thin hollow cone with an angle $\theta \approx 80°$ to the field lines, and with a thickness $\Delta\theta \approx 1°$. The estimates (11.49a, b) would imply such emission for $v_m/c \approx 0.2$ (electron energy $\approx 10 \, \text{keV}$) and $v_c \approx v_m/5$. The fact that the perpendicular-driven cyclotron maser implies just such a weird emission pattern, i.e. emission on the surface of a hollow cone, provides further support for it being the emission mechanism for DAM.

The success of the cyclotron maser mechanism in the interpretation of these planetary radio emissions has led to suggestions for other astrophysical applications. Solar decimetre spike bursts and bright radio emission from some flare stars are likely candidates (e.g. Melrose & Dulk 1982). There is a serious difficulty in these applications: the radiation must escape through the relatively hot ($\approx 10^6 \, \text{K}$) solar or stellar corona where gyromagnetic absorption at $s = 2$ and $s = 3$ is strong. It is not clear how this problem is to be overcome. One suggestion is that the radiation encounters the second-harmonic absorption layer (where ω is equal to twice the local electron cyclotron frequency) at an angle $\theta \approx 0$; second harmonic emission and absorption (being quadrupolar) go to zero as $\sin^2 \theta$ allowing an escape 'window' at small angles. However, the maser emission is generated at a large angle and it is difficult to see how this escape mechanism can operate in a realistic magnetic model. The parallel-driven cyclotron maser can generate emission at small θ, and one might imagine that it could be more favourable in this case. However, the emission from precipitating electrons is directed initially downwards for this mechanism; the rays need to be refracted through $\gtrsim 90°$ to start propagating outwards and through $\approx 180°$ to encounter the second harmonic absorption layer at $\theta \approx 0$. This difficulty with the escape of the radiation leaves a question as to whether the cyclotron maser mechanism is the radiation mechanism for these bright stellar radio emissions.

11.6 Phase bunching and wave trapping

Now let us return to the reactive version of the electron cyclotron instability. As in the unmagnetized case (cf. Chapter 5) we may discuss the effects of phase bunching and wave trapping in terms of a simple model for the interaction between a single particle and a monochromatic wave. First we must generalize the model developed in §5.1 by including the magnetic field.

We generalize (5.8) by allowing the wave to have an arbitrary polarization vector $\mathbf{e}$

$$\mathbf{E} = \mathbf{e}E(t)e^{-i\psi(t,\mathbf{x})} + \text{c.c.}, \tag{11.51}$$

with $E(t) = E_0 G(t)$ as in (5.10). The equation of motion is

$$\frac{d\mathbf{p}}{dt} = q\mathbf{v} \times \mathbf{B}_0 + \frac{q}{\omega}\{(\omega - \mathbf{k}\cdot\mathbf{v})\mathbf{E} + \mathbf{k}\mathbf{v}\cdot\mathbf{E}\}. \tag{11.52}$$

Here it is important to retain relativistic effects. As in (10.12) and (10.14), we write

$$\mathbf{p} = (p_\perp \cos\phi, -\epsilon p_\perp \sin\phi, p_\parallel) \tag{11.53}$$

and

$$\mathbf{X} = \mathbf{x}_0 + (R\sin\phi, \epsilon R\cos\phi, v_\parallel t) \tag{11.54}$$

and use (11.52) to determine how $p_\perp$, $p_\parallel$, γ and ϕ vary with time. (The variations in $p_\perp$, $p_\parallel$ and γ are related through $\gamma^2 m^2 c^2 = p_\perp^2 + p_\parallel^2$.) The phase $\psi(t,\mathbf{x})$ is to be evaluated at the position of the particle. We write

$$\psi(t,\mathbf{X}) = \Psi - k_\perp R\sin\phi, \tag{11.55a}$$

$$\Psi = -\mathbf{k}\cdot\mathbf{x}_0 + (\omega - k_\parallel v_\parallel)t, \tag{11.55b}$$

where $\mathbf{k}$ is assumed in the x–z plane. An expansion in Bessel functions is performed as in (10.18). The resulting expressions are

$$\frac{dp_\perp}{dt} = \sum_{s=-\infty}^{\infty} qE_0 G\left[e^{-i(\Psi-s\phi)}\left\{(1 - N\beta_\parallel\cos\theta)\left(e_x\frac{s}{k_\perp R}J_s + i\epsilon e_y J_s'\right)\right.\right.$$
$$\left.\left. + N\beta_\parallel\sin\theta e_z\frac{s}{k_\perp R}J_s\right\} + \text{c.c.}\right], \tag{11.56a}$$

$$\frac{dp_\parallel}{dt} = \sum_{s=-\infty}^{\infty} qE_0 G\left[e^{-i(\Psi - s\phi)}\left\{\left(1 - \frac{s\Omega}{\omega}\right)e_z J_s + N\beta_\perp \cos\theta \right.\right.$$

$$\left.\left. \times \left(e_x \frac{s}{k_\perp R} J_s + i\epsilon e_y J_s'\right)\right\} + \text{c.c.}\right], \tag{11.56b}$$

$$\frac{d\gamma}{dt} = \sum_{s=-\infty}^{\infty} \frac{qE_0 G}{mc}\left[e^{-i(\Psi - s\phi)}\left\{\beta_\perp\left(e_x \frac{s}{k_\perp R} J_s + i\epsilon e_y J_s'\right) + \beta_\parallel e_z J_s\right\} + \text{c.c.}\right], \tag{11.56c}$$

$$\frac{d\phi}{dt} = \frac{\Omega_0}{\gamma} + \sum_{s=-\infty}^{\infty} \frac{qE_0 G}{p_\perp}\left[ie^{-i(\Psi - s\phi)}\left\{(1 - N\beta_\parallel \cos\theta)\left(-e_x J_s' - i\epsilon e_y \frac{s}{k_\perp R} J_s\right)\right.\right.$$

$$\left.\left. + iN\beta_\perp \sin\theta e_y J_s - N\beta_\parallel \sin\theta e_z J_s'\right\} + \text{c.c.}\right], \tag{11.56d}$$

with $\beta_\parallel = v_\parallel/c$, $\beta_\perp = v_\perp/c$ and where the arguments $k_\perp R$ of the Bessel functions are omitted.

The relations (11.56) apply to any gyromagnetic interaction. Here we are concerned with the small gyroradius limit $k_\perp R \ll 1$ when only the terms $s = 0, \pm 1$ contribute in the sums. Moreover we assume that the frequency mismatch

$$\Delta = \omega - \Omega_0/\gamma - k_\parallel v_\parallel \tag{11.57}$$

from resonance at the fundamental is small, and then only $s = 1$ need be retained in the sums in (11.56). (With $\phi \approx \Omega_0 t/\gamma$ the terms $s \neq 0$ oscillate rapidly and average to zero.) We then find

$$\frac{d\gamma}{dt} = \frac{\epsilon E_0 G\Omega_0}{2B_0 c}\left[e^{-i(\Psi - \phi)}\{\beta_\perp(e_x + i\epsilon e_y)\} + \text{c.c.}\right] \tag{11.58a}$$

$$\frac{d(\Psi - \phi)}{dt} = \Delta + \frac{\epsilon E_0 G\Omega_0}{2B_0 c}$$

$$\times \left[\frac{i}{\gamma\beta_\perp}e^{-i(\Psi - \phi)}\{(1 - N\beta_\parallel \cos\theta)(e_x + i\epsilon e_y) + N\beta_\parallel \sin\theta e_z\} + \text{c.c.}\right] \tag{11.58b}$$

$$\frac{d\Delta}{dt} = \frac{\epsilon E_0 G\Omega_0}{2B_0 c}\left[\frac{\beta_\perp}{\gamma}e^{-i(\Psi - \phi)}\{\omega(1 - N^2 \cos^2\theta) - \Delta\}(e_x + i\epsilon e_y) + \text{c.c.}\right]. \tag{11.58c}$$

The effects discussed in Chapter 5 include (i) systematic phase bunching, (ii) reactive growth treated in terms of the second order rate of change of γ, and (iii) wave trapping. Let us now discuss each of these briefly for the magnetized case using (11.58).

Systematic phase bunching arises from a first order term in the relative phase $(\Psi - \phi)^{(1)}$ which increases linearly with time. Growth or damping is not important in this context, and so we set $\omega_i = 0$. Let ψ_0 be the value of the relative phase at $t = t_0$. Then integrating (11.58c), inserting the result in (11.58b) and integrating again leads to an expression containing the systematic term

$$(\Psi - \phi)^{(1)} \approx -\frac{GE_0\Omega_0}{B_0 c}\frac{\omega\beta_\perp P}{\gamma\Delta}\sin\psi_0(1 - N^2\cos^2\theta)(t - t_0), \qquad (11.59)$$

where we assume that $\omega(1 - N^2\cos^2\theta)$ is much greater than Δ in magnitude. The polarization appears only in the factor

$$P = -\epsilon(e_x + i\epsilon e_y). \qquad (11.60)$$

For circularly polarized waves, or more correctly for $|e_x/e_y| = 1$, one has $P = \sqrt{2}$ or 0 depending on whether the handednesses of the wave and of the spiralling particle are the same or opposite, respectively. We have $P \gtrsim 0$ here. The discussion following (5.24) then implies that systematic phase bunching occurs if $1 - N^2\cos^2\theta$ and Δ have the same sign. It is not difficult to see that the unit term and the $N^2\cos^2\theta$ term in $1 - N^2\cos^2\theta$ arise from the changes in Ω_0/γ and in $k_\parallel v_\parallel$, respectively. They correspond to bunching in ϕ and in Ψ respectively. That is, for $N^2\cos^2\theta < 1$ and $\Delta > 0$ the bunching occurs in azimuthal angle and for $N^2\cos^2\theta > 1$ and $\Delta < 0$ the bunching occurs in z, i.e. along the axis.

Now let us indicate how the growth rate (11.14) may be rederived using this model. Following the arguments in §5.2, we calculate the first order corrections to $(\Psi - \phi)$ and $\beta_\perp$ and use these to find the second order change in γ. From (11.58a) and (11.60) we have

$$\frac{d\gamma^{(2)}}{dt} = -\frac{PE_0 G\Omega_0}{2B_0 c}[e^{-i(\Psi - \phi)}\{-i(\Psi - \phi)^{(1)}\beta_\perp + \beta_\perp^{(1)}\} + \text{c.c.}]. \qquad (11.61)$$

We retain only the terms proportional to G^2 and average over the initial phases. Then we set $n_e m_e c^2 d\gamma^{(2)}/dt$ equal to minus the rate of change of the energy density in the waves. The energy density increases at the rate $2\omega_i\varepsilon_0 E_0^2 G^2/R$, where R is the ratio of electric to total energy; in the following discussion we set $RP^2 = 1$. The resulting identity reduces to

$$1 = \frac{\Delta}{(\Delta^2 + \omega_i^2)}\frac{\omega_p^2\omega\beta_\perp^2}{2\gamma}(1 - N^2\cos^2\theta). \qquad (11.62)$$

Now if Δ and ω_i are the real and imaginary parts of the solution of a cubic equation then we have $|\Delta| = \frac{1}{2}r$ and $|\omega_i| = \sqrt{3}r/2$ with $r^2 = \Delta^2 + \omega_i^2$. It follows that (11.62) reproduces the growth rate (11.14) on setting $|\gamma_{max}| = 2\omega_i$. Moreover (11.62) requires $(1 - N^2\cos^2\theta) > 0$; the solution for $\Delta > 0$, $1 - N^2\cos^2\theta > 0$ corresponds to azimuthal bunching and that for $\Delta < 0$, $1 - N^2\cos^2\theta < 0$ to axial bunching.

The final effect we discuss using (11.58) is wave trapping. On differentiating (11.58b) with respect to time, using (11.58c) and neglecting some small terms, one obtains the pendulum equation (for $G = 1$)

$$\frac{d^2}{dt^2}(\Psi - \phi) = -\frac{PE_0\Omega_0\beta_\perp\omega(1 - N^2\cos^2\theta)}{\gamma B_0 c}\cos(\Psi - \phi). \qquad (11.63)$$

The *bounce frequency* is

$$\omega_T^2 = \left[\frac{PE_0\Omega_0\beta_\perp\omega(1 - N^2\cos^2\theta)}{\gamma B_0 c}\right]. \qquad (11.64)$$

As the wave amplitude increases in the growth of the gyrotron instability, ω_T increases. Saturation of the instability due to trapping occurs for $\omega_T \approx \omega_i$, as in the unmagnetized case. Saturation of the gyrotron instability has been discussed by Sprangle & Drobot (1977).

Exercise set 11

11.1 Show that if one choose the (transverse) right circular polarization vector, cf. (11.17),

$$\mathbf{e} = \frac{1}{\sqrt{2}}(\mathbf{t} + i\mathbf{a})$$

then the result (11.12) applies approximately for small $\sin^2 \theta$.

11.2 (a) Show that the condition for negative absorption at the sth harmonic for the DGH distribution (11.19) may be written in the form (11.35), viz.

$$g_s = s\Omega_0 \frac{s}{j+s} + \frac{V_\perp^2}{V_\parallel^2}(\omega - s\Omega_0 - k_\parallel U) < 0.$$

(b) Noting that the absorption coefficient is of the form

$$\gamma_s \propto g_s \exp\left[-\frac{(\omega - s\Omega_0 - k_\parallel U)^2}{2k_\parallel^2 V_\parallel^2}\right],$$

show that negative absorption near the centre of the line, i.e. for $(\omega - s\Omega_0 - k_\parallel U)^2 \lesssim 2k_\parallel^2 V_\parallel^2$, requires

$$\frac{j}{j+s} + \frac{V_\perp^2}{V_\parallel^2} \gtrsim \frac{s\Omega_0}{|k_\parallel|V_\parallel}.$$

11.3 (a) Suppose a DGH electron distribution (11.19) with number density $n_1 \ll n_e$ causes a small frequency shift $\Delta\omega_M(\mathbf{k})$ in the dispersion relation $\omega = \omega_M(\mathbf{k}) + \Delta\omega_M(\mathbf{k})$ for a mode M near the sth harmonic of the gyrofrequency. Using the hermitian counterpart of (11.31) with (11.32) show that one has

$$\Delta\omega_M \approx \frac{(j+s)!}{V_\perp^2 s! j!} \frac{g_s}{\omega - s\Omega_e - k_\parallel U} \frac{1}{s!}\left(\frac{\lambda}{2}\right)^s R_M \omega_M |e_M^*(\mathbf{k})\cdot\mathbf{v}(\omega_M, \mathbf{k}, s)|^2$$

with g_s defined in Exercise 11.2, and where arguments $(\mathbf{k})$ are omitted for ω_M, R_M, $\mathbf{e}_M$ and $\Delta\omega_M$.

(b) Show that for sufficiently small frequency mismatch $\delta\omega_M = \omega_M - s\Omega_e - k_\parallel U$, the implied quadratic equation for $\Delta\omega_M$ has complex solutions for $g_s < 0$.

(c) Show that the growth rate of the reactive instability in part (b) and the corresponding kinetic instability (Exercise 11.2) pass over continuously into each other, with the growth rates being approximately equal when they are approximately equal to $|k_\parallel|V_\parallel$.

11.4 (a) Starting from the basic integral (11.26), by differentiating with respect to

$k_\perp/\Omega_0$ (with $z = k_\perp v_\perp/\Omega_0$) derive the results

$$\int_0^\infty \frac{dv_\perp v_\perp}{V_\perp^2}\left[\begin{array}{c} v_\perp J_s'(z)J_s(z) \\ v_\perp^2 J_s'^2(z) \end{array}\right]\exp(-v_\perp^2/2V_\perp^2)$$

$$= e^{-\lambda}\left[\begin{array}{c} (\Omega_0/k_\perp)\lambda\{I_s'(\lambda) - I_s(\lambda)\} \\ (\Omega_0/k_\perp)^2[s^2 I_s(\lambda) - 2\lambda^2\{I_s'(\lambda) - I_s(\lambda)\}] \end{array}\right]$$

with $\lambda = k_\perp^2 V_\perp^2/\Omega_0^2$.

Hint: Use Bessel's equation for both the ordinary and modified functions

$$J_s''(z) + \frac{1}{z}J_s'(z) + \left(1 - \frac{s^2}{z^2}\right)J_s(z) = 0,$$

$$I_s''(\lambda) + \frac{1}{\lambda}I_s'(\lambda) - \left(1 + \frac{s^2}{\lambda^2}\right)I_s(\lambda) = 0.$$

(b) Evaluate the dielectric tensor (10.21) in the nonrelativistic approximation for the DGH distribution (11.19) with arbitrary integral $j > 0$. Show that your result may be written in the form, cf. (11.27),

$$K_{il}(\omega, \mathbf{k}) = \delta_{il} - \sum \frac{\omega_p^2}{\omega^2}\left[-A^{(j)}a_{il}(\mathbf{k}) - \frac{\omega^2}{k_\parallel^2 V_\parallel^2}b_i b_l\right.$$

$$+ \sum_{s=-\infty}^\infty \{\phi(y_s) - i\sqrt{\pi}y_s e^{-y_s^2}\}\left[\left(\frac{s\Omega_0}{\omega - s\Omega_0 - k_\parallel U} + \frac{V_\perp^2}{V_\parallel^2}\right)\right.$$

$$\left.\left.\times N_{il}^{(j)}(\omega, \mathbf{k}, s) - \frac{s\Omega_0}{\omega - s\Omega_0 - k_\parallel U}N_{il}^{(j-1)}(\omega, \mathbf{k}, s)\right]\right]$$

with $a_{il}(\mathbf{k})$ defined by (11.28), with

$$A^{(j)} = (j+1)\frac{V_\perp^2}{V_\parallel^2} - 1$$

and with, cf. (11.30),

$$N_{il}^{(j)}(\omega, \mathbf{k}, s) = \frac{(-)^j}{j!}\lambda^{-(j+1)}\left(\frac{\partial}{\partial\lambda^{-1}}\right)^j[\lambda N_{il}(\omega, \mathbf{k}, s)].$$

Hints: Label $f^{(j)}$ with the j-value and establish the identities $(\alpha = \frac{1}{2}V_\perp^2)$

$$f^{(j+1)} = \frac{(-)^j}{j!}\alpha^{j+1}\left(\frac{\partial}{\partial\alpha}\right)^j\left[\frac{f^{(0)}}{\alpha}\right]$$

and

$$\left(\frac{s\Omega_0}{v_\perp}\frac{\partial}{\partial v_\perp} + k_\parallel\frac{\partial}{\partial v_\parallel}\right)f^{(j)} = \frac{s\Omega_0}{V_\perp^2}(f^{(j-1)} - f^{(j)}) - k_\parallel\frac{(v_\parallel - U)}{V_\parallel^2}f^{(j)}.$$

Then write $v_\parallel = (\omega - s\Omega_0)/k_\parallel - (\omega - s\Omega_0 - k_\parallel v_\parallel)/k_\parallel$ and use the sum rules in part (d) of Exercise 10.1.

(c) Show that the terms $s \neq 0$ in part (b) simplify in the limit $\lambda \ll 1$, when one has $N_{il} \propto \lambda^{s-1}$, so that the quantity in square brackets in the expression for

$K_{il}(\omega, \mathbf{k})$ is replaced by, to correct order in λ,

$$\frac{(j+|s|)!}{j!\,|s|!\,V_\perp^2}\left[\left(\frac{s\Omega_0}{\omega-s\Omega_0-k_\parallel U}\right)\left(\frac{|s|}{j+|s|}\right)-\frac{V_\perp^2}{V_\parallel^2}\right]v_i(\omega,\mathbf{k},s)v_l^*(\omega,\mathbf{k},s)$$

with $\mathbf{v}(\omega,\mathbf{k},s)$ given by (11.32).

11.5 (a) Show that the resonance condition $\omega - s\Omega - k_\parallel v_\parallel = 0$ may be rewritten in the form

$$\left(\frac{s^2\Omega_0^2+k_\parallel^2 c^2}{s^2\Omega_0^2}\right)\left(\frac{v_\parallel}{c}-\frac{\omega k_\parallel c}{s^2\Omega_0^2+k_\parallel^2 c^2}\right)^2+\frac{v_\perp^2}{c^2}=\frac{s^2\Omega_0^2+k_\parallel^2 c^2-\omega^2}{s^2\Omega_0^2+k_\parallel^2 c^2},$$

and hence derive the properties (11.42) for the resonance ellipse.

(b) Show that for $k_\parallel^2 c^2 > \omega^2$ the resonance ellipse touches the unit circle $v^2/c^2 = 1$ in velocity space and that the outer section of the ellipse is not a solution of $\omega - s\Omega - k_\parallel v_\parallel = 0$.

(c) Derive an equation analogous to that in part (a) but for $p_\parallel = \gamma m v_\parallel$ and $p_\perp = \gamma m v_\perp$. Show that this equation corresponds to an ellipsoid $(\omega^2 > k_\parallel^2 c^2)$ a paraboloid $(\omega^2 = k_\parallel^2 c^2)$ or an hyperboloid $(k_\parallel^2 c^2 > \omega^2)$ of two sheets in momentum space. Which sheet of the hyperboloid corresponds to solutions of $\omega - s\Omega - k_\parallel v_\parallel = 0$?

11.6 (a) Show, either by direct evaluation of (11.20) or by using the antihermitian part of (11.27) (for $a = 0$, $U = 0$) in (2.67), that in the nonrelativistic approximation (11.18), the gyromagnetic absorption coefficient at the sth harmonic due to thermal electrons is

$$\gamma_M(\mathbf{k},s)=\left(\frac{\pi}{2}\right)^{1/2}\frac{\omega_p^2 A_M(\omega,\theta,s)}{\omega N_M(V_e/c)|\cos\theta|N_M\partial(\omega N_M)/\partial\omega}$$

$$\times\exp\left[-\frac{(\omega-s\Omega_e)^2}{2\omega^2 N_M^2(V_e/c)^2\cos^2\theta}\right]$$

with

$$A(\omega,\theta,s)=\frac{e^{-\lambda}}{1+T^2}\left[\left\{\frac{\omega}{\Omega_e}(K\cos\theta-T\sin\theta)\tan\theta+sT\sec\theta\right\}^2\frac{I_s}{\lambda}\right.$$

$$+2\left\{\frac{\omega}{\Omega_e}(K\cos\theta-T\sin\theta)\tan\theta+sT\sec\theta\right\}(I_s'-I_s)$$

$$+\left.\left(\frac{s^2}{\lambda}+2\lambda\right)I_s-2\lambda I_s'\right]$$

where subscripts M are omitted, as are arguments $\lambda = k_\perp^2 V_e^2/\Omega_e^2$, with $k_\perp = N_M(\omega/c)\sin\theta$, of the modified Bessel functions I_s and their derivatives I_s'.

(b) Show that for the magnetoionic $(M = \sigma = \pm 1)$ waves in the quasi-circular approximation, cf. (10.54), the quantity $A(\omega,\theta,s)$ for $\lambda \ll 1$ may be approximated by

$$A_\sigma(\omega,\theta,s)\approx\frac{s^2}{4}\frac{(\lambda/2)^{s-1}}{s!}(1-\sigma|\cos\theta|)^2.$$

12

Instabilities in warm and in inhomogeneous plasmas

12.1 Longitudinal waves in warm plasmas

The terms 'warm' plasma and 'hot' plasma are used to describe situations where the cold plasma approximation is inadequate. Consider a plasma in which each species α is described by a Maxwellian distribution with temperature $T_\alpha = m_\alpha V_\alpha^2$. The thermal speed V_α appears in the dielectric tensor in two different ways: specifically, if one sets $U = 0$ and $V_\perp = V_\parallel$ in (11.27) et seq. then V_α appear in the argument of the plasma dispersion function

$$y_{\alpha s} = \frac{\omega - s\Omega_\alpha}{\sqrt{2}k_\parallel V_\alpha} \tag{12.1}$$

and in the argument

$$\lambda_\alpha = \frac{k_\perp^2 V_\alpha^2}{\Omega_\alpha^2} \tag{12.2}$$

of the modified Bessel functions. The cold plasma limit corresponds to $y_{\alpha s}^2 \gg 1$ and $\lambda_\alpha \ll 1$ for all α and s. The cold plasma approximation breaks down for species α whenever any of the $y_{\alpha s}$ or λ_α fail to satisfy these conditions.

In §10.4 we have already encountered warm plasma effects. The dielectric tensor (10.74) applies for $\omega^2 \ll 2k_\parallel^2 V_e^2$, i.e. for $y_{e0}^2 \ll 1$. The electron thermal speed then appears through the ion sound speed $v_s = (\omega_{pi}/\omega_p)V_e$. Besides thermal effects due to $v_s \neq 0$ at low frequencies, there are two other important classes of warm plasma waves. One of these consists of *resonant waves*. It is apparent that $y_{\alpha s}^2 \gg 1$ cannot be satisfied for sufficiently large $k_\parallel^2$, and near the resonances in cold plasma theory $k_\parallel^2$ becomes arbitrarily large. As a consequence, thermal corrections to the dispersion of waves near a resonance are necessarily important. This class of resonant waves includes upper and lower hybrid waves, resonant whistlers, and 'parallel' cyclotron resonant waves. The other class involves waves propagating nearly perpendicular to the field lines ($k_\parallel \approx 0$) near the cyclotron harmonics ($\omega \approx s\Omega_\alpha$) with $\lambda_\alpha \gtrsim 1$, called *Bernstein modes* (Bernstein 1958). The Bernstein waves are 'perpendicular' cyclotron waves with longitudinal polarization, as opposed to the 'parallel' cyclotron waves which have circular polarization. The distinction between these two classes of waves is discussed further below.

With the exception of the 'parallel' cyclotron resonant waves, these waves have approximately longitudinal polarization over most of the range of interest. As a first approximation in deriving the wave properties it is reasonable to suppose that they satisfy the longitudinal dispersion equation $K^L(\omega, \mathbf{k}) = 0$. Let us evaluate $K^L(\omega, \mathbf{k})$ using (11.27), retaining $V_\perp \neq V_\parallel$ and $U \neq 0$ for the present. One finds

$$\text{Re}[K^L(\omega, \mathbf{k})] = 1 + \sum_\alpha \frac{\omega_{p\alpha}^2}{k^2 V_{\alpha\perp}^2} \left[1 - \sum_{s=-\infty}^{\infty} \phi(y_{\alpha s}) \left\{ \frac{\omega - k_\parallel U_\alpha}{\omega - s\Omega_\alpha - k_\parallel U_\alpha} + A_\alpha \right\} \right.$$

$$\left. \times I_s(\lambda_\alpha) \exp(-\lambda_\alpha) \right] \tag{12.3a}$$

$$\text{Im}[K^L(\omega, \mathbf{k})] = \sum_\alpha \left(\frac{\pi}{2} \right)^{1/2} \frac{\omega_{p\alpha}^2}{k^2 V_{\alpha\perp}^2 |k_\parallel| V_{\alpha\parallel}} \sum_{s=-\infty}^{\infty} \{\omega - k_\parallel U_\alpha$$

$$+ A_\alpha [\omega - s\Omega_\alpha - k_\parallel U_\alpha]\} I_s(\lambda_\alpha) \exp[-\lambda_\alpha - y_{\alpha s}^2], \tag{12.3b}$$

with $y_{\alpha s} = (\omega - s\Omega_\alpha - k_\parallel U_\alpha)/\sqrt{2} k_\parallel V_{\alpha\parallel}$ and $A_\alpha = V_{\alpha\perp}^2/V_{\alpha\parallel}^2 - 1$. In the case of a thermal plasma ($V_\perp = V_\parallel$, $U = 0$, $A = 0$), (12.3a) simplifies to

$$\text{Re}[K^L(\omega, \mathbf{k})] = 1 + \sum_\alpha \frac{\omega_{p\alpha}^2}{k^2 V_\alpha^2} \left[1 - \sum_{s=-\infty}^{\infty} \phi(y_{\alpha s}) \frac{\omega}{\omega - s\Omega_\alpha} I_s(\lambda_\alpha) \exp(-\lambda_\alpha) \right] \tag{12.4}$$

with $y_{\alpha s}$ now given by (12.1).

First consider the *resonant longitudinal waves in an electronic plasma*. Then only the contribution of the electrons $\alpha = e$ is retained in the sum over α in (12.4). On expanding $\phi(y_{es})$ and $I_s(\lambda_e)$ in powers of $1/y_{es}^2$ and λ_e respectively, one obtains an expansion of $K^L(\omega, \mathbf{k})$ in powers of k^2:

$$\text{Re}[K^L(\omega, \mathbf{k})] = 1 - \frac{\omega_p^2}{\omega^2} \left[\frac{\omega^2 - \Omega_e^2 \cos^2 \theta}{\omega^2 - \Omega_e^2} + \frac{k^2 V_e^2}{\omega^2} F(Y, \theta) + \cdots \right] \tag{12.5}$$

with $Y = \Omega_e/\omega$ and

$$F(Y, \theta) = 3\cos^2 \theta - \frac{\cos^2 \theta \sin^2 \theta}{Y^2} - \frac{\sin^4 \theta}{Y^2(1 - Y^2)}$$

$$+ \frac{(1 + 3Y^2)\sin^2 \theta \cos^2 \theta}{Y^2(1 - Y^2)^3} + \frac{\sin^4 \theta}{Y^2(1 - 4Y^2)}. \tag{12.6}$$

For $k^2 V_e^2/\omega^2 = 0$ the longitudinal dispersion relation $\text{Re}[K^L(\omega, \mathbf{k})] = 0$ reduces to the corresponding condition in the magnetoionic theory, i.e. to $1 - X - Y^2 - XY^2\cos^2 \theta = 0$. The two resonant frequencies $\omega = \omega_\pm(\theta)$ are given by (10.47). Thermal corrections may be included as a perturbation, giving

$$[\omega_\pm^L(k, \theta)]^2 = \omega_\pm^2(\theta) \pm \frac{\omega_p^2 k^2 V_e^2}{\omega_+^2(\theta) - \omega_-^2(\theta)} (1 - Y_\pm^2) F(Y_\pm, \theta) \tag{12.7}$$

with $Y_\pm = \Omega_e/\omega_\pm(\theta)$.

For $\omega_p^2 > \Omega_e^2$ the upper frequency solution in (12.7) corresponds to *generalized*

Langmuir waves. On expanding in Ω_e^2/ω^2, (12.7) gives

$$[\omega_+^L(k,\theta)]^2 = \omega_p^2 + \Omega_e^2 \sin^2\theta + 3k^2 V_e^2, \tag{12.8}$$

which has already been written down in (10.57). For perpendicular propagation these waves are *upper hybrid waves*. For arbitrary Ω_e^2/ω_p^2, (12.7) for $\sin\theta = 1$ gives

$$[\omega_+^L(k,\theta)]^2 = \omega_{UH}^2 + 3k^2 V_e^2 \frac{\omega_p^2}{\omega_{UH}^2 - 4\Omega_e^2} \tag{12.9}$$

with $\omega_{UH}^2 = \omega_p^2 + \Omega_e^2$. Upper hybrid waves may be regarded as a special case of electron Bernstein waves (§12.4).

The lower frequency solution in (12.7) corresponds to *resonant whistler waves* for $\omega_p^2 > \Omega_e^2$. We have $\omega_-^2(\theta) \approx \Omega_e^2 \cos^2\theta$ for $\omega_p^2 \gg \Omega_e^2$ and hence $Y^2 \approx 1/\cos^2\theta$. The resulting expression for the thermal corrections is relatively cumbersome. The thermal corrections are usually not important except near the lower hybrid frequency.

The effect of the ions needs to be included in (12.5) to treat *lower hybrid waves*. Let us assume $\omega^2 \gg \Omega_i^2$, which with $\omega \approx \omega_{LH}$ implies $\omega_{pi}^2 \gg \Omega_i^2$, in which case we may treat the ions as unmagnetized. Their contribution to (12.5) is then

$$\mathrm{Re}[K^{L(i)}(\omega,\mathbf{k})] = -\frac{\omega_{pi}^2}{\omega^2}\left(1 + \frac{3k^2 V_i^2}{\omega^2} + \cdots\right). \tag{12.10}$$

The approximate dispersion relation for lower hybrid waves then implied is [cf. Exercise 12.1]

$$\omega^2 = \omega_{LH}^2\left\{1 + \frac{m_i}{m_e}\cos^2\theta + \frac{k^2 V_e^2}{\Omega_e^2}\left(\frac{3}{4} + \frac{3T_i}{T_e}\right)\right\}. \tag{12.11}$$

At low frequencies the contribution from the sum over s for the electrons may be neglected in (12.4). This is due to setting $\omega/\Omega_e \ll 1$ for $s \neq 0$ and setting $\phi(y_{e0}) \approx 0$ for $s = 0$ due to $y_{e0}^2 = \omega^2/2k_\parallel^2 V_e^2 \ll 1$. Suppose we assume that the ions are cold. Then we have

$$K^L(\omega,\mathbf{k}) = 1 + \frac{1}{k^2\lambda_{De}^2} - \frac{\omega_{pi}^2}{\omega^2}\cos^2\theta - \frac{\omega_{pi}^2}{\omega^2 - \Omega_i^2}\sin^2\theta. \tag{12.12}$$

For $\omega^2 \gg \Omega_i^2$, $K^L = 0$ with (12.12) leads to the dispersion relation $\omega = \omega_s(\mathbf{k})$ for ion-sound waves, cf. (1.5). For arbitrary Ω_i, (12.12) leads to the dispersion relations

$$\omega^2 = \tfrac{1}{2}(\omega_s^2(\mathbf{k}) + \Omega_i^2) \pm \tfrac{1}{2}[(\omega_s^2(\mathbf{k}) + \Omega_i^2)^2 - 4\omega_s^2(\mathbf{k})\Omega_i^2\cos^2\theta]^{1/2}. \tag{12.13}$$

For $\omega_s^2(\mathbf{k}) \gg \Omega_i^2$ the upper frequency solution reduces to the dispersion relation for ion sound waves and the lower frequency solution to

$$\omega \approx \omega_i(\theta) \approx \Omega_i|\cos\theta| \tag{12.14a}$$

which is the dispersion relation for *electrostatic ion cyclotron waves*. In the long wavelength limit, defined here by $\omega_s^2(\mathbf{k}) \ll \Omega_i^2$, (12.14a) is replaced by

$$\omega = \omega_i(\theta) \approx \Omega_i + \frac{k^2 v_s^2 \sin^2\theta}{2\Omega_i}. \tag{12.14b}$$

A dispersion relation for arbitrary λ_i may be derived for the electrostatic ion cyclotron mode in the case $2k_\parallel^2 V_i^2 \ll \Omega_i^2$. The electronic contribution is approximated as in (12.12) and the ionic contribution is reduced by setting $\omega = \Omega_i$ in all terms other than $s = 1$, and then performing the sum over s [cf. Exercise 12.2]. One finds

$$\omega = \Omega_i[1 + g] \tag{12.15}$$

with

$$g = \frac{e^{-\lambda_i}I_1(\lambda_i)}{1 + T_i/T_e - e^{-\lambda_i}I_1(\lambda_i) + \{1 - e^{-\lambda_i}I_0(\lambda_i)\}/\lambda_i}. \tag{12.16}$$

For $\lambda_i \ll 1$ one finds $g \approx \frac{1}{2}\lambda_i(T_e/T_i)$ and (12.15) with (12.16) reproduces (12.14b). For $\lambda_i \gtrsim 1$ the region of validity of (12.15) is restricted to a small range of angles about perpendicular propagation due to the conditions $2k_\parallel^2 V_i^2 \ll \Omega_i^2 \lesssim k_\perp^2 V_i^2$. For $\lambda_i \gg 1$ the asymptotic expansion of the Bessel functions implies $e^{-\lambda}I_s(\lambda) \approx 1/(2\pi\lambda)^{1/2}$.

Electrostatic ion cyclotron waves may be driven unstable due to a *current instability*. This instability is of interest because, unlike the analogous ion sound instability, it does not require $T_e \gg T_i$. The instability may be treated by evaluating the absorption coefficient in terms of $\mathrm{Im}\,[K^L(\omega, \mathbf{k})]$, as given by (12.3b). Assume that the electrons are flowing through the ions and that one has $\lambda_e \ll 1$ and $(\omega - k_\parallel v_d)^2 \ll 2k_\parallel^2 V_e^2$, with $V_{\perp e} = V_{\parallel e}$ and $U_e = v_d$. The condition for growth is $\mathrm{Im}\,[K^L(\omega, \mathbf{k})] < 0$, and this requires that the negative contribution from the electrons (at $s = 0$) overcome the positive contribution from the ions (dominated by $s = 1$). Thus growth requires

$$\frac{\omega_p^2}{V_e^3}(\omega - k_\parallel v_d) + \frac{\omega_{pi}^2}{V_i^3}\omega e^{-\lambda_i}I_1(\lambda_i)\exp\left[-\frac{(\omega - \Omega_i)^2}{2k_\parallel^2 V_i^2}\right] < 0. \tag{12.17}$$

The threshold condition implied by (12.17) may be recast in the form (Exercise 12.3)

$$\frac{v_d}{\sqrt{2}V_i} = \frac{1 + g}{g}\rho\left(1 + \frac{v_s^2 V_e^2}{V_i^3}e^{-\lambda_i}I_1(\lambda_i)e^{-\rho^2}\right) \tag{12.18}$$

with $\rho = (\omega - \Omega_i)/\sqrt{2}k_\parallel V_i$. Kindel & Kennel (1971) found that (12.18) can be satisfied for $v_d/\sqrt{2}V_i \gtrsim 13$ in a hydrogen plasma with $T_e/T_i \approx 1$. The waves produced in this current instability have $k_\perp/k_\parallel \approx 10$ near the threshold.

The important conclusion is that in a plasma with $T_e \approx T_i$ a current can drive electrostatic ion cyclotron waves at large angles unstable. For $T_e \gg T_i$ the ion-cyclotron and ion-sound instabilities may both be possible in principle, and a detailed analysis is required to determine which is the faster growing.

Besides the ion-cyclotron and ion-sound modes there is also an electron-sound mode. This has received relatively little attention because it was first identified as existing only in a plasma with ions much hotter than the electrons (Sizonenko & Stepanov 1967), and this condition is not commonly satisfied, especially in space plasmas. However the electron-sound mode can also exist in a plasma with two electron components, specifically a hot and a cold electron component (Tokar & Gary 1984). This latter requirement is often satisfied in space plasmas, and the possible importance of electron-sound in such plasmas warrants further investigation.

The name electron-sound arises from the form $\omega = |\mathbf{k}| v_{se} |\cos\theta|$ of the dispersion relation at small $|\mathbf{k}|$. This form arises from a balance between the contribution $-\omega_{pc}^2/\omega^2$ from cold (C) electrons and $1/k^2\lambda_{DH}^2$ from the hot (H) component of electrons or ions, giving the electron-sound speed $v_{se} = \omega_{pc}\lambda_{DH} = (\omega_{pc}/\omega_{pH})V_H$, where V_H is the thermal speed of the hot component. The dispersive properties may be treated by reversing the roles of electrons and ions (for hot ions) in (12.12), with λ_{Di} then replaced by λ_{DH} for either hot electrons or hot ions.

12.2 'Parallel' cyclotron waves

The 'parallel' electron cyclotron resonance occurs in the whistler mode for $\omega_p > \Omega_e$ (and in the z-mode for $\omega_p < \Omega_e$) with the frequency $\omega_-(\theta)$, cf. (10.47), approaching Ω_e from below as $\sin\theta$ approaches zero. The analogous 'parallel' ion cyclotron resonances occur in the Alfvén mode as ω approaches Ω_i from below for small $\sin\theta$. As indicated in (10.72) and (10.73), the cold plasma waves for $\sin\theta = 0$ have dispersion relations and polarizations

$$N^2 = R_{\pm}, \quad \mathbf{e} = \frac{1}{\sqrt{2}}(1, \pm i, 0). \tag{12.19a, b}$$

The resonances occur at $\omega = \Omega_e$ for $N^2 = R_+$ and at $\omega = \Omega_i$ for $N^2 = R_-$, and then the handedness of the resonant wave is the same as that of the resonant particles, i.e. electrons and ions respectively. At these resonances the condition (12.1) necessarily breaks down for $s = 1$ for the resonant particles. The 'parallel' cyclotron resonant waves correspond to the thermally modified forms of these parallel cyclotron resonances of cold plasma theory.

The simplest approximation for 'parallel' electron cyclotron waves corresponds to $k_\perp = 0$ (implying $\lambda_e = 0$) and $\omega \approx \Omega_e$ with $\Omega_e^2 \gg 2k_\parallel^2 V_{e\parallel}^2$. The dispersion equation (11.8) for the distribution (11.27) then becomes, retaining only the electronic contribution,

$$\frac{k_\parallel^2 c^2}{\omega^2} = 1 - \frac{\omega_p^2}{\omega^2}\left[\phi\left(\frac{\omega-\Omega_e}{\sqrt{2}k_\parallel V_{e\parallel}}\right)\left\{\frac{\omega}{\omega-\Omega_e} + A_e\right\} - A_e\right] \tag{12.20}$$

The only important damping process is the cyclotron damping by electrons, which is described by (2.67) with

$$\mathrm{Im}\,[K^R(\omega,\mathbf{k})] = \left(\frac{\pi}{2}\right)^{1/2}\frac{\omega_p^2}{\omega^2}\frac{1}{|k_\parallel|V_{e\parallel}}\{\omega + A_e[\omega - \Omega_e]\}\exp[-(\omega-\Omega_e)^2/2k_\parallel^2 V_{e\parallel}^2]. \tag{12.21}$$

A streaming motion is unimportant because it may be eliminated by a change of frame of reference to the rest frame of the electrons. Hence without loss of generality we have set $U_e = 0$ in (12.20) and (12.21). The effect of the anisotropy A_e in (12.20) on the dispersion relation is neglected in the following.

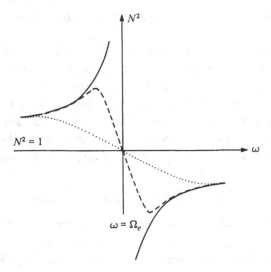

Fig. 12.1 The solid curve represents the cold plasma dispersion equation for electron cyclotron waves, the dashed curve indicates how it is modified by thermal effects and the dotted curve illustrates the case $\omega_p^2/\Omega_e^2 \ll V_{e\parallel}/c$.

The cold plasma dispersion relation corresponds to $(\omega - \Omega_e)^2 \gg 2k_\parallel^2 V_{e\parallel}^2$ in (12.20) when the ϕ-function may be approximated by unity. As ω increases towards Ω_e the argument $y_{e1} = (\omega - \Omega_e)/\sqrt{2}k\, V_{e\parallel}$ decreases in magnitude, and ϕ approaches a maximum just greater than unity for $|y_{e1}| \approx 1$, cf. Figure 2.1. A further decrease in $|\omega - \Omega_e|$ then causes ϕ to decrease. Using the expansions (2.40) and (2.41) for ϕ, we have

$$\frac{k_\parallel^2 c^2}{\omega^2} = 1 - \frac{\omega_p^2}{\omega(\omega - \Omega_e)} \begin{cases} 1 + \dfrac{k_\parallel^2 V_{e\parallel}^2}{(\omega - \Omega_e)^2} & y_{e1}^2 \gg 1 \\[3mm] \dfrac{(\omega - \Omega_e)^2}{k_\parallel^2 V_{e\parallel}^2} & y_{e1}^2 \ll 1. \end{cases} \tag{12.22}$$

The solution for $k_\parallel^2 c^2/\omega^2$ varies as illustrated in Figure 12.1; it has a maximum at $(\omega - \Omega_e)^2 \approx k_\parallel^2 V_{e\parallel}^2$ with

$$|k_\parallel|_{\max} \approx \left(\frac{\omega_p^2 \Omega_e}{V_{e\parallel} c^2}\right)^{1/3} \tag{12.23}$$

and $k_\parallel^2 c^2/\omega^2$ then decreases and passes through unity at $\omega = \Omega_e$.

The discussion so far has concerned the region $\omega < \Omega_e$ where the cold plasma resonance occurs. For $\omega > \Omega_e$ the cold plasma result implies a region of evanescence with a real solution re-appearing at the x-mode cutoff. For $(k_\parallel)_{\max} c/\Omega_e \gg 1$, the curve for $k_\parallel^2 c^2/\omega^2$ as a function of ω decreases for ω above the value where the maximum (12.23) appears, passing through unity at $\omega = \Omega_e$, then becoming negative, reaching a maximum negative value, with $|k_\parallel| = (k_\parallel)_{\max}$ given by (12.23), then increasing and passing through zero at the cold plasma cutoff. As $V_{e\parallel}$ is increased the maximum value (12.23) decreases and the effect of the cyclotron resonance on the dispersion curve is reduced (Figure 12.1).

At sufficiently low plasma densities the electron cyclotron resonance can be completely washed out. This occurs when the region where $k_\parallel^2 c^2 / \omega^2$ is negative disappears, i.e. when $(k_\parallel)_{max} c / \Omega_e$ is less than unity. There is then one continuous curve joining the z-mode to the x-mode. For $|k_\parallel|_{max} c / \Omega_e \ll 1$ there is only a small bump on the dispersion curve for $\omega \approx \Omega_e$, as illustrated in Figure 12.1. Thus for

$$\frac{\omega_p^2}{\Omega_e^2} \ll \frac{V_{e\parallel}}{c} \tag{12.24}$$

the refractive index does not deviate much from unity near $\omega = \Omega_e$. It is then reasonable to ignore the dispersive effects of the electron gas and to treat the waves as transverse waves in vacuo, at least to a first approximation. The limiting case (12.24) is relevant to the discussion of the gyrotron (§11.2) for example.

Electron cyclotron waves can be driven unstable by a temperature anisotropy. Growth requires $\mathrm{Im}\, K^R > 0$, and for $\omega < \Omega_e$ this requires $A_e > 1$, i.e. $T_{e\perp} > T_{e\parallel}$, according to (12.3b). The temperature must exceed a threshold value $\omega / (\Omega_e - \omega)$, i.e. one requires

$$A_e = \frac{T_{e\perp}}{T_{e\parallel}} - 1 > \frac{\omega}{\Omega_e - \omega}. \tag{12.25}$$

Except for very large anisotropies, (12.25) restricts growth to well below the region $|\omega - \Omega_e| \lesssim |k_\parallel| V_{e\parallel}$ where thermal modifications are important. Thus for moderate values of A_e (12.25) requires $(\omega - \Omega_e)^2 \gg k_\parallel^2 V_{e\parallel}^2$, and the exponential factor in (12.21) is very small and the growth is very weak. The growth rate is large only when $\Omega_e - \omega$ is of order or less than the value

$$(\Omega_e - \omega)_{max} \approx \Omega_e \left(\frac{\omega_p^2}{\Omega_e^2} \frac{V_{e\parallel}^2}{c^2} \right)^{1/3}$$

where the maximum (12.23) in $k_\parallel$ occurs. The required anisotropy is then

$$A_e \gtrsim \left(\frac{\omega_p^2}{\Omega_e^2} \frac{V_{e\parallel}^2}{c^2} \right)^{-1/3}. \tag{12.26}$$

In most applications growth of 'parallel' electron and ion cyclotron waves is due to a suprathermal component of electrons. Similarly ion cyclotron waves tend to be driven unstable by suprathermal anisotropic ions.

12.3 Ion cyclotron waves in multi-ion plasmas

In a thermal plasma with a single ion species, *ion cyclotron waves* may be treated in a way closely analogous to the foregoing treatment of electron cyclotron waves. In place of (12.20) one has

$$\frac{k_\parallel^2 c^2}{\omega^2} = 1 - \sum_i \frac{\omega_{pi}^2}{\omega^2} \left[\frac{\omega}{\Omega_i} + \phi\left(\frac{\omega - \Omega_i}{\sqrt{2} k_\parallel V_{i\parallel}} \right) \left\{ \frac{\omega}{\omega - \Omega_i} + A_i \right\} - A_i \right] \tag{12.27}$$

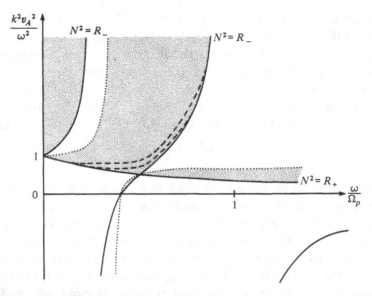

Fig. 12.2 The dispersion curves are plotted for a plasma with $v_A^2 \ll c^2$ consisting of 25% He$^+$ and 75% protons. The solid curves are for parallel propagation and the dotted curves for perpendicular propagation. The curves for an intermediate angle lie in the shaded regions; dashed curves illustrate their form schematically.

where the electronic term has been evaluated for $\omega \ll \Omega_e$ and rewritten using the condition (10.67) for charge neutrality. The relevant polarization is now that with the minus sign in (12.19b). The anti-hermitian part of the dielectric tensor is of the form (12.21), with e replaced by i, for each ionic species. Once again thermal effects modify the dispersion relation near $\omega = \Omega_i$. The dispersion curves are similar to those for electron cyclotron waves (Figure 12.1) with $|k_\parallel|_{max}$ given by replacing e by i in (12.23).

Several different effects are associated with the presence of two or more species of ion in a plasma, and these effects are important in both space and laboratory plasmas.

The dispersion curves for the cold plasma modes are plotted for a two-ion plasma in Figure 12.2. For parallel propagation the dispersion relations reduce to $N^2 = R_+$ and $N^2 = R_-$ with $R_\pm$ defined in (10.26). The mode $N^2 = R_-$ has a resonance at the lower of the two ion-cyclotron frequencies, followed by a region of evanescence, a cutoff, and another resonance at the higher of the two cyclotron frequencies. Note that there is a 'crossover', where the refractive indices of the modes $N^2 = R_+$ and $N^2 = R_-$ are equal. Now the refractive indices of the two cold plasma modes are different (i.e. they never cross) except for the singular points where $\sin \theta$ and DP vanish simultaneously; only then does F, as given by (10.32), vanish. The singular point $\sin \theta = 0$, $P = 0$ is between the o-mode and the z-mode, as illustrated in Figure 10.3. The singular point $\sin \theta = 0$ and $D = 0$ can occur in a two-ion plasma (and several such points can occur in a multi-ion plasma). More generally $D = 0$ is the condition for the crossover, that is $D = 0$ corresponds to $R_+ = R_-$. As in the o-z-mode coupling in Figure 10.3, for $\sin\theta \neq 0$ the two curves do not cross. (The dispersion curves for $\sin\theta \neq 0$ are also illustrated in Figure 12.2.) The Alfvén mode has the resonance at the lower gyrofrequency, it reappears at the cutoff and then becomes the magnetoacoustic mode; the magnetoacoustic mode at low frequency passes continuously through the lower resonance, changes its sense of polarization (from opposite to that of ions to the same sense as ions) at the crossover frequency (cf. Exercise 12.4) and has a resonance at the higher cyclotron frequency.

The resonances occur at the cyclotron frequencies (in the cold plasma approximation) only for $\sin\theta = 0$. More generally the resonances occur where A/C is zero; using (10.30) with (10.26), this condition becomes

$$\frac{S\sin^2\theta + P\cos^2\theta}{PR_+R_-} = 0. \tag{12.28}$$

Provided θ is not too close to $\pi/2$, (12.28) with $S = \frac{1}{2}(R_+ + R_-)$ implies that the resonances occur close to the infinities of

$$R_- \approx 1 + \sum_i \frac{\omega_{pi}^2}{\Omega_i^2} \frac{1}{1 - \omega/\Omega_i},$$

which follows from (10.26) on assuming $\omega \ll \Omega_e$ and using (10.67). As $\sin\theta$ is increased away from zero, the resonances remain close to the cyclotron frequencies, shifting very slightly to lower frequencies. As θ approaches $\pi/2$ the resonances implied by (12.28) move rapidly from $R_- \approx \infty$ to $S \approx 0$. In a two-ion plasma (labels 1 and 2), one has

$$S \approx 1 + \frac{\omega_{p1}^2}{\Omega_1^2} \frac{1 - \omega^2/\omega_{ii}^2}{(1 - \omega^2/\Omega_1^2)(1 - \omega^2/\Omega_2^2)} \tag{12.29a}$$

$$\omega_{ii}^2 = \frac{\omega_{p1}^2\Omega_2^2 + \omega_{p2}^2\Omega_1^2}{\omega_{p1}^2 + \omega_{p2}^2}. \tag{12.29b}$$

For $\omega_{p1}^2 \gg \Omega_1^2$ (implying $v_A^2 \ll c^2$ in practice) the zero of S occurs at the ion-ion hybrid frequency ω_{ii} for $\theta = \pi/2$. For θ sufficiently close to $\pi/2$ the resonance is just above ω_{ii}. In Figure 12.2 the resonance at $\omega = \Omega_1$ for $\sin\theta = 0$ moves to $\omega = \omega_{ii}$ for $\theta = \pi/2$. The resonance at $\omega = \Omega_2$ for $\sin\theta = 0$ moves to lower frequencies, and effectively disappears for $\theta = \pi/2$ due to the Alfvén mode then disappearing.

Thermal effects may be included for parallel propagation using (12.27). As for electron cyclotron waves in Figure 12.1, the dispersion curve no longer has a resonance and there is a section where N^2 decreases with increasing ω. As a consequence of these effects an additional crossover appears just above the lower cyclotron frequency in Figure 12.2. For sufficiently hot ions and sufficiently small density ratio $\omega_{p2}^2/\omega_{p1}^2$ of the second species ($\Omega_2 < \Omega_1$) the deviation in the dispersion curve for $\omega \approx \Omega_2$ can become small. Then modified curve $N^2 = R_-$ lies above $N^2 = R_+$ so that no crossover occur. Thus, the resonance associated with minor ionic species can be washed out entirely by thermal effects.

Multi-ion effects can be important in both laboratory and space plasmas, as the following examples illustrate. In tokamak plasmas a mixture of ions, e.g. deuterium and tritium, offers the possibility of radio frequency (RF) heating through the ion-ion resonance. More generally RF energy has access to the interior of the plasma only through specific resonances, such as the lower hybrid, the 'Alfvén resonance' (Ott, Wersinger & Bonoli 1978) and the ion-ion resonance offers an alternative (Stix 1975) to these. At these resonances a wave in the magnetoacoustic mode (in the exterior of the plasma) couples into a damped mode, e.g. an ion Bernstein wave.

In space plasmas the dominant ionic component is protons, with an admixture of $\approx 10\%$ of Helium and $\lesssim 1\%$ of other ions. In the solar wind the minor ionic components have properties which may possibly be explained in terms of interactions with the ion cyclotron waves. The minor ions have flow speeds, temperatures and temperature anisotropies which differ from protons and from one minor species to another. Despite extensive theoretical work (e.g. McKenzie, Ip & Axford 1979) no entirely satisfactory explanation has emerged. One idea involves the 'magnetic beach' effect (Stix 1962) in which the decrease in B with radial distance from the Sun causes ω/Ω_i to increase for outward propagating Alfvén waves. Sufficiently high frequency waves eventually encounter an ion-cyclotron resonance and the first resonance encountered is that for the ion with the lowest cyclotron frequency. The cold plasma resonance for ions with very low abundances are washed out by thermal effects, but it is not clear that this is the case for Helium, cf. Exercise 12.6. The interaction of the various ionic species with the waves can lead to strongly species-dependent acceleration, with protons being the least affected. However, as already remarked, the observed properties have yet to be explained satisfactorily either in terms of this or any other mechanism.

Another application in which the ion cyclotron waves are observed directly involves ultra low frequency (ULF) waves ($< 10\,$Hz) in the terrestrial magnetosphere. These waves have been studied simultaneously from the ground and from geo-stationary spacecraft (e.g. Young et al. 1981), which provide information on those waves which can reach the ground and those which cannot. The data show that the Helium abundance in the source region is important in determining the properties of the waves and of their growth, which is attributed to a temperature anisotropy. In this application the interpretation of the data has reached a relatively high degree of sophistication.

12.4 The Bernstein modes

The Bernstein modes are perpendicular cyclotron modes. They are usually treated assuming that the waves are longitudinal and at strictly perpendicular propagation; also the particles are usually assumed to have nonrelativistic distributions. Then there is a Bernstein mode associated with each harmonic of the cyclotron frequency of each species of particle.

Consider the longitudinal part of the dielectric tensor for non-relativistic Maxwellian distributions, as given by (12.4). Perpendicular propagation ($k_\parallel = 0$) implies $y_{\alpha s} = \infty$ in (12.1), and hence $\phi(y_{\alpha s}) = 1$ in (12.4). The properties of the modified Bessel functions, specifically

$$\sum_{s=-\infty}^{\infty} e^{-\lambda} I_s(\lambda) = 1, \quad I_{-s}(\lambda) = I_s(\lambda), \tag{12.30a,b}$$

then allow one to simplify (12.4) to

$$\mathrm{Re}[K^L(\omega, \mathbf{k})] = 1 - \sum_\alpha \sum_{s=1}^{\infty} \frac{2s^2 \omega_{p\alpha}^2}{\omega^2 - s^2 \Omega_\alpha^2} \frac{I_s(\lambda_\alpha) \exp[-\lambda_\alpha]}{\lambda_\alpha} \tag{12.31}$$

When treating the electron Bernstein waves one retains only the electronic contribution $\alpha = e$ in (12.31). For the ion modes one can usually assume $\lambda_e \ll 1$ and then only $s = 1$ contributes for the electrons, and the electronic contribution may be approximated by ω_p^2/Ω_e^2. The detailed properties of the electron and ion modes are closely analogous to each other. In the following discussion we concentrate on the electron modes.

The form of the dispersion relation, for ω^2 as a function of $\lambda_e = k_\perp^2 V_e^2/\Omega_e^2$ say, for the nth harmonic Bernstein mode may be determined by assuming $\omega^2 \approx n^2 \Omega_e^2$ in (12.31) and evaluating the sum in the limits $\lambda_e \ll 1$ and $\lambda_e \gg 1$ using

$$e^{-\lambda} I_s(\lambda) \approx \begin{cases} \dfrac{1}{s!} \left(\dfrac{\lambda}{2} \right)^s & \lambda \ll 1 \\[2ex] \dfrac{1}{(2\pi\lambda)^{1/2}} & \lambda \gg 1 \end{cases} \tag{12.32}$$

One finds

$$\frac{\omega^2 - n^2 \Omega_e^2}{n^2 \Omega_e^2} \approx \begin{cases} \dfrac{\omega_p^2 (n^2 - 1)}{(n^2 - 1)\Omega_e^2 - \omega_p^2} \dfrac{1}{n!} \left(\dfrac{\lambda_e}{2} \right)^{n-1} \\[2ex] \dfrac{2\omega_p^2}{\Omega_e^2 \lambda_e (2\pi\lambda_e)^{1/2}} \end{cases} \tag{12.33}$$

Let us introduce the upper hybrid frequency $\omega_{UH}^2 = \omega_p^2 + \Omega_e^2$ and note that the sign of the right hand side for $\lambda_e \ll 1$ is negative for $n^2 < \omega_{UH}^2/\Omega_e^2$ and positive for $n^2 > \omega_{UH}^2/\Omega_e^2$. For $\lambda_e \gg 1$ the dispersion curves approach the cyclotron harmonics from above. Moreover the dispersion curves cannot cross the lines $\omega^2 = n^2 \Omega_e^2$. It is then not difficult to see that for $n^2 < \omega_{UH}^2/\Omega_e^2$ the dispersion curves start at $\omega^2 = n^2 \Omega_e^2$ at $\lambda_e = 0$ and decrease to $\omega^2 = (n-1)^2 \Omega_e^2$ at $\lambda_e = \infty$, and that for $n^2 > \omega_{UH}^2/\Omega_e^2$ the dispersion curve starting at $\omega^2 = n^2 \Omega_e^2$ at $\lambda_e = 0$ must rise to a maximum and decrease back to $\omega^2 = n^2 \Omega_e^2$ at $\lambda_e = \infty$.

The dispersion curves for the electron Bernstein modes are illustrated in Figure 12.3. At small k the longitudinal approximation breaks down in the vicinity of the upper hybrid frequency. As shown, following Puri, Leuterer & Tutter (1973), when the case of perpendicular propagation is treated without making the longitudinal approximation, the exact dispersion curve in the vicinity of the upper hybrid

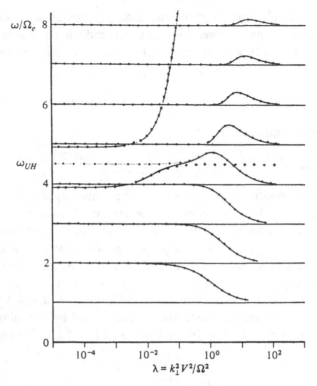

Fig. 12.3 The dispersion curves for the Bernstein modes are illustrated for perpendicular propagation in a plasma with the fully electromagnetic (———), cold plasma ($\times$———$\times$) and longitudinal ($\bullet$———$\bullet$) solutions in the upper hybrid band. (After Puri *et al.* 1973).

frequency follows the cold plasma (z-mode) curve for $\lambda_e \lesssim 1$ and then the Bernstein mode curve for $\lambda_e \gtrsim 1$ with a small region of approximate overlap. An implication is that the Bernstein mode curve starting at the upper hybrid frequency is non-physical for $\lambda_e \lesssim 1$ and that the relevant waves should then be regarded as z-mode waves. This point has already been made in connection with (10.57).

For $k_\parallel \neq 0$ the wave properties for the Bernstein modes are further complicated due to the dispersion equation including two different transcendental functions ($I_s(\lambda)$ and $\phi(y_{as})$). Numerical methods are required to make much progress, and the more so when relativistic effects, electromagnetic effects and non-Maxwellian effects (required to drive any instability) are included. The $k_\parallel$-dependence is particularly relevant because it determines the parallel component of the group velocity, through $v_{g\parallel} = \partial\omega/\partial k_\parallel$. Growth of the waves is favoured not only by a large temporal growth rate, but also by a low value of $v_{g\parallel}$ leading to a large spatial growth rate.

A certain class of magnetospheric emissions is interpreted in terms of the Bernstein modes. These are referred to as '$n + \frac{1}{2}$ – waves' due to their location roughly midway between harmonics of the electron cyclotron frequency. Detailed calculations have been made of the wave properties including the growth rate, for a mixture of hot and cold electrons, with the hot distribution function of the form

$$f_H \propto \Delta \exp\left(-\frac{v^2}{2V_e^2}\right) + \frac{1-\Delta}{1-\beta}\left[\exp\left(-\frac{v_\perp^2}{2V_e^2}\right) - \exp\left(-\frac{v_\perp^2}{2\beta V_e^2}\right)\right]\exp\left(-\frac{v_\parallel^2}{2V_e^2}\right)$$

(12.34)

where Δ and β are free parameters (e.g. Ashour-Abdalla & Kennel 1978). The theory appears capable of accounting for the detailed features of the observational data. The various parameters in the theory influence the wave properties and wave growth in quite a complicated way.

From an observational viewpoint an $(n + \frac{1}{2})$-band in the magnetosphere can be greatly enhanced when it coincides with the upper hybrid frequency. The theory mentioned above can be extended to account for this observation (e.g. Kurth, Craven, Frank & Gurnett 1979). A related enhancement was predicted in connection with solar radio bursts by Zheleznyakov & Zlotnik (1975) who argued that the growth rate is enhanced when the upper hybrid frequency coincides with a cyclotron harmonic; they referred to this as a 'double resonance'. The application of Bernstein waves to the interpretation of solar radio bursts has been directed primarily towards fine structures in decimetre-wave bursts (e.g. Kuijpers 1980b), where one probably has $\omega_p \approx \Omega_e$. The double resonance is relevant for $\omega_p > \Omega_e$ and may lead to observable effects for metre-wave bursts from regions with $\omega_p \gg \Omega_e$.

As has already been remarked, ion Bernstein waves have properties similar to those of electron Bernstein waves in the sense that the dispersion curves for perpendicular propagation differ from those of the electron Bernstein waves only by scaling factors. The lower hybrid frequency plays a role for the ion waves analogous to that of the upper hybrid frequency for the electron waves. The electrostatic ion cyclotron wave, cf. (12.14a, b) may be regarded as the fundamental ion Bernstein wave.

12.5 Drift motions and the drift kinetic equation

An important class of plasma instabilities is that involving drift waves. Both the wave properties and the wave growth of drift waves are associated with gradients in the plasma or with an external force. These effects lead to real or apparent drift motions across the magnetic field lines and it is these motions which lead to the distinctive character of drift waves. Drift waves are discussed in §12.7. Here we discuss the drift motions, then we introduce the drift-kinetic equation.

At low frequencies the motion of a particle in a magnetic field may be described in terms of the *guiding-centre approximation*. The idea is that if changes over a gyroperiod and a gyroradius are small, then one can average over a gyroperiod and describe the motion of the particle in terms of the way its centre of gyration drifts.

The most familiar result of the guiding-centre approximation is that the magnetic moment of the particle is an adiabatic invariant. The so-called first adiabatic

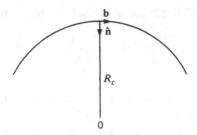

Fig. 12.4 The radius of curvature R_c and unit vector $\hat{\mathbf{n}}$ are defined for a curved magnetic field line by identifying the circle (centre 0 radius R_c) which is locally tangent to the magnetic field line.

invariant is

$$M = \frac{p_\perp^2}{2B} = m\gamma\mu, \qquad (12.35)$$

where $\mu = p_\perp v_\perp/2B$ is the magnetic moment. The drift velocity $\mathbf{v}_D$ of the particle in the presence of an electric field $\mathbf{E}$ and an external force $\mathbf{F}_{ext}$ may be described in terms of an effective force $\mathbf{F}$

$$\mathbf{F} = qE_\parallel \mathbf{b} - \mu\,\mathrm{grad}\,B - p_\parallel v_\parallel(\mathbf{b}\cdot\mathrm{grad})\mathbf{b} - m\frac{d\mathbf{v}_E}{dt} + \mathbf{F}_{ext} \qquad (12.36)$$

where

$$\mathbf{v}_E = \frac{(\mathbf{E} + v_\parallel \mathbf{b}\times\mathbf{B})\times\mathbf{B}}{B^2} \qquad (12.37)$$

is the electric drift. Here $\mathbf{b}$ is a unit vector along the mean (time-averaged) field, and $v_\parallel$ and $v_\perp$ refer to components relative to the mean field. The term involving

$$(\mathbf{b}\cdot\mathrm{grad})\mathbf{b} = \frac{\hat{\mathbf{n}}}{R_c} \qquad (12.38)$$

is more readily interpreted in terms of the radius R_c of curvature of the field lines and the unit normal $\hat{\mathbf{n}}$ towards the centre of curvature (Figure 12.4). The drift velocity is then given by

$$\mathbf{v}_D = \left(v_\parallel + \frac{\mu}{q}\mathbf{b}\cdot\mathrm{curl}\,\mathbf{b}\right)\mathbf{b} + \mathbf{v}_E + \frac{\mathbf{F}\times\mathbf{b}}{qB}. \qquad (12.39)$$

Besides the electric drift, the drifts included in the final term in (12.39) with (12.36) are the grad B drift, the curvature drift, the polarization drift and the external force drift, respectively. The energy of the particle changes at the rate

$$\dot{\varepsilon} = (q\mathbf{E} + \mathbf{F}_{ext})\cdot\mathbf{v}_D - \mu\mathbf{b}\cdot\mathrm{curl}(q\mathbf{E} + \mathbf{F}_{ext}). \qquad (12.40)$$

Drift-kinetic equation

The Vlasov equation describes the evolution of a distribution function $f(\mathbf{p}, t, \mathbf{x})$ in the 6-dimensional phase space $(\mathbf{p}, \mathbf{x})$. Suppose we make the guiding-centre approximation. An individual particle is then described by the parallel momentum of its guiding centre and the position vector, $\mathbf{x}_D$ say, of the guiding centre. Let $f_D(p_{\parallel}, t, \mathbf{x}_D)$ be the density of the particles in the 4-dimensional phase space $(p_{\parallel}, \mathbf{x}_D)$. Continuity in phase requires that f_D satisfy the *drift-kinetic equation*

$$\frac{\partial f_D}{\partial t} + \frac{\partial}{\partial \mathbf{x}_D} \cdot (\mathbf{v}_D f_D) + \frac{\partial}{\partial p_{\parallel}} (F_{\parallel} f_D) = 0 \tag{12.41}$$

where $\mathbf{v}_D$ is the drift velocity and $F_{\parallel} = \mathbf{b} \cdot \mathbf{F}$ is the parallel force operating on the guiding centre, as given by (12.36).

The role of the adiabatic invariant M in the drift-kinetic equation is that of an external parameter, rather than that of a coordinate in the phase space. Although f_D is implicitly a function of M, there is no term in the drift-kinetic equation involving a derivative with respect to M. In normalizing f_D we need to take account of the dependence on M. A suitable normalization would be

$$n = \int dp_{\parallel} \, dM f_D \tag{12.42}$$

where n is the number density.

The drift kinetic equation (12.41) would appear to be a continuity equation in a phase space consisting of one momentum and three space coordinates. However formally a phase space involves N coordinates and the N conjugate momenta. In fact, we may interpret the four dimensions in the drift-kinetic equation as two coordinates and two momenta. Referring to the discussion leading to (10.92) above, the gyration motion is described in terms of y and p_y, with p_x interpreted as describing the y-coordinate of the centre of gyration. Adiabatic invariants in general are essentially action angle variables, with

$$J_q = \oint dq \, p, \tag{12.43}$$

defined as the integral around a closed orbit of $dq \, p$, where q and p are a coordinate and its canonical momentum. When the relevant variables (here $q = y$ and $p = p_y$) are then eliminated the dimensionality of our mathematical description of the system is reduced from $N = 3$ to $N = 2$. Thus the $2N = 4$ dimensions in the drift kinetic equation are x, p_x, z, p_z with p_x interpreted as stated and rewritten as the 'coordinate' $Y = -p_x/qB$.

To emphasize this point further, suppose the particles are in a magnetic trap so that their bounce motion is also periodic. The quantity

$$J = \oint dz \, p_z \tag{12.44}$$

is called the *second adiabatic invariant*. The dimensionality of our system is then reduced to $N = 1$, the corresponding phase space has $2N = 2$ dimensions. In magnetospheric physics the resulting continuity equation is often written in terms of parameters L and ϕ, where L is the radial distance in the equatorial plane, in units of the planetary radius, to the gyrocentre of the particle and where ϕ is azimuthal angle around the planet at its magnetic equator. These are not actually canonically conjugate; the action angle variable conjugate to ϕ is, apart from a constant factor, the magnetic flux Φ_M enclosed by the line $0 \lesssim \phi \lesssim 2\pi$ encircling the Earth. However Φ_M is a function of L ($\Phi_M \propto L^{-2}$ in a dipole field) and hence one may write the equation in terms of L. In the relevant equation both M and J are free parameters.

In a system in which the bounce motion is periodic, there is the possibility of an additional type of wave-particle resonance: a *bounce-resonance*. In this case the frequency of the wave is an integral multiple of the bounce frequency. Effects associated with bounce resonance have been studied in the magnetospheric context, e.g. Roberts (1969).

Historically the drift-kinetic equation was used in an early treatment (in 1961) of drift waves by Rudakov and Sagdeev. The drift-kinetic equation is useful in treating some, but not all drift instabilities. It is limited by the fact that finite gyroradii are excluded. We have averaged over the spiralling motion and so have lost all information on the gyroradius of a particle, effectively having set it equal to zero. A more general treatment using kinetic theory is required when effects associated with non-zero gyroradii are important. An advantage of the drift-kinetic equation is that it allows real drift motions to be included simply.

12.6 Dielectric tensor for an inhomogeneous plasma

The calculation of the dielectric tensor for an inhomogeneous plasma is rather lengthy and tedious. Rather than present a detailed kinetic-theory calculation here, we use the drift-kinetic equation to determine the form of the result in the small gyroradius limit. The effect of non-zero gyroradii are important for many applications involving drift waves, and we simply cite the relevant generalizations.

In the guiding-centre approximation the current density $\mathbf{J}$ may be calculated from

$$\mathbf{J} = \frac{\partial \mathbf{P}}{\partial t} + \text{curl } \mathbf{M}, \tag{12.45}$$

where the polarization $\mathbf{P}$ and the magnetization $\mathbf{M}$ are the electric and magnetic, respectively, dipole moments per unit volume. Thus we have

$$\mathbf{J} = \sum \int dp_\parallel dM [q\mathbf{v}_D f_D - \text{curl} \{\mu \mathbf{b} f_D\}] \tag{12.46}$$

where the sum is over all species of particle. In (12.46) we use the fact that the magnetic moment per particle is $\boldsymbol{\mu} = -\mu\mathbf{b}$, where the minus reflects the fact that plasmas are diamagnetic.

Let us assume that the gradient is along the y-axis, that the waves are longitudinal and that $\mathbf{k}$ is in the x–z plane. Then for a plane wave we have

$$\mathbf{E} = -\text{grad } \Phi = -i(k_\perp, 0, k_\parallel)\Phi. \tag{12.47}$$

The relevant perpendicular components of $\mathbf{v}_D$ are the electric and polarization drifts:

$$\mathbf{v}_{D\perp} = \mathbf{v}_E + \frac{m}{qB^2}\frac{d\mathbf{E}}{dt}$$

$$= -i\left(0, \frac{k_\perp}{B}, 0\right)\Phi - \frac{m}{qB^2}(\omega - k_\parallel v_\parallel)(k_\perp, 0, 0)\Phi, \qquad (12.48)$$

where we use $d\mathbf{E}/dt = -i(\omega - k_\parallel v_\parallel)\mathbf{E}$.

On linearizing (12.41), the perturbed value $f_D^{(1)}$ is determined by

$$-i(\omega - k_\parallel v_\parallel)f_D^{(1)} + v_{Dy}\frac{\partial f_D^{(0)}}{\partial y} + i\mathbf{k}\cdot\mathbf{v}_{D\perp}f_D^{(0)} + qE_\parallel\frac{\partial f_D^{(0)}}{\partial p_\parallel} = 0, \qquad (12.49)$$

where $f_D^{(0)}$ is the unperturbed distribution function. The longitudinal part of the dielectric tensor may then be evaluated using

$$K^L(\omega, \mathbf{k}) = 1 - \frac{\mathbf{k}\cdot\mathbf{J}}{\varepsilon_0 \omega k^2 \Phi}. \qquad (12.50)$$

For a thermal plasma, with non-relativistic Maxwellian distributions, explicit evaluation leads to

$$K^L(\omega, \mathbf{k}, \mathbf{x}) = 1 + \sum_\alpha \frac{\omega_{p\alpha}^2}{\Omega_\alpha^2}\frac{k_\perp^2}{k^2} + \sum_\alpha \frac{\omega_{p\alpha}^2}{k^2 V_\alpha^2}\left[1 - \frac{\omega - \omega_{D\alpha}}{\omega}\{\phi(y_\alpha) - i\sqrt{\pi}y_\alpha e^{-y_\alpha^2}\}\right] \qquad (12.51)$$

with $y_\alpha = \omega/\sqrt{2}k_\parallel V_\alpha$ and with

$$\omega_{D\alpha} = \mathbf{k}\cdot\mathbf{V}_{D\alpha}, \qquad \mathbf{V}_{D\alpha} = -\epsilon_\alpha\frac{V_\alpha^2}{\Omega_\alpha}\mathbf{G}_\alpha \times \mathbf{b}, \qquad (12.52a, b)$$

$$\mathbf{G}_\alpha = \text{grad}(\ln f_{D\alpha}). \qquad (12.52c)$$

The frequency $\omega_{D\alpha}$ is the *drift frequency* for species α and $\mathbf{V}_{D\alpha}$ is the *diamagnetic drift velocity* for species α. The diamagnetic drift is an apparent motion of the particles associated with the gradient in the distribution of the guiding centres and/or the size of the gyroradii. This point is illustrated in Figure 12.5. A simplified analytic treatment of the diamagnetic drift is given in Exercise 12.7.

Although (12.51) suffices for some purposes for treating drift waves in thermal plasmas, there are important examples where the small gyroradius and/or the longitudinal approximation break down. A derivation of a general form of the dielectric tensor in an inhomogeneous plasma has been given by Mikhailovskii (1967, 1974b). The following results are obtained as a result of the general derivation.

The longitudinal part of the dielectric tensor for a thermal plasma must reduce to (12.3a), with $A_\alpha = 0$ and $U_\alpha = 0$, in the absence of inhomogeneity, and to (12.51) in the small gyroradius limit. The result quoted by Mikhailovskii (1974b) is equivalent to

$$K^L(\omega, \mathbf{k}, \mathbf{x}) = 1 + \sum_\alpha \chi^{L(\alpha)}(\omega, \mathbf{k}, \mathbf{x}), \qquad (12.53a)$$

(a)

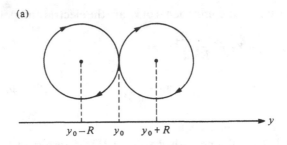

(b)

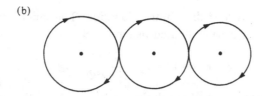

Fig. 12.5(a) In a homogeneous plasma a gradient (along the y-axis) in the number density of particles leads to a net local imbalance in the flux of particles along the x-axis due to an incomplete cancellation of those moving down with their gyrocentres on the left and those moving up with their gyrocentres on the right.
(b) A similar imbalance results from gradients in B or T, which affect the gyroradii for thermal particles and the change in R with y again leads to a lack of cancellation between upgoing and downgoing particle fluxes.

with

$$\chi^{L(\alpha)}(\omega, \mathbf{k}, \mathbf{x}) = \frac{\omega_{p\alpha}^2}{k^2 V_\alpha^2}\left[1 - \sum_{s=-\infty}^{\infty}\left\{1 - \hat{D}_y^{(\alpha)}\left(1 + \frac{s\omega}{\lambda_\alpha \Omega_\alpha}\right)\right\}\right.$$
$$\left. \times \left(\frac{\omega}{\omega - s\Omega_\alpha}\right)\{\phi(y_{\alpha s}) - i\sqrt{\pi}y_{\alpha s}e^{-y_{\alpha s}^2}\}e^{-\lambda_\alpha}I_s(\lambda_\alpha)\right], \quad (12.53b)$$

where we introduce the differential operator

$$\hat{D}_y^{(\alpha)} = \frac{\epsilon_\alpha k_\perp \cos\psi\, V_\alpha^2}{\omega \Omega_\alpha}\left[\frac{\partial \ln n_\alpha}{\partial y} + \frac{\partial V_\alpha^2}{\partial y}\frac{\partial}{\partial V_\alpha^2}\right] \qquad (12.54)$$

with $k_y = k_\perp \sin\psi$ now allowed to be non-zero. For most purposes one may replace $\hat{D}_y^{(\alpha)}$ by $\omega_{D\alpha}/\omega$, and we do so in the following.

When the longitudinal approximation is relaxed one needs to estimate K_{xx}, $K_{xz} = K_{zx}$ and K_{zz} separately. We may assume $y_{\alpha s}^2 \gg 1$ except for $s = 0$, and then (11.27) for $A_\alpha = 0$ and $U_\alpha = 0$ may be used to derive relevant approximations for a homogeneous plasma. Assuming modifications by factors $(\omega - \omega_{D\alpha})/\omega$ to take account of inhomogeneity one finds

$$K_{xx}(\omega, \mathbf{k}, \mathbf{x}) = K_\perp(\omega, \mathbf{k}, \mathbf{x}) = 1 + \sum_\alpha \chi_\perp^{(\alpha)}(\omega, \mathbf{k}, \mathbf{x}), \qquad (12.55a)$$

$$K_{xz}(\omega, \mathbf{k}, \mathbf{x}) = K_{zx}(\omega, \mathbf{k}, \mathbf{x}) = 0, \tag{12.55b}$$

$$K_{zz}(\omega, \mathbf{k}, \mathbf{x}) = K_{\parallel}(\omega, \mathbf{k}, \mathbf{x}) = 1 + \sum_{\alpha} \chi_{\parallel}^{(\alpha)}(\omega, \mathbf{k}, \mathbf{x}), \tag{12.55c}$$

with (Mikhailovskii 1967)

$$\chi_{\perp}^{(\alpha)}(\omega, \mathbf{k}, \mathbf{x}) = \frac{\omega_{p\alpha}^2}{k_{\perp}^2 V_{\alpha}^2} \frac{\omega - \omega_{D\alpha}}{\omega} \{1 - e^{-\lambda_{\alpha}} I_0(\lambda_{\alpha})\}, \tag{12.56a}$$

$$\chi_{\parallel}^{(\alpha)}(\omega, \mathbf{k}, \mathbf{x}) = \frac{\omega_{p\alpha}^2}{k_{\parallel}^2 V_{\alpha}^2} \left[\frac{\omega - \omega_{D\alpha}}{\omega} \{1 - \phi(y_{\alpha 0}) + i\sqrt{\pi} y_{\alpha 0} e^{-y_{\alpha 0}^2}\} e^{-\lambda_{\alpha}} I_0(\lambda_{\alpha}) \right]. \tag{12.56b}$$

In evaluating

$$\chi^{L(\alpha)} = \frac{k_{\perp}^2 \chi_{\perp}^{(\alpha)} + k_{\parallel}^2 \chi_{\parallel}^{(\alpha)}}{k^2} \tag{12.57}$$

the result (12.53) is reproduced (for $\hat{D}_y^{(\alpha)}$ replaced by $\omega_{D\alpha}/\omega$) only when one neglects the contribution from $\chi_{\perp}^{(\alpha)}$; the term proportional to $k_{\perp}^2/k^2$ in (12.51) arises from retaining this term in the limit $\lambda_{\alpha} \ll 1$.

The approximations (12.51), (12.53) and (12.55) with (12.56) are used in the discussion of drift waves in the next Section.

[The reader might note that the approximations (12.51), (12.53) and (12.57) with (12.56) are not mutually consistent. It should be emphasized that in (12.56) the sum over s for $\omega \ll |s\Omega_{\alpha}|$ has been performed using sum rules for the modified Bessel functions. However this fact accounts only partially for the differences, and the remaining differences may be attributed to the somewhat artificial inclusion of the effect of the inhomogeneity through the factor $(\omega - \omega_{D\alpha})/\omega$, which is well justified only for $s = 0$ in (12.53).]

12.7 Drift waves

The existence of drift motions lead to modifications of wave properties, compared with a plasma with no drifts, at frequencies less than or of order the drift frequencies $\omega_{D\alpha}$. The modified wave modes are called *drift modes*. The drift modes associated with the diamagnetic drift, that is those associated with inhomogeneity in the plasma, are of particular importance: plasmas are inevitably inhomogeneous and unstable drift modes cannot be avoided. As a consequence the instability of drift waves due to gradients in the plasma is sometimes referred to as a universal instability. In practice many drift waves have small or zero $k_{\parallel}$ and can be stabilized through a shear motion in the magnetic field; this effectively limits the wavelength $k_{\parallel}^{-1}$ along the magnetic field. However, there are drift waves which exist in sheared fields, and the onset of instability can be inhibited but not eliminated by a shear.

There is a wide variety of drift modes, including modified versions of Alfvén, ion-sound, ion-cyclotron and lower-hybrid waves in a collisionless plasma, and of MHD modes in a collisional plasma. Although instabilities involving these waves can be either reactive or kinetic the instabilities of interest here are kinetic and are due to resonant electrons (or ions) at $s = 0$. For a Maxwellian distribution this resonance leads to Landau damping in a homogeneous plasma. However when the factor $(\omega - \omega_{D\alpha})/\omega$, e.g. in (12.51), is negative the resonance leads to negative absorption.

Let is start with the example of low frequency waves. This case is somewhat more complicated than others considered below because the waves cannot be assumed longitudinal, e.g. Alfvén waves certainly are not longitudinal. Assuming the inhomogeneity to be along the y-axis the simplest approximate dispersion equation we can use is one which arises from the wave equation by assuming $\mathbf{E}$ and $\mathbf{k}$ to be in the x–z plane. This case includes Alfvén waves and ion sound waves in the homogeneous limit. The relevant dispersion equation (from the determinant of the 2×2 matrix Λ_{ij} for $i, j = x, z$) is

$$\frac{k^2 c^2}{\omega^2} K^L - K_\perp K_\parallel = 0 \tag{12.58}$$

where we take account of (12.55). Then using (12.56) and (12.57),with $\omega_{pi}^2/\Omega_i^2 = c^2/v_A^2$ and $\omega_{pi}^2 V_e^2/\omega_p^2 = v_s^2$, we obtain the following dispersion equation for $\lambda_e \ll 1$ and $\sqrt{2} k_\parallel V_i \ll \omega \ll \sqrt{2} k_\parallel V_e$:

$$\left\{ \frac{\omega - \omega_{De}}{\omega} - \frac{k_\parallel^2 v_s^2}{\omega^2} \frac{\omega - \omega_{Di}}{\omega} e^{-\lambda_i} I_0(\lambda_i) \right\} \left\{ \frac{\omega - \omega_{Di}}{\omega} - \frac{k_\parallel^2 v_A^2}{\omega^2} \frac{\lambda_i}{1 - e^{-\lambda_i} I_0(\lambda_i)} \right\}$$

$$= \lambda_i \frac{T_e}{T_i} \frac{k_\parallel^2 v_A^2}{\omega^2} \left(\frac{\omega - \omega_{Di}}{\omega} \right). \tag{12.59}$$

In the limit $\lambda_i \ll 1$, (12.59) factorizes into $\omega^2 = k_\parallel^2 v_s^2$ and $\omega^2 = k_\parallel^2 v_A^2$, corresponding to ion-sound and Alfvén waves respectively.

Two simplifications occur if we assume $T_e = T_i$. First we then have $\omega_{De} = -\omega_{Di}$ and, second, $\omega^2 \gg 2k_\parallel^2 v_i^2$ then implies $\omega^2 \gg 2k_\parallel^2 v_s^2$ so that the term proportional to $k_\parallel^2 v_s^2/\omega^2$ in (12.59) may be neglected. The resulting equation is

$$(\omega - \omega_{De}) \left\{ \omega(\omega + \omega_{De}) - k_\parallel^2 v_A^2 \frac{\lambda_i}{1 - e^{-\lambda_i} I_0(\lambda_i)} \right\} = \lambda_i k_\parallel^2 v_A^2 (\omega + \omega_{De}). \tag{12.60}$$

For small λ_i this has three solutions

$$\omega = \omega_{De}, \ -\frac{\omega_{De}}{2} \pm \left\{ \left(\frac{\omega_{De}}{2} \right)^2 + k_\parallel^2 v_A^2 \right\}^{1/2}. \tag{12.61}$$

The solution $\omega = \omega_{De}$ and one of the others necessarily cross. At this crossover we have a double solution of the dispersion equation.

The double root for $\lambda_i \ll 1$ is important because the maximum growth occurs in its vicinity. This may be seen from the expression (2.65) for the absorption coefficient $\gamma_M \propto (\text{Im} K^M) (\partial \Lambda/\partial \omega)^{-1}$. The condition for a double root is that $\Lambda = 0$ and $\partial \Lambda/\partial \omega = 0$ be satisfied simultaneously. Thus near a double root $\partial \Lambda/\partial \omega$ is small and the absorption coefficient is large as a consequence. In the present case $\text{Im} K^M$ may be identified as the imaginary part of K^L with only the electronic contribution for $s = 0$ retained in (12.53). For $\lambda_e \ll 1$, $y_{e0}^2 \ll 1$, the maximum growth rate is for $k_\parallel^2 v_A^2 \approx 2\omega_{De}^2$ and is

$$|\gamma|_{\max} = \frac{\sqrt{\pi}}{3} |\omega_{De}| \frac{v_A}{V_e} \lambda_i^{1/2}. \tag{12.62}$$

As a second example of drift waves let us consider ion-sound-like waves. These may be treated using the longitudinal approximation. The limit $\lambda_e \ll 1$ with $\lambda_i \gtrsim 1$ and $\omega \ll \Omega_i$ may be treated by setting the leading terms in curly brackets in (12.59) equal to zero. Rather than discuss this low-frequency limit, let us consider the opposite limit $\omega \gg \Omega_i$, $\lambda_i \gg 1$ when the ions may be regarded as unmagnetized. Note that we have (for singly charged ions)

$$\frac{\lambda_i}{\lambda_e} = \frac{m_i}{m_e} \frac{T_i}{T_e}, \tag{12.63}$$

and hence $\lambda_i \gg 1$ and $\lambda_e \lesssim 1$ are compatible. Then assuming $\omega \ll \Omega_e$, $\omega \ll \sqrt{2} k_\parallel V_e$, only the term $s = 0$ contributes for the electrons in (12.53b). Then we have $\chi^{L(e)} \approx 1/k^2 \lambda_{De}^2$ and $\chi^{L(i)} \approx -\omega_{pi}^2/\omega^2$ and hence

$$\mathrm{Re}[K^L(\omega, \mathbf{k}, \mathbf{x})] = 1 + \frac{1}{k^2 \lambda_{De}^2} - \frac{\omega_{pi}^2}{\omega^2}. \tag{12.64}$$

The dispersion relation for ion sound waves is then the same as in a homogeneous plasma, cf. (1.5). However, the imaginary part of K^L for $\omega \ll \sqrt{2} k_\parallel V_e$ is, from (12.53b),

$$\mathrm{Im} K^L = \left(\frac{\pi}{2}\right)^{1/2} \frac{\omega - \omega_{De}}{k^2 \lambda_{De}^2 |k_\parallel| V_e} e^{-\lambda_e} I_0(\lambda_e). \tag{12.65}$$

The important qualitative point is that for $\omega_{De} > \omega = \omega_s(\mathbf{k})$ the sign of the absorption coefficient (2.45) with (12.65) becomes negative. Thus an inhomogeneity can cause ion sound waves to grow. Note that a large gradient is required. Specifically the gradient needs to be over a characteristic length

$$L \lesssim \frac{k_\perp}{k_\parallel} \frac{V_e}{\Omega_e} \left(\frac{m_i}{m_e}\right)^{1/2}. \tag{12.66}$$

Another related example is that of ion cyclotron waves. This case is intermediate between the low frequency limit $\omega \ll \Omega_i$ in (12.59) and the high frequency limit $\omega \gg \Omega_i$ in (12.64). As in the example of ion sound waves the electronic contribution to $\mathrm{Re} K^L$ may be approximated by $1/k^2 \lambda_{De}^2$. For the ionic contribution we use (12.53b), assume $k_\parallel$ is sufficiently small that we have $y_{is}^2 \gg 1$ for all relevant s. The drift frequency for the ions appears in a factor $(\omega - \omega_{Di})/\omega$, which we ignore. The justification for this is as follows. The growing waves have $\omega - \omega_{De} < 0$. Now ω_{Di} is equal to $-\omega_{De}$ for $T_e = T_i$ and ω_{Di} and ω_{De} have opposite signs more generally. Hence for $\omega \approx \omega_{De} \approx s\Omega_i$, $(\omega - \omega_{Di})/\omega$ is of order $(s + 1)/s$. Actually, approximating the operator in (12.53b) by $(\omega - \omega_{D\alpha})/\omega$ is not well justified for $s \neq 0$ but the conclusion that it introduces a correction of the form of a multiplicative factor or order unity remains valid. The factor does not affect the qualitative form of the dispersion relation (12.15) for ion cyclotron waves, or the dispersion relations for the ion Bernstein modes. Thus, as in the case of ion sound waves, the only significant effect of the inhomogeneity is in the absorption coefficient. For $\omega \approx s\Omega_i \lesssim \omega_{De}$ (12.65) implies growth of ion cyclotron waves ($s = 1$) or ion Bernstein waves due to the inhomogeneity.

As a final example let us consider the lower-hybrid drift instability. This is of interest in connection with anomalous transport in laboratory plasmas (e.g. Davidson & Gladd 1975), with observed magnetospheric turbulence in the regions of high density gradients (the plasmapause and the plasma sheet) (e.g. Kintner & Gurnett 1978) and with certain ionospheric irregularities (e.g. Huba & Ossakow 1981). In treating this instability let us assume the waves are longitudinal, that the electrons are described by (12.51) with $y_e \gg 1$ and that the ions are unmagnetized and drifting with velocity U_i. Then we have

$$\operatorname{Re}K^L \approx 1 + \frac{\omega_p^2}{\Omega_e^2} \frac{k_\perp^2}{k^2} + \frac{1}{k^2 \lambda_{De}^2} \frac{\omega_{De}}{\omega} + \frac{1}{k^2 \lambda_{Di}^2}[1 - \phi(y_i)] \qquad (12.67)$$

with $y_i = (\omega - \mathbf{k} \cdot \mathbf{U}_i)/\sqrt{2}kV_i$. For $k_\perp^2 \approx k^2$, $\mathbf{k} \cdot \mathbf{v}_1 \ll \omega$ and $\omega \gg \sqrt{2}kV_i$, $\operatorname{Re}K^L = 0$ implies $\omega^2 = \omega_{LH}^2$.

The following simplified treatment of the *lower-hybrid-drift instability* is based on the analysis by Davidson & Gladd (1975). We assume that the ions are drifting with speed U_i such that $U_i + V_{Di}$ is zero. Then transforming to the frame in which the ions are at rest and the electrons are drifting with velocity $-U_i = V_{Di}$, we have

$$\operatorname{Re}K^L = 1 + \frac{\omega_p^2}{\Omega_e^2} + \frac{1}{k^2 \lambda_{De}^2} \frac{\omega_{De}}{\omega - \mathbf{k} \cdot \mathbf{V}_{Di}} + \frac{1}{k^2 \lambda_{Di}^2}. \qquad (12.68)$$

Solutions to $\operatorname{Re}K^L = 0$ exist for $\omega < \mathbf{k} \cdot \mathbf{V}_{Di}$. Landau 'damping' by thermal ions then leads to growth ($\operatorname{Im}K^L > 0$, $(\partial/\partial\omega)\operatorname{Re}K^L < 0$). Maximum growth for $T_e \ll T_i$ occurs for $\omega = \mathbf{k} \cdot \mathbf{V}_{Di}/2$ with

$$|\gamma|_{\max} = \tfrac{1}{2}\left(\frac{\pi}{2}\right)^{1/2}\left(\frac{V_{Di}}{V_i}\right)^2 \omega_{LH}. \qquad (12.69)$$

Under other conditions $|\gamma|_{\max}$ can be of order ω_{LH}.

This discussion of drift waves is far from complete. There are also drift waves associated with actual forces (e.g. gravity) and real drifts (e.g. the curvature drift). The interested reader is referred to the reviews by Krall (1968) and Mikhailovaskii (1967) and to the book by Mikhailovskii (1974b) for more detailed discussions.

Exercise set 12

12.1 Derive the dispersion relation (12.11) for lower hybrid waves as follows. Assume the electronic contribution is determined by (11.27) with (11.30) with $A_e = 0$ and $U_e = 0$ and expand in $\lambda_e = k_\perp^2 V_e^2/\Omega_e^2$, ω^2/Ω_e^2 and $k_\parallel^2/k_\perp^2$. Then add the contribution (12.10) of the ions and the electronic contribution from (11.27) to determine the dispersion equation $S = \operatorname{Re}K_{xx} = 0$. Note that in (12.11) one assumes $\omega_{LH}^2 = \omega_{pi}^2 \Omega_e^2/\omega_p^2$.

12.2 Show that if one makes the approximation $y_s^2 \gg 1$ for all $s \neq 1$ then in the ionic contribution to K^L, as given by (12.3a), the sum over s may be evaluated as follows for $\omega \approx \Omega_i$

$$\sum_{s=-\infty}^{\infty} \phi(y_s) \frac{\omega}{\omega - s\Omega} e^{-\lambda} I_s(\lambda) \approx \frac{\omega}{\omega - \Omega} \phi(y_1) e^{-\lambda} I_1(\lambda) - \sum_{s \neq 1} \frac{1}{s-1} e^{-\lambda} I_s(\lambda)$$

$$\sum_{s \neq 1} \frac{1}{s-1} e^{-\lambda} I_s(\lambda) = -\frac{1}{\lambda} \{1 - e^{-\lambda} I_0(\lambda)\}.$$

Hence derive (12.15) with (12.16)

Hint: Note the relations

$$2\frac{s}{\lambda} I_s(\lambda) = I_{s-1}(\lambda) - I_{s+1}(\lambda), \quad I_{-s}(\lambda) = I_s(\lambda)$$

$$\sum_{s=-\infty}^{\infty} e^{-\lambda} I_s(\lambda) = 1.$$

12.3 Derive (12.18) by replacing the inequality (12.17) by an equality and using (12.15).

12.4 (a) Show that the handednesses of the two cold plasma modes interchange as D passes through zero. (*Hint:* consider the sign $T_{\pm}$ as given by (10.36).)

   (b) Evaluate the crossover frequency $\omega = \omega_{cr}$, at which D vanishes, in a hydrogen plasma with 10% admixture of the He^+ ions when the plasma frequency is equal to 1 GHz. Express ω_{cr} as a fraction of the proton gyrofrequency. How does ω_{cr} vary as ω_p varies?

12.5 Include thermal corrections in the dispersion for Alfvén waves by the following procedure.

   (i) Show that to first order in λ_i for $\omega \ll \Omega_i$, (11.27) with (11.30) for $A_i = 0$ and $U_i = 0$ implies the ionic contribution

$$K_{xx}^{(i)} \approx \frac{c^2}{v_A^2}\left(1 - \frac{3}{4}\lambda_i\right).$$

   (ii) Justify the approximations $K_{yy} \approx c^2/v_A^2$ and $K_{zz} \approx 1/k^2 \lambda_{De}^2 \cos^2\theta$ by determining their ranges of validity.

   (iii) Hence show that the dispersion equation (10.76) leads to the dispersion relation

$$N^2 \cos^2\theta \approx \frac{c^2}{v_A^2}\left\{1 - \lambda_i\left(\frac{3}{4} + \frac{T_e}{T_i}\right)\right\}$$

   for $\cos^2\theta \ll 1$. Such waves are sometimes called *kinetic Alfvén waves*.

12.6 Consider the resonance at the He^+ gyrofrequency in a plasma with 10% He^+ to 90% H^+ by number.

   (a) Argue that the counterpart of (12.24) for the resonance to be washed out by thermal effects is

$$|k_{\parallel}|_{max} = \left(\frac{\omega_{pHe}^2 \Omega_{He}}{c^2 V_{He}}\right)^{1/3} < \frac{\Omega_{He}}{v_A}.$$

(b) Show that for the He^+/H^+ plasma under consideration, the He^+ resonance is washed out for

$$V_{He} \gtrsim 0.4 v_A.$$

12.7 Suppose a pressure gradient is balanced by a $\mathbf{J} \times \mathbf{B}$ force:

$$\text{grad } P = \mathbf{J} \times \mathbf{B}.$$

Show that the implied current, written in the form

$$\mathbf{J} = \sum_\alpha q_\alpha n_\alpha \mathbf{V}_{D\alpha}$$

with

$$P = \sum_\alpha n_\alpha T_\alpha$$

implies the diamagnetic drifts as given by (12.52) for $T_\alpha = \text{constant}$.

13

Instabilities due to anisotropic fast particles

13.1 Resonant scattering

The scattering rate due to Coulomb interactions between a fast particle (speed v) and thermal particles decreases with increasing v as v^{-3}. Thus one might expect that fast particles are scattered very ineffectively. The reverse is the case in the low density (low β) plasmas in the magnetosphere, and in space and astrophysical plasmas generally. Fast particles are scattered very efficiently due to resonant interaction with low frequency waves, called *resonant scattering*.

The evidence which led to the initial development of the theory of resonant scattering came from the properties of the trapped particles in the magnetosphere. By the early 1960's it was clear that for the stability of the distributions of trapped magnetospheric particles (in the terrestrial 'radiation' or 'van Allen' belts) to be consistent with the observations of precipitation of these particles, both the fast electrons and the fast ions must be scattered very efficiently. The development of the theory of resonant scattering led to a satisfactory qualitative and semi-quantitative explanation for these magnetospheric observations. Resonant scattering also offered ways of resolving serious difficulties connected with the scattering and acceleration of fast particles in astrophysical plasmas. The most obvious of these concerns the confinement of galactic cosmic rays (§13.4). Another serious difficulty was with the acceleration of fast particles: early theories for the acceleration required (either implicitly or explicitly) very efficient scattering. The problems of scattering and acceleration are discussed in §§13.4 & 13.5.

The scattering of fast electrons and fast ions in the magnetosphere is attributed to whistlers and to ion cyclotron waves, respectively, and the scattering of relativistic electrons and faster ions is attributed to Alfvén and/or magnetoacoustic waves. In §11.1 it is pointed out that the Doppler condition (11.1) can be satisfied for $|s\Omega| \approx |k_\parallel v_\parallel| \gg \omega$, and this case is the one of relevance in resonant scattering. In particular only $s = \pm 1$ need be considered for qualitative and semi-quantitative purposes, and we make this simplifying assumption here. (Resonances at $|s| > 1$ are not necessarily negligible, but their inclusion leads to no qualitatively different effects than for $|s| = 1$.) For $|s\Omega| \approx |k_\parallel v_\parallel| \gg \omega$ the change in the properties of the particle in an interaction are predominantly in pitch angle α. This may be seen by considering the changes $\Delta\alpha$ and $\Delta\varepsilon$ on emission of a wave quantum. Using (10.91) with $\alpha = \arctan p_\perp/p_\parallel$ one finds

$$\Delta\alpha = \frac{\hbar(\omega\cos\alpha - k_\parallel v)}{pv\sin\alpha}, \quad \Delta\varepsilon = \hbar\omega, \quad \Delta p = \frac{\hbar\omega}{v}. \tag{13.1a,b,c}$$

Assuming that $\cos\alpha$ and $\sin\alpha$ are not close to zero, one has $\Delta\alpha \approx -(\Delta p/p)(k_\parallel v_\parallel/\omega)$, implying $\Delta\alpha \gg \Delta p/p$ for $k_\parallel v_\parallel \gg \omega$. Hence the pitch angle of the particle changes much more rapidly than does p or its energy ε. One says that the resonant interaction leads predominantly to *pitch-angle scattering*.

There are three signs involved: s, $k_\parallel/|k_\parallel|$ and $v_\parallel/|v_\parallel|$. For resonance to occur the product of these must be negative

$$s\frac{k_\parallel}{|k_\parallel|}\frac{v_\parallel}{|v_\parallel|} = -1. \tag{13.2}$$

For electrons interacting with whistlers and ions interacting with ion cyclotron waves one effectively requires $s = 1$. The reason arises from the form (10.23) of $\mathbf{V}(\mathbf{k}, \mathbf{p}; s)$ and of the polarization of these waves. The interaction involves $|\mathbf{e}^* \cdot \mathbf{V}(\mathbf{k}, \mathbf{p}; s)|^2$ which is proportional to the factor P^2 with P defined by (11.60) for $s = 1$ and by

$$P = -\epsilon(e_x + i\epsilon s e_y) \tag{13.3}$$

for $s = \pm 1$. For parallel propagation whistlers have $e_y/e_x = i$ and ion cyclotron waves have $e_y/e_x = -i$, and these waves interacting with electrons and ions respectively lead to $P \neq 0$ only for $s = 1$. While interaction at $s = -1$ is not forbidden for oblique propagation, it can be ignored for qualitative purposes. Then (13.2) implies that the resonant particles and waves must be propagating in opposite directions along the field lines ($k_\parallel v_\parallel < 0$) in order to interact.

Alfvén (A) waves and magnetoacoustic (M) waves have polarization vectors $\mathbf{e}_A = (1, 0, 0)$ and $\mathbf{e}_M = (0, i, 0)$, and (12.3) implies $P \neq 0$ for either sign of ϵ or s. The restrictions to $s > 0$ and to $k_\parallel v_\parallel < 0$ do not apply to interactions with these waves.

The conditions $\Omega \approx |k_\parallel v_\parallel| \gg \omega$ lead to restrictions on the parallel velocity $v_\parallel$ of the resonant particles. For simplicity let us consider the case of parallel propagation. For whistlers at $\omega \ll \Omega_e$ and Alfvén and magnetoacoustic waves at $\omega \ll \Omega_i$, we have $|k_\parallel| \approx (\omega_p/\Omega_e)(\omega\Omega_e)^{1/2}/c \ll \Omega_e/43 v_A$ and $|k_\parallel| \approx \omega/v_A \ll \Omega_i/v_A$, respectively, where we write $\Omega_e/\omega_p = 43 v_A/c$ for a hydrogen plasma. It then follows that we require

$$|p_\parallel| > 43 m_e v_A = \frac{m_i v_A}{43} \tag{13.4a}$$

for electrons to resonate with whistlers and

$$|p_\parallel| > m_i v_A \tag{13.4b}$$

for ions or electrons to resonate with Alfvén or magnetoacoustic waves. These conditions are relaxed near the cyclotron resonances where $|k_\parallel|$ becomes larger than the values assumed above. The value of $|k_\parallel|_{\max}$ given by (12.23) for electron cyclotron waves (ECW) and the corresponding limit for ion cyclotron waves (ICW) in a thermal plasma require $|p_\parallel| > |p_\parallel|_{\min}$ with

$$|p_\parallel|_{\min} = \begin{cases} 43 m_e v_A \left(\dfrac{V_e}{43 v_A} \right)^{1/3} & \text{for ECW} \\[2ex] m_i v_A \left(\dfrac{V_i}{v_A} \right)^{1/3} & \text{for ICW} \end{cases} \tag{13.5}$$

Between the limits implied by (13.4) and (13.5) the relevant waves must be regarded as ECW and ICW, i.e. $k_\parallel$ must be determined by the dispersion relations taking thermal effects into account.

Pitch-angle scattering may be treated using the quasilinear equation (10.92). Retaining only the term describing diffusion in pitch angle, we have

$$\frac{df}{dt} = \frac{1}{\sin \alpha} \frac{\partial}{\partial x} \left[\sin \alpha \, D \, \frac{\partial f}{\partial \alpha} \right] \tag{13.6}$$

with

$$D = \sum_{s=-\infty}^{\infty} \frac{1}{p^2 \sin^2 \alpha} \int \frac{d^3 k}{(2\pi)^3} \hbar^2 w_M(s, \mathbf{p}, \mathbf{k}) N_M(\mathbf{k}) k_\parallel^2. \tag{13.7}$$

We now make the approximation (10.23) to $\mathbf{V}(\mathbf{k}, \mathbf{p}; s)$, retain only $s = \pm 1$, use the properties (10.60) for whistlers and use the properties

$$\omega_A = |k_\parallel| v_A, \quad R_A^{\cdot} = \frac{v_A^2}{2c^2}, \quad \mathbf{e}_A = (1, 0, 0) \tag{13.8a}$$

$$\omega_M = k v_A, \quad R_M = \frac{v_A^2}{2c^2}, \quad \mathbf{e}_M = (0, i, 0) \tag{13.8b}$$

for the other two modes. The probabilities in (13.7) are then given by

$$w_{\dot W}(s, \mathbf{p}, \mathbf{k}) \approx \frac{\pi q^2}{4\varepsilon_0} \frac{\Omega_e}{\hbar \omega_p^2} \frac{v \sin^2 \alpha}{|\cos \alpha|} \frac{(1 + s|\cos \theta|)^2}{\cos^2 \theta} \delta(k + s\Omega_e / \gamma v \cos \alpha \cos \theta) \tag{13.9}$$

for electrons interacting with whistlers, and

$$w_{A,M}(s, \mathbf{p}, \mathbf{k}) = \frac{\pi q^2}{4\varepsilon_0} \frac{v_A^2}{c^2} \frac{v \sin^2 \alpha}{\hbar \omega |\cos \alpha \cos \theta|} \delta(k + s\Omega_0 / \gamma v \cos \alpha \cos \theta) \tag{13.10}$$

for either Alfvén or magnetoacoustic waves interacting with either fast electrons $(p > m_i v_A)$ or ions.

13.2 Whistler waves in the magnetosphere: quasilinear theory

In the magnetosphere whistler waves occur in the VLF (very low frequency) band from 3 to 30 kHz. The three most prominent forms of emission are called 'hiss', 'chorus' and 'discrete VLF emissions'. Both hiss and chorus involve resonant or nearly resonant whistlers, i.e. waves near $\omega = \omega_-(\theta) \approx \Omega_e |\cos \theta|$. The likely generation mechanisms involve Cerenkov and cyclotron instabilities respectively. The discrete emissions have the more familiar properties of whistlers, e.g. as described by (10.60). In the discussion in this Section and in §13.3 we concentrate on possible generation mechanisms for these waves, and discuss the observational data on them only where necessary to place the theory in context. In this Section we present quasilinear theories for the interaction between whistlers and electrons with a loss-cone distribution, and for the interaction of resonant whistlers (hiss) with streaming electrons. Discrete VLF emissions are discussed in §13.3.

The absorption coefficient (10.90) may be evaluated for whistlers using the expression (13.9) for the probability. It is convenient to introduce the resonant momentum

$$p_R = \frac{m_e \Omega_e}{k|\cos \alpha \cos \theta|}. \tag{13.11}$$

Then, after summing over $s = \pm 1$, one finds

$$\gamma_W(k, \theta) = -\frac{\pi^2 k}{2n_e} \int_{-1}^{+1} d\cos \alpha \, \sin^2 \alpha \left[p^3 v \left\{ -\frac{\cos \theta}{\sin \alpha} \frac{\partial f}{\partial \alpha} + \frac{43 v_A}{v|\cos \alpha|} p \frac{\partial f}{\partial p} \right\} \right.$$
$$\left. \times \left\{ (1 + \cos^2 \theta) - 2 \cos \theta \frac{\cos \alpha}{|\cos \alpha|} \right\} \right]_{p = p_R}. \tag{13.12}$$

The p-derivative leads to a stabilizing contribution for $\partial f/\partial p < 0$, which is the case in nearly all applications. The α-derivative leads to a destabilizing contribution if f is an increasing function of $\sin \alpha$. (Recall that we require $\cos \theta \cos \alpha < 0$.) Thus an instability can be driven by a loss-cone distribution.

Let us assume a step-function loss-cone distribution, specifically

$$f(p, \alpha) = f_0(p)\{H(\alpha - \alpha_0) + H(\pi - \alpha_0 - \alpha)\} \tag{13.13}$$

where α_0 is the loss-cone angle. Then waves at $\theta > \pi/2$ may be driven unstable by the contribution from $\partial f/\partial \alpha = f_0 \delta(\alpha - \alpha_0)$, and likewise waves at $\theta < \pi/2$ may be driven unstable by $\partial f/\partial \alpha = -f_0 \delta(\alpha - \pi + \alpha_0)$. For a power-law distribution $f_0(p) \propto p^{-a}$ one finds (cf. Exercise 13.1)

$$\gamma_W(k) \approx -\frac{n_1}{n_e} \Omega_e \left\{ \alpha_0^2 - a \left(\frac{43 v_A}{v} \right)^2 \right\} \tag{13.14}$$

with k and v related by $kv \approx \Omega_e/\gamma$, and where n_1 is the integrated number density of particles with speed $\gtrsim v$. Growth occurs only for $\alpha_0 > 43 v_A/v$, which requires $v > 43 v_A$, in accord with (13.4a).

The accepted qualitative interpretation of the population of energetic electrons in the Earth's magnetosphere is as follows. The electrons enter from the solar wind and diffuse inwards. The inward diffusion violates the third adiabatic invariant Φ_M but conserves M and J, cf. (12.35) and (12.44). Now the magnetic field B and the length of the field lines vary with the radial distance (the L-value) of the field lines as L^{-3} and L respectively, and hence $M = $ constant and $J = $ constant imply $p^2 \sin^2 \alpha \propto L^{-3}$ and $p \cos \alpha \propto L^{-1}$. As a consequence both p and $\sin \alpha$ increase as the electrons diffuse to smaller L. The number density of the particles also increases. (In diffusive equilibrium, which is not achieved, f would be a constant and $n_1 \propto p_\perp^2 p_\parallel f$ would vary as L^{-4}.) The loss-cone angle α_0 also increases as the electrons diffuse inwards. Eventually γ_W exceeds the effective loss rate for whistlers, and the whistlers then grow. (The effective loss rate is due to the propagation time t_W for whistlers across the source region and growth becomes effective for $\gamma_W t_W \gg 1$.)

Once the whistlers have been generated they scatter the electrons causing them to diffuse in pitch angle, in accord with (13.6). If the loss cone is empty or nearly empty then $\partial f/\partial \alpha$ is large for $\alpha \approx \alpha_0$ and diffusion is rapid there. Electrons diffuse into the loss cone and can then precipitate. However, if the diffusion is fast enough the loss cone can be filled, or nearly filled, and a typical electron diffuses out again before it precipitates. Let the magnetic flux tube be of length l. Then including a loss rate $v/l = v$, (13.6) for $\alpha \ll 1$ reduces to

$$\frac{1}{\alpha}\frac{\partial}{\partial \alpha}\left\{\alpha D \frac{\partial f}{\partial \alpha}\right\} - vf = 0 \qquad (13.15)$$

inside the loss cone, i.e. for $\alpha < \alpha_0$. Now (13.7) with (13.9) implies that D is not a sensitive function of α for $\alpha \ll 1$, and hence we take D independent of α. Then the relevant solution of (13.15) is

$$f \propto I_0(\alpha(v/D)^{1/2})/I_0(\alpha_0(v/D)^{1/2}). \qquad (13.16)$$

Kennel (1969) defined the following two cases

strong diffusion: $D \gg v\alpha_0^2$

weak diffusion: $D \ll v\alpha_0^2$.

In weak diffusion the loss cone is nearly empty while in strong diffusion it is nearly full (cf. Exercise 13.2).

The foregoing theory provides a satisfactory qualitative and semi-quantitative interpretation of the distribution of energetic electrons in the magnetosphere. However it is by no means clear that the theory is correct in detail. The strongest argument against the theory is that the whistlers observed in connection with precipitating electrons are discrete VLF emissions. These waves are phase-coherent and narrowband, and one cannot treat the effect of such a wave on the electrons using quasilinear theory. However it may be that one can justify using quasilinear theory, and specifically (13.15), as follows. Each discrete VLF emission causes particles with an initial $\mathbf{p}$ to change by some amount $\Delta \mathbf{p}$. After many such interactions the net effect may well be reasonably described in terms of diffusion in momentum space. Quasilinear theory is then applicable, although the diffusion coefficient may differ in detail from that given in §10.6.

VLH hiss

As has already been mentioned VLF hiss is thought to be generated by streaming electrons. It is straightforward to calculate the growth rate due to interaction at $s = 0$, e.g. as outlined in Exercise 13.3. However treatment of the quasilinear relaxation is not quite as straight forward. We discuss only the quasilinear relaxation.

For interactions at $s = 0$ the only contribution to the quasilinar equation (10.91) is

$$\frac{\partial f}{\partial t} = \frac{\partial}{\partial p_\parallel}\left[D_{\parallel\parallel}\frac{\partial f}{\partial p_\parallel}\right]. \qquad (13.17)$$

Suppose the waves are resonant whistlers at $\omega = \Omega_e|\cos\theta|$. The resonance condition $\omega - k_\parallel v_\parallel = 0$ implies growth for a streaming distribution only for $\omega/k_\parallel < U_e$, where U_e is the streaming speed. Hence only waves with $k > \Omega_e/U_e$ should be generated. It then follows that the diffusion coefficient in (13.17) is of the form (cf. Exercise 13.4)

$$D_{\parallel\parallel} = \begin{cases} D_0\left(\dfrac{U_e}{v_\parallel}\right)^3 & v_\parallel < U_e \\ 0 & v_\parallel > U_e \end{cases}. \qquad (13.18)$$

It is of interest to examine how the quasilinear relaxation suppresses the instability. The rate of change of U_e and of $V_\parallel^2$ may be found by setting $M = v_\parallel$ and $(v_\parallel - U_e)^2$ in

$$\frac{d\bar{M}}{dt} = \int d^3\mathbf{p}\, M\frac{\partial f}{\partial t}. \qquad (13.19)$$

Assuming $f \propto \exp[-(v_\parallel - U_e)^2/2V_\parallel^2]$, (13.17) and (13.18) in (13.19) imply that the rate of decrease of U_e and the rate of increase of $V_\parallel$ are comparable in magnitude. For $V_\parallel \ll U_e$ initially the main effect is to increase the velocity spread. The growth rate ($\propto 1/V_\parallel^2$) then decreases tending to suppress the instability. Suppression occurs when the rate of increase in $V_\parallel^2$, i.e. $d(\ln V_\parallel^2)/dt$, becomes comparable with the growth rate, and this occurs when roughly $V_\parallel/U_e$ of the energy in the stream has been transferred to the waves.

Qualitatively, this instability and its saturation is similar to that of Langmuir waves due to a streaming instability in an unmagnetized plasma.

13.3 Discrete VLF emissions

Discrete VLF emissions in the terrestrial magnetosphere are narrowband and phase coherent. They can drift to either higher or lower frequencies. Different discrete emissions can interact with each other in quite specific ways. Some of the spectral features are illustrated schematically in Figure 13.1. These discrete emissions correlate with the electron precipitation, and it seems that the whistlers which control the scattering of electrons into the loss cone, as discussed in §13.2, are these discrete emissions. Perhaps the most intriguing characteristic of these emissions is that they can be triggered artificially. This was discovered accidentally through observation of the effects of morse code transmissions in the VLF range. It was found that morse code dashes, of duration 150 ms, can trigger emissions but that morse code dots, of duration 50 ms, usually do not ('dot-dash anomaly'). It was later found that not only morse code emissions (≈ 1 MW for $\gtrsim 100$ ms) but also low-

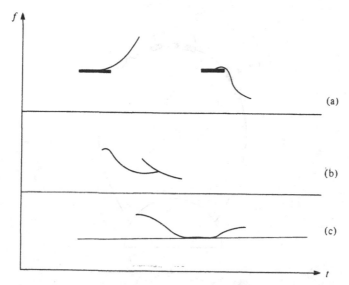

Fig. 13.1 Some schematic examples of VLF emissions. (a) The triggered signal starts about 100 ms after the onset of the triggering morse code signal (solid line) appearing first at a slightly higher frequency ($\lesssim 0.5$ kHz) and then drifting to higher or lower frequencies. (b) Cutoff at another signal. (c) Entrainment of a signal by a steady background signal.

power transmissions (≈ 100 W for $\gtrsim 1$ s) can trigger emissions. The properties of discrete VLF emissions and of theories for their interpretation have been reviewed by Matsumoto (1979).

The basic interpretation of discrete VLF emission is embodied in a phenomenological model developed initially by Helliwell (1967). After summarizing this model, we discuss theories for triggered VLF emissions, concentrating on the growth mechanism. It seems that the growth mechanism is not a standard kinetic or reactive instability, and because of this there is considerable interest from a formal viewpoint in identifying the instability mechanism in detail. At present there is no satisfactory widely-accepted mechanism.

Helliwell's model for discrete VLF emissions is illustrated in Figure 13.2. Recall that the interaction between electrons and whistlers occurs when the particle and the wave are travelling in opposite directions. Helliwell separated the region of interaction into two parts; let us call them the bunching region (BR) and the growth region (GR). As electrons enter the BR they are assumed to have a frequency mismatch

$$\Delta = \omega - \Omega_e - k_\parallel v_\parallel$$

which is moderately small and negative. The whistlers (with $N^2 \gg 1$ and with $\cos \theta = 1$ by assumption) then cause the electrons to bunch, as discussed in §10.6. There is little exchange of energy in the BR. The electrons propagate to the GR where Δ is very small and energy exchange becomes effective. The whistlers grow

(a)

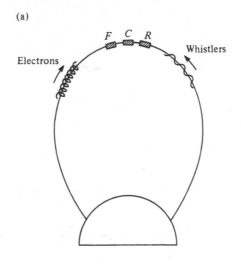

(b)

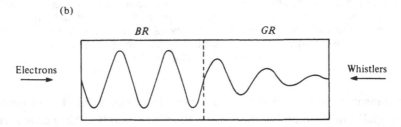

Fig. 13.2 Helliwell's model for discrete VLF emission.
(a) Location of the interaction region when the emission is falling (*F*), constant
(*C*) and rising (*R*) in frequency.
(b) Electrons propagating from the left are bunched by the waves in the *BR*, then
contribute to the amplification of the wave in the *GR*. A frequency drift results
from a spatial drift of the *BR/GR* region due to the spatial variation of Ω_e.

and they propagate back towards the BR to cause bunching of newly entering
electrons.

An important ingredient in the model is that in the GR one has both $\Delta = 0$ and
$d\Delta = 0$ where '*d*' denotes differentials due to the variation with distance along the
field line and perhaps also due to a drift of the GR itself along the field lines. Helliwell
(1967) referred to the conditions $\Delta = 0$ and $d\Delta = 0$ as the 'consistent wave condition'
and Nunn (1974) referred to an equivalent condition as the 'second order resonance'.
This condition defines those regions where the distance or time available for the
wave-particle interaction is a maximum, or, put another way, where the effect of
inhomogeneity in limiting the resonant interaction is minimized.

Growth in discrete VLF emissions is exponential, at least initially. Helliwell &
Inan (1982) extended the phenomenological model to include a gain factor (due to an
unspecified mechanism) in each feedback cycle, and with this extension the model

can account for most of the observed features of discrete VLF emissions. The growth mechanism is evidently some form of enhanced amplification. That is, the triggering signal enhances the value of the growth rate. It is likely that the electron distribution is close to marginal instability, due to a loss-cone anisotropy or other feature, in that the growth rate for the kinetic instability is positive but $\gamma_W t_W$ is less than about unity; growth becomes effective only when the triggering signal enhances the value of γ_W. Two ideas for enhanced growth have been considered: trapping and phase bunching.

Trapping of electrons by whistlers may be treated using the theory developed in §11.6. If we assume parallel propagation for the whistlers then the bounce frequency (11.64) becomes

$$\omega_T = \left(\Omega_e k v_\perp \frac{B_W}{B} \right)^{1/2} \tag{13.24}$$

where $B_W (= N_W E_0/\sqrt{2}c)$ is the magnetic amplitude of the whistlers.

The idea that trapping can cause enhanced growth is as follows. Consider a trapped particle at $v_\parallel = v_R = (\omega - \Omega_e)/k_\parallel$ and an untrapped particle at a neighbouring value of $v_\parallel$. As they propagate along the field line the trapped particle has $v_\parallel$ changing so that it remains equal to the changing value of v_R. The untrapped particle has $v_\parallel$ changing in accord with $v_\perp^2/B = \text{constant}$ and $v_\parallel^2 + v_\perp^2 = \text{constant}$. Hence the distribution function becomes distorted. A distortion which increases the value of $\partial f/\partial v_\parallel$ can enhance the growth rate for instabilities, such as the whistler instability, driven by a parallel gradient.

There is a strong observational argument against trapping being the cause of the enhanced growth. If triggering were related directly to a process involving trapping one would expect there to be a threshold on the wave amplitude below which the enhanced growth is ineffective. This is because trapping occurs only on a timescale $> \omega_T^{-1} \propto B_W^{-1/2}$, and trapping must occur on a timescale less than the propagation time across the source region. This threshold on B_W translates into a threshold on the power required from a transmitter before amplification should occur. The large variation (100 W to 1 MW) in power which is known to cause triggering argues against a trapping mechanism, and this argument is strengthened considerably by estimations of ω_T for the low-power transmission seemingly being inconsistent with trapping on the required timescale.

Trapping is certainly important in discrete VLF emissions. As in some other applications it is probably the saturation mechanism for the instability rather than an essential ingredient in the growth mechanism.

Phase bunching can cause only a minor enhancement in the growth rate for a reactive instability, and causes no enhancement at all in the growth rate for a kinetic instability. In the reactive instability, one averages over relative initial phases in passing from (11.61) to (11.62), and the average of a factor $\cos^2(\psi - \phi)$ is unity for complete initial phase bunching and one half for random initial phases. Kinetic growth is independent of the relative phase of the particles and waves and is unaffected by any phase bunching.

A specific mechanism for enhanced growth due to phase bunching (proposed by Winglee (1985)) involves transitory growth, as defined in §5.5. The idea is an adaption of one due to Sprangle & Smith (1980) for enhanced growth in the gyrotron. In the BR the frequency mismatch is small enough to allow effective phase bunching but not so small that the energy exchange is effective. Then Δ decreases in magnitude as the electrons pass from the BR to the GR. As they enter the GR they can cause transitory growth on a timescale $\sim \bar{\Delta}^{-1}$, where $\bar{\Delta}$ is the average of Δ over the electron distribution. Provided we have $\omega_i \gg |\bar{\Delta}|$ the growth occurs at a faster rate than phase mixing, and it then follows that the transient contribution $K_{ij}^{(TR)}(\omega, \mathbf{k})$ is a slowly varying function of t.

Assuming a DGH distribution (11.19), the growth rate, due to transitory growth, is found to be (cf. Exercise 13.5)

$$|\gamma_W| = \frac{n_1}{n_e} \frac{(j+1)V_\perp^2 k_\parallel^2}{|\Delta_0|} (\Omega_e - \omega) t \cos(\Delta_0 t) \exp\left[-\tfrac{1}{2}(k_\parallel V_\parallel t)^2\right] \qquad (13.25)$$

with $\Delta_0 = \omega - \Omega_e - k_\parallel U$, and where n_1 is the number density of electrons in the DGH distribution. Transitory growth persists for a time

$$t \lesssim \min\left[\frac{1}{|k_\parallel|V_\parallel}, \frac{1}{|\Delta_0|}\right]. \qquad (13.26)$$

The attractive features of this suggestion are that it allows growth without a threshold (unlike the trapping mechanism) and that it incorporates the phase-bunching mechanism of Helliwell's model in a natural way. The main difficulty is the required narrow spread in $v_\parallel$.

The interpretation of the growth mechanism in discrete VLF emissions remains a challenging problem. It is clear from an observational point of view that phase-bunching is an essential feature, but it is not clear how this is to be incorporated in a theory for growth. Transitory growth is one specific idea for including phase bunching in instability theory; the important point is that pre-phase-bunched particles cannot be treated using a time-asymptotic theory. It is likely that further study of VLF emissions, and other similar phenomena, will lead to a deeper understanding of this aspect of instability theory.

13.4 Scattering of streaming cosmic rays

Resonant scattering is an ingredient in current ideas on the propagation and acceleration of cosmic rays. An important question is the source of the waves required to scatter the particles. It is thought that the waves are generated by the anisotropic particles themselves. Before discussing the instability involved, let us briefly summarize the background astrophysical results relating to streaming cosmic rays.

The central problem in the propagation of galactic cosmic rays can be summarized as follows (e.g. Wentzel 1974). The cosmic rays evidently fill the galactic disc which is about 100 parsecs or 300 light years thick. Hence a typical relativistic particle propagating freely would cross the disc in ≈ 300 years. However evidence from spallation products imply a confinement time $\approx 10^7$ years in the galactic disc, and this is supported by measurement of the anisotropy of cosmic rays which implies that they are streaming very slowly. Thus the cosmic rays appear to diffuse through the galactic disc at a mean speed $\lesssim 10^{-4}c$. This diffusion can be explained, in principle, in terms of pitch-angle scattering causing spatial diffusion along the field lines.

In both the applications to streaming and to acceleration the resonant waves can be generated by the anisotropic particles. Consider the absorption coefficient derived from (10.86) using (13.10),

$$\gamma_{A,M}(\omega) = \frac{2\pi^2}{\varepsilon_0} \frac{q^2 v_A}{kc} \int_{-1}^{+1} d\cos\alpha \frac{\sin^2\alpha}{|\cos\alpha|}$$
$$\times \left[\frac{p^3 c}{\varepsilon} \left\{ \frac{\cos\theta}{|\cos\theta|} \frac{1}{\sin\alpha} \frac{\partial}{\partial\alpha} - \frac{v_A}{v} p \frac{\partial}{\partial p} \right\} f(p,\alpha) \right]_{p=p_R}, \qquad (13.27)$$

where for simplicity we assume parallel propagation for the waves ($\sin\theta = 0$), and where the resonant momentum is given by $p_R = m_i v_A \Omega_i / \omega |\cos\alpha|$.

In treating the generation of the waves by streaming cosmic rays let us assume a distribution of the form

$$f(p,\alpha) = f_0(p)\left(1 + \frac{3v_{CR}}{v}\cos\alpha\right) \qquad (13.28)$$

where v_{CR} is the streaming speed of cosmic rays with speed v. Data on cosmic rays imply

$$f_0(p) = K\left(\frac{p}{p_0}\right)^{-a} \qquad (13.29)$$

with $a = 4.6$ and $Kk_0^3 \approx 2.5 \times 10^{-7} \, \text{m}^{-3}$ for $p_0 = m_p c$. Explicit evaluation then gives

$$\gamma_{A,M}(\omega) = -\frac{3\pi}{2} \frac{a-3}{a(a-2)} \frac{q^2 v_A n(p_0)}{\varepsilon_0 k_0 p_0 c^2} \left(\frac{\omega}{k_0 v_A}\right)^{a-3} \left\{ \frac{\cos\theta}{|\cos\theta|} v_{CR} - \frac{a}{3} v_A \right\} \qquad (13.30)$$

with $k_0 p_0 = |q| B$ and where $n(p_0)$ is the number density of cosmic rays with $p > p_0$. Growth is possible for streaming in excess of a threshold speed $v_{CR} > a v_A/3$. The growing waves are propagating in the streaming direction ($v_{CR} \cos\theta > 0$), and one may regard the waves as carrying off forward momentum of the particles.

In the case of streaming it is relatively simple to calculate how v_{CR} decreases due to the effect of scattering by the waves. One has

$$\frac{dv_{CR}}{dt} = v \int_{-1}^{+1} d\cos\alpha \, \cos\alpha \frac{df}{dt} \Big/ \int_{-1}^{+1} d\cos\alpha \, f \qquad (13.31)$$

with df/dt given by (13.6). If we retain an additional term from (10.92), specifically the term involving $D_{\alpha p}$, then one finds (cf. Exercise 13.6)

$$\frac{dv_{CR}}{dt} = v_s \left(v_{CR} - \frac{a}{3} v_A \right) \qquad (13.32)$$

with

$$v_s = 3 \int_{-1}^{+1} d\cos\alpha \quad D_{\alpha\alpha} \sin^2\alpha. \qquad (13.33)$$

Explicit evaluation of $D_{\alpha\alpha}$ gives

$$D_{\alpha\alpha} = \frac{\pi q^2}{4\varepsilon_0 \varepsilon p |\cos\alpha|} W(k_R) \qquad (13.34)$$

where

$$W_{A,M} = \int dk_{\parallel} W(k_{\parallel}) \qquad (13.35)$$

is the energy density in the resonant waves and with $k_R p_R = k_0 p_0$.

Although it seems likely that the scattering of cosmic rays is due to resonant scattering in some form, a serious difficulty remains: the growth rate is too small. Suppose v_{CR} is of order $v_A \approx 10^{-2}$ c; then the growth time implied by (13.30) is of order $(p_R/p_0)^{1.6}$ years with $p_0 c = 10^9$ eV. It follows that for $pc \gtrsim 10^{13}$ eV the growth time is even longer than the inferred confinement time of the cosmic rays to the galactic disc. For even quite modest cosmic ray energies ($\lesssim 10^{12}$ eV) there is simply not enough time for the cosmic rays to generate the resonant waves. One way of minimizing this difficulty is to appeal to localized scattering regions so that the cosmic rays are confined by encountering such 'scattering centres' along any prospective escape path. How the resonant waves are generated in these scattering centres still remains a problem.

13.5 Small amplitude Fermi acceleration

The acceleration of cosmic rays and other energetic particles is a long-standing astrophysical problem which is still not adequately understood; it may be that a variety of different acceleration mechanisms contributes. Here we are concerned with only two mechanisms: small-amplitude Fermi acceleration and diffusive acceleration at shock fronts. Both mechanisms were discussed in the astrophysical literature in the 1940's but their efficiency was found to be too low. This efficiency is greatly enhanced in the presence of resonant scattering. 'Small amplitude Fermi acceleration' is essentially the damping of MHD turbulence by fast particles, with the energy in the MHD turbulence being transferred to the fast particles. Provided they are scattered efficiently the cosmic rays act like a viscous low-density gas, and their viscous drag on the wave motion may be regarded as causing the damping of the waves.

Now let us discuss the instability involved in the acceleration process. Compression or rarefaction of the magnetic field due to the MHD turbulence causes the particles to develop an anisotropy which is an even function of $\cos \theta$. Formally one has

$$\frac{\partial f}{\partial t} = -\frac{1}{p_\perp} \frac{\partial}{\partial p_\perp} \left[p_\perp \left\{ \frac{dp_\perp}{dt} f \right\} \right] \tag{13.36}$$

with $p_\perp^2/B = $ constant implying

$$\frac{dp_\perp}{dt} = \frac{p_\perp}{2B} \frac{dB}{dt}. \tag{13.37}$$

In the presence of scattering the anisotropy is modified from that implied by (13.36). Rather than use (13.36) let us suppose the anisotropy is kept small due to the scattering, and let us expand in Legendre polynomials. We write

$$f(p, \alpha) = f_0(p)\{1 - AP_2(\cos \alpha)\}, \tag{13.38}$$

where A is an anisotropy factor. Semi-quantitatively the resulting form of the absorption coefficient is similar to (13.30) with v_{CR} replaced by Av, and growth can occur provided the anisotropy exceeds a threshold value of order v_A/v. A qualitative difference is that growth requires that the product of the signs of A, $\cos \theta$ and $\cos \alpha$ be negative, corresponding to $As > 0$. That is, for $A > 0$ (i.e. $\langle p_\perp^2 \rangle > 2\langle p_\parallel^2 \rangle$) growth occurs at $s = 1$, and for $A < 0$ growth occurs at $s = -1$.

A point which is not widely recognized is that it is only the magnetoacoustic component in the MHD turbulence which is effective in the acceleration. (Evolution of the turbulent spectrum can convert energy in Alfvén waves into magneto-acoustic waves however.) This point becomes obvious when one treats the acceleration in terms of damping at $s = 0$ of the magnetoacoustic waves (Achterberg 1981). Damping at $s = 0$ of Alfvén waves is very weak, being zero in the approximation $e_A = (1, 0, 0)$, cf. (10.79a). We consider only the damping of the magneto-acoustic component by the energetic particles. The relevant absorption coefficient follows from (10.90) with the wave properties given by (13.8b)

$$\gamma_M(\mathbf{k}) = -\frac{\pi q^2 v_A^2}{\varepsilon_0 \omega c^2} \int d^3\mathbf{p} \, v_\perp^2 J_0'^2(k_\perp v_\perp/\Omega) k_\parallel \frac{\partial f(\mathbf{p})}{\partial p_\parallel} \delta(\omega - k_\parallel v_\parallel). \tag{13.39}$$

The particle distribution, in the absence of scattering, evolves according to

$$\frac{df(\mathbf{p})}{dt} = \frac{\partial}{\partial p_\parallel} \left[D_{\parallel\parallel} \frac{\partial f(\mathbf{p})}{\partial p_\parallel} \right] \tag{13.40}$$

with

$$D_{\parallel\parallel} = \frac{\pi q^2 v_A^2}{\varepsilon_0 c^2} \int \frac{d^3\mathbf{k}}{(2\pi)^3} W_M(\mathbf{k}) \frac{k_\parallel^2}{\omega^2} v_\perp^2 J_0'^2(k_\perp v_\perp/\Omega) \delta(\omega - k_\parallel v_\parallel). \tag{13.41}$$

Explicit evaluation of $D_{\parallel\parallel}$ for $k_\perp v_\perp \ll \Omega$ gives

$$D_{\parallel\parallel} = \frac{\pi^2 v_A \omega}{B^2/2\mu_0} \frac{p_\perp^2 v_\perp^2}{|v_\parallel|^3} \int dk \, k^2 W_M(k, \theta_R) \left(1 - \frac{v_A^2}{v_\parallel^2}\right) \tag{13.42}$$

with $\cos \theta_R = v_A/v_\parallel$.

The damping of the MHD turbulence by the particles causes the particles to diffuse in $p_\parallel$ at constant $p_\perp$. Particles gain energy only by gaining in $|p_\parallel|$. As $p_\parallel$ increases $\cos\theta_R = v_A/v_\parallel$ decreases and hence the particles can interact with waves only in a decreasing range of angles about $\theta = \pi/2$. An initially isotropic distribution of particles becomes increasingly anisotropic (with $\langle p_\parallel^2 \rangle > \frac{1}{2}\langle p_\perp^2 \rangle$) and this has the effect of decreasing the efficiency of the energy transfer. Thus in the absence of resonant scattering the efficiency of the acceleration is limited by the anisotropy which develops.

Let us suppose that the particles generate resonant waves when their anisotropy $|A|$ exceeds the threshold value $\approx v_A/v$. The resulting pitch-angle scattering then restricts the anisotropy to of order v_A/v. The evolution of the distribution of particles may then be described by an equation

$$\frac{df_0(p)}{dt} = \frac{1}{p^2}\frac{\partial}{\partial p}\left\{p^2 D_{pp}\frac{\partial f_0(p)}{\partial p}\right\} \tag{13.43}$$

obtained by averaging (13.40) over pitch angles. One finds

$$\begin{aligned}
D_{pp} &= \frac{1}{2}\int_{-1}^{+1} d\cos\alpha\,\cos^2\alpha\,D_{\parallel\parallel}\\
&= \frac{\pi}{4}\frac{v_A^2}{B^2/2\mu_0}\frac{p^2}{v}\int\frac{d^3k}{(2\pi)^3}\frac{W_M(\mathbf{k})k_\perp^2}{|k_\parallel|}\left(1 - \frac{\omega^2}{k_\parallel^2 v^2}\right)^2\\
&\approx \frac{\pi}{8}\frac{v_A p^2 \bar\omega}{v}\frac{W_M}{B^2/2\mu_0}\ln\frac{v}{v_A}
\end{aligned} \tag{13.44}$$

where in the final expression the waves are assumed isotropic with a mean frequency $\bar\omega$ and an energy density W_M.

The mean rate of energy gain

$$\left\langle\frac{d\varepsilon}{dt}\right\rangle = \int d^3\mathbf{p}\,\varepsilon\frac{df}{dt}\bigg/\int d^3\mathbf{p}\,f \tag{13.45}$$

is given by

$$\begin{aligned}
\left\langle\frac{d\varepsilon}{dt}\right\rangle &= \frac{1}{p^2}\frac{\partial}{\partial p}(p^2 v D_{pp})\\
&\approx 4\eta\bar\omega p v_A\left(\frac{\delta B}{B}\right)^2
\end{aligned} \tag{13.46}$$

where η is a factor of order unity and $\delta B/B$ is the relative change in the magnetic field due to the turbulence. The result (13.46) is well known in astrophysical applications; it corresponds to small amplitude Fermi acceleration.

Although it is accepted that MHD turbulence may damp by transferring effectively all its energy to fast particles in astrophysical plasmas, it is not clear that this is the dominant acceleration mechanism for the fast particles. Acceleration at shock fronts seems more favourable in many applications.

13.6 Acceleration at shock fronts

There has long been indirect evidence that fast particles are accelerated by shocks in the solar corona and in other astrophysical plasmas, and there is now direct detailed data on such acceleration for the bow shock of the Earth and for interplanetary shocks. Acceleration occurs in a variety of forms. At the bow shock one of the prominent components involves ion beams; these are evidently accelerated by an electric field in the bow shock. Energetic electrons are also produced and these generate second harmonic plasma emission as they flow back upstream. At interplanetary shocks already suprathermal particles are further accelerated as they cross the shock. This type of acceleration, called shock-drift acceleration, may be treated by calculating the energy change when a particle encounters a shock and is either reflected or transmitted (e.g. the reviews by Toptyghin 1980 and Drury 1983).

Here we concentrate on diffusive acceleration at shock fronts. The idea is that if fast particles are scattered on either side of a shock front, then they can diffuse back and forth crossing the shock many times, and every time they are reflected they gain energy. To see this suppose the particles cross the shock from downstream to upstream, cf. Figure 13.3; they gain energy because they are reflected from scattering centres flowing into the shock. If the particles cross the shock from upstream to downstream the scattering centres again appear to be flowing towards the upstream plasma and reflection causes an energy gain. Besides being an efficient mechanism for acceleration, a relatively simple theory for this acceleration leads naturally to a power-law momentum distribution $f(p) \propto p^{-a}$ of the form which occurs widely in astrophysical sources.

The acceleration is called 'diffusive' because the accelerated particles diffuse across the shock front due to resonant scattering. Before discussing how the resonant waves are generated let us consider this spatial diffusion.

Let $D_\parallel$ be the spatial diffusion coefficient along the field lines. Diffusion is then described by

$$\frac{df}{dt} = \frac{\partial}{\partial z}\left(D_\parallel \frac{\partial f}{\partial z}\right).$$
(13.47)

The equation (13.6) describing the pitch-angle scattering may be written

$$\left(\frac{\partial}{\partial t} + v\cos\alpha \frac{\partial}{\partial z}\right)f = \frac{1}{\sin\alpha}\frac{\partial}{\partial\alpha}\left(\sin\alpha\, D \frac{\partial f}{\partial\alpha}\right)$$
(13.48)

where the term $\partial f/\partial z$ is non-zero whenever there is a gradient in the distribution of fast particles. Let us expand f in Legendre polynomials and keep only the first two terms,

$$f(p,\alpha) = f_0(p) + f_1(p)\cos\alpha + \cdots.$$
(13.49)

Now on averaging (13.48) over $\cos\alpha$, and averaging (13.48) times $\cos\alpha$ over

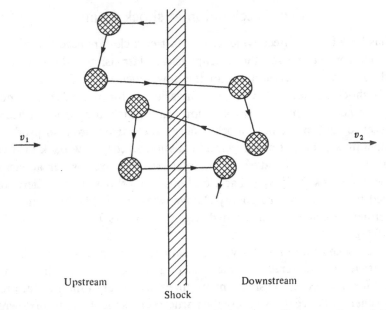

Fig. 13.3 Diffusive shock acceleration may be interpreted in terms of a model in which a fast $(v \gg v_1, v_2)$ particle bounces off scattering centres in the upstream and the downstream plasma. Viewed from one side of the shock the scattering centres on the other side of the shock are approached at a speed $|\mathbf{v}_1 - \mathbf{v}_2|$. Hence each time the particle crosses the shock it is reflected head-on and gains energy. A large energy gain can result after many shock crossings.

$\cos \alpha$, one obtains

$$\frac{\partial f_0}{\partial t} + \frac{v}{3} \frac{\partial f_1}{\partial z} = 0, \tag{13.50a}$$

$$\frac{\partial f_1}{\partial t} + \frac{v}{3} \frac{\partial f_0}{\partial z} = \tfrac{1}{2} f_1 \int_{-1}^{+1} d \cos \alpha \, D \sin^2 \alpha. \tag{13.50b}$$

Ignoring the term $\partial f_1/\partial t$ in comparison with $\partial f_0/\partial t$, one obtains (13.47) with

$$D_{\parallel} = \frac{v^2}{9} \left[\frac{1}{2} \int_{-1}^{+1} d \cos \alpha \, D \sin^2 \alpha \right]^{-1}. \tag{13.51}$$

The instability which generates the resonant waves is due to a gradient-induced anisotropy. From (13.50b), again with the term $\partial f_1/\partial t$ neglected, one finds

$$f_1 = -\frac{3D_{\parallel}}{v} \frac{\partial f_0}{\partial z}. \tag{13.52}$$

On substituting (13.52) in (13.49) and thence in (13.27), one finds that the resonant waves grow for

$$\frac{\cos \theta}{|\cos \theta|} 3D_{\parallel} \frac{\partial f_0}{\partial z} - v_A p \frac{\partial f_0}{\partial p} < 0. \tag{13.53}$$

Spatial diffusion tends to smooth out the density gradient thereby reducing $\partial f_0/\partial z$ and hence the growth rate. The spatial diffusion rate should adjust so that it is stationary in a frame fixed relative to the shock.

The power-law spectrum is predicted as follows. Firstly suppose that the distribution reaches a stationary state. Let the particles be streaming with a velocity u, which is a function of z. Then in a stationary state we have [cf. Exercise 13.7]

$$u\frac{\partial f_0}{\partial z} - \frac{1}{3}p\frac{\partial f_0}{\partial p}\frac{\partial u}{\partial z} = \frac{\partial}{\partial z}D_\parallel\frac{\partial f_0}{\partial z}. \tag{13.54}$$

Now let us assume

$$u = \begin{cases} v_1 & z < 0 \text{ (upstream)} \\ v_2 & z > 0 \text{ (downstream)} \end{cases} \tag{13.55}$$

so that we have $\partial u/\partial z = (v_1 - v_2)\delta(z)$. Let $f_\pm$ be the values of z at $z = \pm\infty$. The form of the solution of (13.54) for $z \neq 0$, and hence $\partial u/\partial z = 0$, is

$$f_0 = A + B\exp\left[\int dz\frac{u}{D_\parallel}\right]. \tag{13.56}$$

For $z > 0$ we have $u/D_\parallel \neq 0$ and then to avoid a divergent solution we require $B = 0$ and $A = f_+$ in (13.56). Hence the solution is

$$f = \begin{cases} f_- + (f_+ - f_-)\exp\left[v_1\int_0^z\frac{dz}{D_\parallel}\right] & z < 0 \\ f_+ & z > 0 \end{cases} \tag{13.57}$$

Now on integrating (13.54) across $z = 0$ one finds

$$-v_1(f_+ - f_-) = \frac{1}{3}(v_1 - v_2)p\frac{\partial f_+}{\partial p}, \tag{13.58}$$

which integrates to

$$f_+(p) = ap^{-a}\int_0^p dp'p'^{(a-1)}f_-(p') \tag{13.59}$$

with

$$a = \frac{3v_1}{v_2 - v_1}. \tag{13.60}$$

In particular if a δ-function distribution is injected ahead of the shock, i.e. $f_-(p) \propto \delta(p - p_0)$, then (13.59) implies $f_+(p) \propto p^{-a}$ for $p > p_0$ behind the shock.

The implied power law index q is related to the Mach number of the shock. From (8.69) we have $v_1/v_2 = n_2/n_1$ and then a result quoted in Exercise 8.7 implies

$$a = \frac{3r}{r-1}, \qquad r = \frac{(\Gamma+1)M^2}{2 + (\Gamma-1)M^2}, \tag{13.61a, b}$$

and hence

$$a = \frac{3(\Gamma+1)M^2}{2(M^2-1)}. \tag{13.62}$$

For $M^2 \gg 1$ the value of a is close to 4. Thus this mechanism provides a natural explanation for the value $a \approx 4.6$ for cosmic rays and for similar values in a variety of other astrophysical contexts.

Exercise set 13

13.1 Evaluate the absorption coefficient (13.12) for the distribution (13.13) in the case $\cos\theta < 0$, $f_0(p) \propto p^{-a}$ with $a > 4$. Hence show that the condition for growth of whistlers is

$$\sin^2 \alpha_L \cos \alpha_L > \frac{a}{a-3}\left(\frac{\omega}{\Omega_e|\cos\theta|}\right)^{1/2}\left\{1 - \frac{a-3}{a-1}\cos^2\alpha_L\right\}$$

where the refractive index for the whistlers is taken to be $N = \omega_p/(\omega\Omega_e|\cos\theta|)^{1/2}$.

13.2 (a) Show that the equations

$$\frac{1}{\alpha}\frac{\partial}{\partial\alpha}\left\{\alpha D\frac{\partial f}{\partial\alpha}\right\} - vf = 0 \qquad\qquad \alpha < \alpha_L \ll 1$$

$$\frac{1}{\sin\alpha}\frac{\partial}{\partial\alpha}\left\{\sin\alpha D\frac{\partial f}{\partial\alpha}\right\} \propto \delta(\alpha - \pi/2) \qquad \alpha_L < \alpha \lesssim \pi/2$$

in the case where D does not depend on α have the solution

$$f \propto \begin{cases} \phi_0 + \ln\left\{(\tan\alpha/2)/(\tan\alpha_L/2)\right\} & \alpha_L < \alpha < \pi/2 \\ I_0(\alpha(v/D)^{1/2})/I_0(\alpha_L(v/D)^{1/2}) & \alpha < \alpha_L \end{cases}$$

(b) Plot the solution for $\alpha < \alpha_L$ in the case $\alpha_L^2 v \gg D$.

13.3 Show that the absorption coefficient at $s = 0$ for whistlers due to the distribution

$$f(\mathbf{p}) = \frac{n_1}{(2\pi)V^3}\exp\left[-\frac{(\mathbf{v}-\mathbf{U})^2}{2V^2}\right]$$

with $\mathbf{U}$ along the magnetic field lines is given by

$$\gamma_W(\mathbf{k}) = \left(\frac{\pi}{2}\right)^{1/2}\frac{n_1}{n_e}\frac{\Omega_e^2}{kV|\cos\theta|}\frac{\omega(\omega - \mathbf{k}\cdot\mathbf{U})}{k^2V^2}\exp\left[-\frac{(\omega - \mathbf{k}\cdot\mathbf{U})^2}{2k^2V^2\cos^2\theta}\right].$$

13.4 Show that the diffusion coefficient $D_{\parallel\parallel}$ in (13.17) for whistlers at $s = 0$ for nonrelativistic particles is given by

$$D_{\parallel\parallel} = \frac{\pi e^2\Omega_e^2}{\varepsilon_0\omega_p^2 v_\parallel^2}\int\frac{d^3k}{(2\pi)^3}\frac{\omega^2}{k_\parallel^2}\delta(\omega - k_\parallel v_\parallel)W(\mathbf{k}).$$

Hence derive the form (13.18) in the case where $W(\mathbf{k})$ is zero for $\omega/k_\parallel > U$.

13.5 The projection of the transient response onto the right hand polarization vector for parallel propagation for nonrelativistic electrons is

$$K^{R(TR)}(\omega, \mathbf{k}) = -\frac{e^2}{\varepsilon_0 \omega^2} \int d^3\mathbf{p} \left\{ \frac{\omega - k_\parallel v_\parallel}{v_\perp} \frac{\partial}{\partial p_\perp} \left(\frac{v_\perp^2 f}{\Delta} \right) + k_\parallel \frac{\partial}{\partial p_\parallel} \left(\frac{v_\perp^2 f}{\Delta} \right) \right\} \exp\left[i \Delta t \right]$$

where we assume $k_\perp v_\perp \ll \Omega_e$ and write $\Delta = \omega - \Omega_e - k_\parallel v_\parallel$. Evaluate $K^{R(TR)}$ for the DGH distribution (11.19), and hence derive the absorption coefficient (13.25).

13.6 Show that the diffusion coefficient $D_{\alpha p}$, e.g. in (10.89), for Alfvén waves is given by

$$D_{\alpha p} = -\frac{\cos \theta}{|\cos \theta|} \frac{v_A}{v} p \sin \alpha \, D_{\alpha \alpha},$$

with $D_{\alpha \alpha}$ given by (13.34). Hence (i) derive (13.32) in the case where the waves are confined to the forward hemisphere ($v_{CR} \cos \theta > 0$), and (ii) show that the term $a v_A / 3$ in (13.32) is absent if $W(-k_\parallel) = W(k_\parallel)$ is an even function of $\cos \theta$.

13.7 Consider scattering of particles in a medium in which the fluid velocity $\mathbf{u}$ is a function of position $\mathbf{x}$. Suppose that the distribution $f(\mathbf{p}, t, \mathbf{x})$ of particles evolves according to

$$\frac{df(\mathbf{p}, t, \mathbf{x})}{dt} = \left[\frac{df}{dt} \right]_{\text{scatt}},$$

where the right hand member describes scattering, cf. (13.6), and where on the left hand side one has

$$\frac{d}{dt} = \frac{\partial}{\partial t} + \dot{\mathbf{x}} \cdot \frac{\partial}{\partial \mathbf{x}} + \dot{\mathbf{p}} \cdot \frac{\partial}{\partial \mathbf{p}}.$$

Assume that $\dot{\mathbf{x}}$ and $\dot{\mathbf{p}}$ are determined by Hamilton's equations

$$\dot{\mathbf{x}} = \frac{\partial H}{\partial \mathbf{p}}, \qquad \dot{\mathbf{p}} = -\frac{\partial H}{\partial \mathbf{x}}$$

and assume, for nonrelativistic particles,

$$H = \frac{(\mathbf{p} + m\mathbf{u})^2}{2m}.$$

(a) Show that the isotropic part f_0 of f satisfies

$$\frac{\partial f_0}{\partial t} + \mathbf{u} \cdot \frac{\partial f_0}{\partial \mathbf{x}} - \frac{1}{3} (\text{div } \mathbf{u}) p \frac{\partial f_0}{\partial p} = \left[\frac{df}{dt} \right]_{\text{scatt}}.$$

(b) Hence derive equation (13.54).

Appendix A

The plasma dispersion function

The plasma dispersion function may be defined by, cf. (2.30),

$$\bar{\phi}(z) = -\frac{z}{\sqrt{\pi}} \int_{-\infty}^{\infty} \frac{dt\, e^{-t^2}}{t - z} \tag{A.1}$$

for $z = x + iy$ with $y > 0$. In the limit $iy = i0$ one has

$$\bar{\phi}(x + i0) = \phi(x) - i\sqrt{\pi}\, x e^{-x^2}, \tag{A.2}$$

$$\phi(x) = 2x e^{-x^2} \int_0^x dt\, e^{t^2}. \tag{A.3}$$

Related functions are that of Fried & Conte (1961)

$$Z(z) = \frac{1}{\sqrt{\pi}} \int_{-\infty}^{\infty} \frac{dt\, e^{-t^2}}{t - z} = -\frac{1}{z}\bar{\phi}(z) \tag{A.4}$$

and that of Faddeyeva & Terent'ev (1954)

$$w(z) = e^{-z^2}\left(1 + \frac{2i}{\sqrt{\pi}} \int_0^z dt\, e^{t^2}\right) = \frac{i}{\sqrt{\pi} z}\bar{\phi}(z). \tag{A.5}$$

The function $\bar{\phi}(z)$ has the following properties:

$$\bar{\phi}(z^*) = [\phi(-z)]^*, \tag{A.6}$$

$$\frac{d\bar{\phi}(z)}{dz} = \frac{\bar{\phi}(z)}{z} + 2z\{1 - \bar{\phi}(z)\}, \quad \bar{\phi}(0) = 0. \tag{A.7}$$

For $|z|^2 \ll 1$ one has

$$\bar{\phi}(z) = 2z^2 - \tfrac{4}{3}z^4 + \cdots - i\sqrt{\pi} z e^{-z^2} \tag{A.8}$$

and for $|z|^2 \gg 1$ one has

$$\bar{\phi}(z) = 1 + \frac{1}{2z^2} + \frac{3}{4z^4} + \cdots - i\sigma\sqrt{\pi} z e^{-z^2} \tag{A.9a}$$

with

$$\sigma = \begin{cases} 0 & y > |x| \\ 1 & |y| > |x| \\ 2 & y < -|x| \end{cases} \tag{A.9b}$$

Appendix B
Bessel functions

For integral arguments $s = 0, \pm 1, \pm 2 \cdots$ the *ordinary Bessel functions* $J_s(z)$ and the *modified Bessel functions* $I_s(\lambda)$ have the following properties:—

Differential equation

$$J_s''(z) + \frac{1}{z} J_s'(z) + \left(1 - \frac{s^2}{z^2} \right) J_s(z) = 0 \tag{B.1}$$

$$I_s''(\lambda) + \frac{1}{\lambda} I_s'(\lambda) - \left(1 + \frac{s^2}{\lambda^2} \right) I_s(\lambda) = 0. \tag{B.2}$$

Recurrence relations

$$J_{s-1}(z) + J_{s+1}(z) = 2 \frac{s}{z} J_s(z) \tag{B.3a}$$

$$J_{s-1}(z) - J_{s+1}(z) = 2 J_s'(z) \tag{B.3b}$$

$$I_{s-1}(\lambda) + I_{s+1}(\lambda) = 2 I_s'(z) \tag{B.4a}$$

$$I_{s-1}(\lambda) - I_{s+1}(\lambda) = 2 \frac{s}{\lambda} I_s(\lambda). \tag{B.4b}$$

Symmetry properties

$$J_{-s}(z) = (-)^s J_s(z) \tag{B.5}$$

$$J_s(-z) = (-)^s J_s(z) \tag{B.6}$$

$$I_{-s}(\lambda) = I_s(\lambda) \tag{B.7}$$

$$I_s(-\lambda) = (-)^s I_s(\lambda). \tag{B.8}$$

Power series expansions

$$J_s(z) = \sum_{k=0}^{\infty} \frac{(-)^k}{k!(k+s)!} \left(\frac{z}{2} \right)^{2k+s} \tag{B.9}$$

$$I_s(\lambda) = \sum_{k=0}^{\infty} \frac{1}{k!(k+s)!} \left(\frac{\lambda}{2} \right)^{2k+s}. \tag{B.10}$$

Generating functions

$$\exp\left[\frac{1}{2}\left(t-\frac{1}{t}\right)z\right] = \sum_{s=-\infty}^{\infty} t^s J_s(z) \tag{B.11}$$

$$\exp\left[\frac{1}{2}\left(t+\frac{1}{t}\right)\lambda\right] = \sum_{s=-\infty}^{\infty} t^s I_s(\lambda). \tag{B.12}$$

Infinite sums

$$\sum_{s=-\infty}^{\infty} J_s^2(z) = 1, \quad \sum_{s=-\infty}^{\infty} J_s'^2(z) = \tfrac{1}{2}, \quad \sum_{s=-\infty}^{\infty} \frac{s^2}{z^2} J_s^2(z) = \tfrac{1}{2} \quad \text{(B.13a,b,c)}$$

$$\sum_{s=-\infty}^{\infty} e^{-\lambda} I_s(\lambda) = 1, \quad \sum_{s=-\infty}^{\infty} e^{-\lambda} I_s'(\lambda) = 1, \quad \sum_{s=-\infty}^{\infty} e^{-\lambda}\frac{s^2}{\lambda} I_s(\lambda) = 1. \quad \text{(B.14a,b,c)}$$

Asymptotic expansions

$$J_s(z) = \left(\frac{2}{\pi z}\right)^{1/2}\left\{\left(1 - \frac{(s^2-\frac{1}{4})(s^2-9/4)}{2(2z)^2} + \cdots\right)\cos\left(z + \frac{s\pi}{2} - \frac{\pi}{4}\right)\right\}$$

$$-\left(\frac{s^2-\frac{1}{4}}{2z} + \cdots\right)\sin\left(z + \frac{s\pi}{2} - \frac{\pi}{4}\right)\right\} \tag{B.15}$$

$$I_s(\lambda) = \frac{e^\lambda}{(2\pi\lambda)^{1/2}}\left(1 - \frac{s^2-\frac{1}{4}}{2\lambda} + \cdots\right). \tag{B.16}$$

Interrelation

$$I_s(\lambda) = i^{-s} J_s(i\lambda). \tag{B.17}$$

Appendix C
Collision frequencies

The effect on suprathermal particles of species α of Coulomb interactions with particles of species β may be described in terms of collision frequencies $v^{(\alpha,\beta)}$. For an initially monoenergetic beam of particles of species α interacting with isotropic thermal particles of species β, four collision frequencies may be defined: these are for slowing down

$$\frac{d\mathbf{v}}{dt} = -v_s^{(\alpha,\beta)}\mathbf{v}, \tag{C.1}$$

and for energy loss

$$\frac{d}{dt}(\tfrac{1}{2}m_\alpha v^2) = -v_e^{(\alpha,\beta)}\tfrac{1}{2}m_\alpha v^2, \tag{C.2}$$

plus two diffusion rates $v_\perp^{(\alpha,\beta)}$ and $v_\parallel^{(\alpha,\beta)}$ describing the diffusive increase in the components $(\mathbf{v} - \langle\mathbf{v}\rangle)_\perp^2$ and $(\mathbf{v} - \langle\mathbf{v}\rangle)_\parallel^2$ where $\perp$ and $\parallel$ refer to the direction of the mean velocity $\langle\mathbf{v}\rangle$, i.e. of the initial direction. These collision frequencies may be expressed in terms of the function

$$\Phi(x) = \frac{2}{\sqrt{\pi}} \int_0^x dt \; e^{-t^2} \tag{C.3}$$

and the related function

$$G(x) = \frac{1}{2x^2}\left\{\Phi(x) - x\frac{d\Phi(x)}{dx}\right\}. \tag{C.4}$$

These functions are to be evaluated at $x = x_\beta$ with

$$x_\beta = \frac{v}{\sqrt{2}V_\beta} \tag{C.5}$$

where $V_\beta = (T_\beta/m_\beta)^{1/2}$ is the thermal speed for species β.

The collision frequencies are given by (e.g. Trubnikov 1965)

$$v_s^{(\alpha,\beta)} = 2\left(1 + \frac{m_\alpha}{m_\beta}\right)v_0^{(\alpha,\beta)}x_\beta^2 G(x_\beta) \tag{C.6}$$

$$v_E^{(\alpha,\beta)} = 2v_s^{(\alpha,\beta)} - v_\perp^{(\alpha,\beta)} - v_\parallel^{(\alpha,\beta)} \tag{C.7}$$

$$v_\perp^{(\alpha,\beta)} = 2v_0^{(\alpha,\beta)}[\Phi(x_\beta) - G(x_\beta)] \tag{C.8}$$

$$v_\parallel^{(\alpha,\beta)} = 2v_0^{(\alpha,\beta)}G(x_\beta) \tag{C.9}$$

with

$$v_0^{(\alpha,\beta)} = \frac{q_\alpha^2 q_\beta^2 n_\beta}{4\pi\varepsilon_0^2 m_\alpha^2 v^3}\ln\Lambda^{(\alpha,\beta)}. \tag{C.10}$$

The Coulomb logarithms $\ln\Lambda^{(\alpha,\beta)}$ for thermal particles of species α and β are given by
electron–electron collisions

$$\ln\Lambda^{(e,e)} = \begin{cases} 16.0 - \frac{1}{2}\ln n_e + \frac{3}{2}\ln T_e & T_e \lesssim 7 \times 10^4 \, K \\ 21.6 - \frac{1}{2}\ln n_e + \ln T_e & T_e \gtrsim 7 \times 10^4 \, K \end{cases} \tag{C.11}$$

electron-ion collisions

$$\ln\Lambda^{(e,i)} = \ln\Lambda^{(i,e)} = \begin{cases} 16.0 - \frac{1}{2}\ln n_e + \frac{3}{2}\ln T_e - \ln Z_i & T_e \lesssim 1.4 \times 10^5 \, K \\ 22.0 - \frac{1}{2}\ln n_e + \ln T_e & T_e \gtrsim 1.4 \times 10^5 \, K \end{cases} \tag{C.12}$$

The rate at which temperture equilibrium between the electrons and the ions is approached is determined by

$$\frac{dT_e}{dt} = v_{eq}^{(e,i)}(T_i - T_e) \tag{C.13}$$

with

$$v_{eq}^{(e,i)} = \frac{e^2 q_i^2 n_e \ln\Lambda^{(e,i)}}{3(2\pi)^{1/2}\pi m_e m_i \varepsilon_0^2 (V_e^2 + V_i^2)^{3/2}} \tag{C.14}$$

Appendix D
Transport coefficients

Braginskii (1965) evaluated the transport coefficients defined in §9.3 for an electron–ion plasma using the Landau form for the collision term in (9.15):

$$C^{(\alpha,\beta)} = -\frac{q_\alpha^2 q_\beta^2 \ln \Lambda^{(\alpha,\beta)}}{8\pi\varepsilon_0^2} \frac{\partial}{\partial p_i} \int d^3\mathbf{p}' \left\{ f^{(\alpha)}(\mathbf{p}) \frac{\partial f^{(\beta)}(\mathbf{p}')}{\partial p_j'} - f^{(\beta)}(\mathbf{p}') \frac{\partial f^{(\alpha)}(\mathbf{p})}{\partial p_j} \right\}$$

$$\times \frac{\{(\mathbf{v} - \mathbf{v}')^2 \delta_{ij} - (\mathbf{v} - \mathbf{v}')_i(\mathbf{v} - \mathbf{v}')_j\}}{|\mathbf{v} - \mathbf{v}'|^2}. \tag{D.1}$$

The Coulomb logarithm $\ln \Lambda^{(\alpha,\beta)}$ is slightly different for different species, cf. (C.11 & 12) but for most purposes the value (9.24) suffices. Braginskii's results for an electron–ion plasma may be summarized as follows.

Characteristic collision times τ_e and τ_i for the electrons and ions are defined by

$$\tau_e = 3\left(\frac{\pi}{2}\right)^{1/2} \frac{1}{v_s} \approx 2.8 \times 10^5 \frac{T_e^{3/2}}{n_e \ln \Lambda} \tag{D.2}$$

and (for protons)

$$\tau_i = 3\pi^{1/2}\left(\frac{4\pi n_e \lambda_{Di}^3}{\omega_{pi} \ln \Lambda}\right) \approx 1.7 \times 10^7 \frac{T_i^{3/2}}{n_e \ln \Lambda} \tag{D.3}$$

where n_e is per cubic metre and temperatures are in kelvins. The momentum transfer rate is written

$$\mathbf{R}^{(e,i)} = -\mathbf{R}^{(i,e)} = en_e\left(\frac{\mathbf{J}_\parallel}{\sigma_\parallel} + \frac{\mathbf{J}_\perp}{\sigma_\perp} - \frac{0.71}{e}(\mathrm{grad}\, T_e)_\parallel - \frac{3}{2}\frac{1}{e\Omega_e \tau_e}\mathbf{b} \times \mathrm{grad}\, T_e\right) \tag{D.4}$$

with $\Omega_e = eB/m_e$ the electron cyclotron frequency, and with

$$\sigma_\parallel = 2.0\sigma_\perp, \quad \sigma_\perp = \varepsilon_0 \omega_p^2 \tau_e. \tag{D.5a,b}$$

In (D.4), $\parallel$ and $\perp$ refer to components relative to $\mathbf{b} = \mathbf{B}/B$. The electron and ion heat fluxes are given, respectively, by

$$\mathbf{q}^{(e)} = \frac{0.71 T_e}{e}\mathbf{J}_\parallel + \frac{3}{2}\frac{T_e}{e\Omega_e \tau_e}\mathbf{b} \times \mathbf{J}_\parallel - \kappa_\parallel^e(\mathrm{grad}\, T_e)_\parallel - \kappa_\perp^e(\mathrm{grad}\, T_e)_\perp - \frac{5n_e T_e}{2m_e\Omega_e}\mathbf{b} \times \mathrm{grad}\, T_e \tag{D.6}$$

and

$$\mathbf{q}^{(i)} = -\kappa_\parallel^i (\text{grad } T_i)_\parallel - \kappa_\perp^i (\text{grad } T_i)_\perp + \frac{5n_i T_i}{2m_i \Omega_i} \mathbf{b} \times \text{grad } T_i \qquad (\text{D.7})$$

with $\Omega_i = Z_i eB/m_i$ the ion cyclotron frequency. The thermal conductivities are given by

$$\kappa_\parallel^e = 3.16 \frac{n_e T_e \tau_e}{m_e}, \quad \kappa_\perp^e = 4.66 \frac{n_e T_e}{m_e \Omega_e^2 \tau_e}, \qquad (\text{D.8a,b})$$

$$\kappa_\parallel^i = 3.9 \frac{n_i T_i \tau_i}{m_i}, \quad \kappa_\perp^i = 2 \frac{n_i T_i}{m_i \Omega_i^2 \tau_i}, \qquad (\text{D.9a,b})$$

These results apply for $\Omega_e \tau_e \gg 1$ and $\Omega_i \tau_i \gg 1$. In the opposite limit $\Omega_i \tau_i \ll 1$ one replaces $\kappa_\perp^i$ by $\kappa_\parallel^i$ and ignores the $\mathbf{b} \times$ term in (9.32), and for $\Omega_e \tau_e \ll 1$ one similarly replaces $\kappa_\perp^e$ by $\kappa_\parallel^e$ and $\sigma_\perp$ by $\sigma_\parallel$ and ignores the $\mathbf{b} \times$ terms in (D.4) and (D.6).

For completeness we quote the results for the viscosity. The rate-of-strain tensor

$$W_{ij}^{(\alpha)} = \frac{\partial v_i^{(\alpha)}}{\partial x_j} + \frac{\partial v_j^{(\alpha)}}{\partial x_i} - \tfrac{2}{3}\delta_{ij} \text{div } \mathbf{v}^{(\alpha)} \qquad (\text{D.10})$$

is symmetric and traceless, as is the stress tensor $\Pi_{ij}^{(\alpha)}$. Each therefore has five different components, and five different viscosities are required in general. With the z-axis along $\mathbf{b}$ Braginskii found, for each species

$$\Pi_{zz} = -\eta_0 W_{zz},$$

$$\Pi_{xx} = -\frac{\eta_0}{2}(W_{xx} + W_{yy}) - \frac{\eta_1}{2}(W_{xx} - W_{yy}) - \eta_3 W_{xy},$$

$$\Pi_{yy} = -\frac{\eta_0}{2}(W_{xx} + W_{yy}) - \frac{\eta_1}{2}(W_{xx} - W_{yy}) + \eta_3 W_{xy}, \qquad (\text{D.11})$$

$$\Pi_{xy} = -\eta_1 W_{xy} + \frac{\eta_3}{2}(W_{xx} - W_{yy}),$$

$$\Pi_{xz} = -\eta_2 W_{xz} - \eta_4 W_{yz},$$

$$\Pi_{yz} = -\eta_2 W_{yz} + \eta_4 W_{xy}$$

with

$$\eta_0^e = 0.73 n_e T_e \tau_e, \quad \eta_1^e = 0.51 \frac{n_e T_e}{\Omega_e^2 \tau_e},$$

$$\eta_2^e = 4\eta_1^e, \quad \eta_3^e = -\frac{n_e T_e}{2\Omega_e}, \quad \eta_4^e = 2\eta_3^e;$$

$$\eta_0^i = 0.96 n_i T_i \tau_i, \quad \eta_1^i = \frac{3}{10} \frac{n_i T_i}{\Omega_i^2 \tau_i}, \qquad (\text{D.12})$$

$$\eta_2^i = 4\eta_1^i, \quad \eta_3^i = \frac{n_i T_i}{2\Omega_i}, \quad \eta_4^i = 2\eta_3^i.$$

Another important result is the rate of entropy generation $\dot{S}_\alpha$ for the electron and ions:

$$T_e \dot{S}_e = \frac{J_\parallel^2}{\sigma_\parallel} + \frac{J_\perp^2}{\sigma_\perp} + \frac{\kappa_\parallel^e}{T_e}(\operatorname{grad} T_e)_\parallel^2 + \frac{\kappa_\perp^e}{T_e}(\operatorname{grad} T_e)_\perp^2 + \frac{1}{2}\sum_{N=0}^{2} \eta_N^e (W_N^e)_{ij}(W_N^e)_{ij}, \quad \text{(D.13)}$$

$$T_i \dot{S}_i = \frac{\kappa_\parallel^i}{T_i}(\operatorname{grad} T_i)_\parallel^2 + \frac{\kappa_\perp^i}{T_i}(\operatorname{grad} T_i)_\perp^2 + \frac{1}{2}\sum_{N=0}^{2} \eta_N^i (W_N^i)_{rs}(W_N^i)_{rs}, \quad \text{(D.14)}$$

(N.B. i is not a tensor index in (D.14)), with

$$(W_0)_{rs} = \begin{pmatrix} \tfrac{1}{2}(W_{xx}+W_{yy}) & 0 & 0 \\ & \tfrac{1}{2}(W_{xx}+W_{yy}) & 0 \\ 0 & 0 & W_{zz} \end{pmatrix},$$

$$(W_1)_{rs} = \begin{pmatrix} \tfrac{1}{2}(W_{xx}-W_{yy}) & W_{xy} & 0 \\ W_{xy} & \tfrac{1}{2}(W_{xx}-W_{yy}) & 0 \\ 0 & 0 & 0 \end{pmatrix}, \quad \text{(D.15)}$$

$$(W_2)_{rs} = \begin{pmatrix} 0 & 0 & W_{xz} \\ 0 & 0 & W_{yz} \\ W_{xz} & W_{yz} & 0 \end{pmatrix}.$$

Bibliographical notes

The following references to text books and monographs is intended to supplement references cited in the text and to provide further background reading.

General references on plasma instabilities

A detailed, thorough but uncritical review of the literature on plasma instabilities is contained in the three volume series

Cap, F.F. (1976). *Handbook on Plasma Instabilities, Volume 1*. Academic Press: New York.

Cap, F.F. (1978). *Handbook on Plasma Instabilities, Volume 2*. Academic Press: New York.

Cap, F.F. (1982). *Handbook on Plasma Instabilities, Volume 3*. Academic Press: New York.

A systematic treatment is given in the two-volume set

Mikhailovskii, A.B. (1974a). *Theory of Plasma Instabilities, Volume 1, Instabilities of a Homogeneous Plasma*. Consultants Bureau: New York.

Mikhailovskii, A.B. (1976b). *Theory of Plasma Instabilities, Volume 2, Instabilities of an Inhomogeneous Plasma*. Consultants Bureau: New York.

Other books devoted primarily to plasma instabilities include

Briggs, R.J. (1964). *Electron-Stream Interaction with Plasmas*. M.I.T. Press: Cambridge, Mass.

Hasegawa, A. (1975). *Plasma Instabilities and Nonlinear Effects*. Springer-Verlag: Berlin.

Lominadze, D.G. (1981). *Cyclotron Waves in Plasma*. Pergamon Press: Oxford.

PART I: Introduction to plasma theory

For background reading on plasma physics there are numerous introductory texts, including.

Boyd, T.J.M. & Sanderson, J.J. (1969). *Plasma Dynamics*. Nelson: London.

Chandrasekhar, S. (1960). *Plasma Physics*. Univ. of Chicago Press: Chicago.

Chen, F.F. (1974). *An Introduction to Plasma Physics*. Plenum Press: New York.

Clemmow, P.C. & Dougherty, J.P. (1969). *Electrodynamics of Particles and Plasmas*. Addison-Wesley: Reading, Mass.

Krall, N.A. & Trivelpiece, A.W. (1973). *Principles of Plasma Physics*. McGraw-Hill: New York.

Schmidt, G. (1966). *Physics of High Temperature Plasmas*. Academic Press: New York.

The notations used here for *tensors* and for the *electromagnetic field* are essentially those used by

Jackson, W.D. (1975). *Classical Electrodynamics*. John Wiley & Sons: New York.

Panofsky, W.K.H. & Phillips, M. (1962). *Classical Electricity and Magnetism*. Addison-Wesley: Reading, Mass.

The use of *complex variables* and the *Landau prescription* are described in the plasma physics text books described above. The use of the Laplace transform rather than the Fourier transform is described in detail by

Montgomery, D.C. & Tidman, D.A. (1964). *Plasma Kinetic Theory*. McGraw Hill: New York.

The terminology for *waves in plasmas* is based on that used by

Stix, T.H. (1962). *The Theory of Plasma Waves*. McGraw Hill: New York.

Other books giving systematic discussions of waves in plasmas include

Akhiezer, A.I., Akhiezer, I.A., Polovin, R.V., Sitenko, A.G. & Stepanov, K.N. (1967). *Collective Oscillations in a Plasma*. M.I.T. Press: Cambridge, Mass.

Allis, W.P., Buchsbaum, S.S. & Bers, A. (1963). *Waves in Anisotropic Plasmas*. M.I.T. Press: Cambridge, Mass.

Denisse, J.F. & Delcroix, J.L. (1962). *Plasma Waves*. Interscience: New York.

The notation used for solving the wave equation and for treating weakly damped waves (cf. §2.6) is basically that of

Sitenko, A.G. (1967). *Electromagnetic Fluctuations in Plasma*. Academic Press: New York.

PART II Instabilities in unmagnetized plasmas

Further examples of reactive and kinetic instabilities are treated in more advanced text books on plasma physics, e.g.

Akhiezer, A.I., Akhiezer, I.A., Polovin, R.V., Sitenko, A.G. & Stepanov, K.N. (1975a & b). *Plasma Eectrodynamics, Volume 1: Linear Theory & Volume 2: Nonlinear Theory and Fluctuations*. Pergamon Press: Oxford.

Drummond, J.E. (1961). *Plasma Physics*. McGraw-Hill: New York.

Ichimaru, S. (1973). *Basic Principles of Plasma Plysics*. W.A. Benjamin Inc.: Reading, Mass.

Liftshitz, E.M., & Pitaevskii, L.P. (1981). *Physical Kinetics*. Pergamon Press: Oxford.

Wave trapping is discussed in books on nonlinear plasma theory, e.g.

Davidson, R.C. (1972). *Methods in Nonlinear Plasma Theory*. Academic Press: New York.

Sagdeev, R.Z. & Galeev, A.A. (1969). *Nonlinear Plasma Theory*. W.A. Benjamin: New York.

Wave turbulence theory is also discussed in the foregoing books on nonlinear plasma theory. The treatment given here is based on the approach adopted by Tsytovich:

Tsytovich, V.N. (1970). *Nonlinear Effects in Plasma*. Plenum Press: New York.

Tsytovich, V.N. (1977). *An Introduction to the Theory of Plasma Turbulence*. Pergamon Press: Oxford.

An older book on weak turbulence is

Kadomstev, B.B. (1965). *Plasma Turbulence*. Academic Press: New York.

Books in which the statistical aspects are discussed in more detail include

Klimontovich, Yu.L. (1967). *The Statistical Theory of Non-Equilibrium Processes in a Plasma*. Pergamon Press: Oxford.

Montgomery, D.C. (1971). *Theory of Unmagnetized Plasma*. Gordon and Breach: New York.

The theory of *strong plasma turbulence* is still under development and the reader is referred to the reviews cited in the text. An introduction to some aspects of the theory is included in the book:

Nicholson, D.R. (1983). *Introduction to Plasma Theory*. John Wiley & Sons: New York.

Part III Collision-dominated magnetized plasmas

Application of MHD theory to plasmas is discussed in most text books on plasma
  physics. The book
Thompson, W.B. (1962). *An Introduction to Plasma Physics*. Pergamon Press: Oxford.
includes a thorough introduction, including a discussion of MHD shock waves, cf. also
Anderson, J.E. (1963). *Magnetohydrodynamic Shock Waves*. M.I.T. Press: Cambridge,
  Mass.
  The standard text on MHD instabilities is
Chandrasekhar, S. (1961). *Hydrodynamic and Hydromagnetic Stability*. Oxford University
  Press: Oxford.
MHD instabilities in plasma machines are discussed in detail in
  Bateman, G. (1978). *MHD Instabilities*. M.I.T. Press: Cambridge, Mass.
  There is an extensive literature concerning the application of MHD Theory to
astrophysical plasmas. Some older books which remain of interest include.
Alfvén, H., & Fälthammar, C.-G. (1963). *Cosmical Electrodynamics*. Oxford University
  Press: Oxford.
Cowling, T.G. (1957). *Magnetohydrodynamics*. Interscience: New York.
Dungey, J.W. (1958). *Cosmic Electrodynamics*. Cambridge University Press: London.
  Collisional effects in plasmas are treated in
Spitzer, L., Jr. (1956). *Physics of Fully Ionized Gases*. Interscience: New York.

Part IV Instabilities in magnetized collisionless plasmas

Besides the books listed under Part I, *waves in magnetized plasmas*, with particular
  emphasis on the *magnetoionic theory* are discussed in
Budden, K.G. (1961). *Radio Waves in the Ionosphere*. Cambridge University Press.
Ginzburg, V.L. (1970). *The Propagation of Electromagnetic Wavs in Plasmas*. Pergamon
  Press: Oxford·
Radcliffe, J.A. (1959). *The Magneto-ionic Theory and its Application to the Ionosphere*.
  Cambridge University Press.
  The *kinetic equations* (§10.5) in the presence of a magnetic field were written down in a
form similar to that used here by
Tsytovich, V.N. (1972). *An Introduction to the Theory of Plasma Turbulence*. Pergamon
  Press: Oxford.
  Plasma Instabilities in astrophysical plasmas have been discussed in
Kaplan, S.A. & Tsytovich, V.N. (1973). *Plasma Astrophysics*. Pergamon Press: Oxford.
Melrose, D.B. (1980a & b). *Plasma Astrophysics, Volume 1, National Processes in
  Diffuse Magnetized Plasmas, & Volume 2, Astrophysical Applications*. Gordon and
  Breach: New York.
Zheleznyakov, V.V. (1970). *Radio Emission from the Sun and Planets*. Pergamon Press:
  Oxford.
Other books in which the theory of *solar radio emission* is discussed are
Krüger, A. (1979). *Introduction to Solar Radio Astronomy and Radio Physics*. D Reidel,
  Dordrecht.

McLean, D.J. & Labrum, N.R. (1985). *Solar Radio Physics.* Cambridge University Press.

The physics of *solar flares* is discussed in detail by

Švestka, Z. (1976). *Solar Flares.* D. Reidel: Dordrecht.

A recent detailed review is in

Sturrock, P.A. (1980). *Solar Flares.* Colorado Associated University Press: Boulder, Colorado.

A detailed description of many aspects of plasma physics in the solar-terrestrial environment is given in the three-volume series

Parker, E.N., Kennel, C.F. & Lanzerotti, L.J. (1979). *Solar System Plasma Physics, I. Solar and Solar Wind Plasma Physics.* North Holland: Amsterdam.

Kennel, C.F. Lanzerotti, L.J. & Parker, E.N. (1979). *Solar System Plasma Physics, II. Magnetospheres.* North Holland: Amsterdam.

Lanzerotti, L.J., Kennel, C.F. & Parker, E.N. (1979). *Solar System Plasma Physics, III. Solar System Plasma Processes.* North Holland: Amsterdam.

Relevant books specifically on magnetospheric plasma physics include

Lyons, L.R. & Williams, D.J. (1984). *Quantitative Aspects of Magnetospheric Physics.* D. Reidel: Dordrecht.

Schulz, M., & Lanzerotti, L.J. (1974). *Particle Diffusion in the Radiation Belts.* Springer-Verlag: Berlin.

References

Achterberg, A. (1981). On the nature of small amplitude Fermi acceleration. *Astron. Astrophys.* **97**, 259.

Alfvén, H. (1942). On the existence of electromagnetic-hydromagnetic waves. *Arkiv. Mat. Astron, Fysik. 29B(2)*.

Appleton, E.V. (1932). Wireless studies of the ionosphere. *J. Inst. Elec. Engrs.* **71**, 642.

Ashour-Abdalla, M. & Kennel, C.F. (1978). Nonconvective and convective electron cyclotron harmonic instabilities *J Geophys. Res.* **83**, 1531.

Aström, E.O. (1950). On waves in an ionized gas. *Arkiv. Fysik.* **2**, 443.

Bardwell, S. & Goldman, M.V. (1976). Three-dimensional Langmuir wave instabilities in type III solar radio bursts. *Astrophys. J* 209, 912.

Bateman, G. (1978). *MHD instabilities.* MIT Press: Cambridge Mass.

Bekefi, G. (1966). *Radiation Processes in Plasmas.* John Wiley & Sons: New York.

Benson, R.F., Calvert, W. & Klumpar, D.M. (1980). Simultaneous wave and particle observations in the auroral kilometric radiation source region. *Geophys. Res. Lett.* **7**, 959.

Bernstein, I.B. (1958). Waves in a plasma in a magnetic field *Phys. Rev.* **109**, 10.

Biskamp, D. (1982). Effects of secondary tearing instability on the coalescence of magnetic islands. *Phys. Lett.* **87A**, 357.

Bohm, D. & Gross, E.P. (1949a&b). The theory of plasma oscillations: A. Origin of medium-like behavior; B. Excitation and damping of oscillations. *Phys. Rev.* **75**, 1851 & 1864.

Braginskii, S.I. (1965). Transport processes in a plasma. *Rev. Plasma Phys.* **1**, 205.

Briggs, R.J. (1964). *Electron-Stream Interaction with Plasmas.* M.I.T. Press: Cambridge Mass.

Buneman, O. (1958). Instability, turbulence and conductivity in a current carrying plasma. *Phys. Rev. Lett.* **1**, 8.

Buneman, O. (1958). Dissipation of currents in ionized media. *Phys. Rev.* **115**, 503.

Cap. F.F. (1976). *Handbook on Plasma Instabilities Volume 1.* Academic Press: New York.

Cap, F.F. (1978). *Handbook on Plasma Instabilities Volume 2.* Academic Press: New York.

Chen, L. & Hasegawa, A. (1974). Plasma heating by spatial resonance of Alfvén waves. *Phys. Fluids* **17**, 1399.

Coppi, B. (1983). Magnetic reconnection driven by velocity space instabilities. *Astrophys. J.* **273**, L101.

Croley, D.R., Jr., Mizera, P.F. & Fennell, J.F. (1978). Signature of a parallel electric field in ion and electron distributions in velocity space. *J. Geophys. Res.* **83**, 2701.

Davidson, R.C. & Gladd, N.T. (1975). Anomalous transport properties associated with the lower-hybrid drift instability. *Phys. Fluids* **18**, 1327.

Dessler, A.J. (1983). *Physics of the Jovian Magnetosphere.* Cambridge University Press.

Dory, R.A., Guest, G.E. & Harris. E.G. (1965). Unstable electrostatic plasma waves propagating perpendicular to a magnetic field. *Phys. Rev. Lett.* **14**, 131.

Drury, L. O'C. (1983). An introduction to the theory of diffuse shock acceleration of energetic particles in tenuous plasmas. *Reports on Prog. in Phys.* **46**, 973.

Dulk, G.A. (1967). Apparent changes in the rotation rate of Jupiter. *Icarus* **7**, 173.

Faddeyeva, V.N. & Terent'ev, N.M. (1954). *Tables of Values of the Probability Integral for Complex Arguments.* GITTL: Moscow.

Ferrari, A., Trussoni, E., & Zaninetti, L. (1981). Magnetohydrodynamic Kelvin–Helmholtz instabilities in astrophysics—II cylindrical boundary layer vortex sheet approximation. *Mon. Not. Roy. Astron. Soc.* **196**, 1051.

Forslund, D.W. (1970). Instabilities associated with the heat conduction in the solar wind and their consequences. *J. Geophys. Res.* **75**, 70.

Fried, B.D. & Conte, S.D. (1961). *The Plasma Dispersion Function.* Academic Press: New York.

Furth, H.P., Killeen, J. & Rosenbluth, M.N. (1963). Finite-resistivity instabilities of a sheet pinch. *Phys. Fluids* **6**, 459.

Gaponov, A.V. (1959). Addendum. *Izv. VUZ. Radiofiz.* **2**, 836, Interaction between electron fluxes and electromagnetic waves in wave guides. *Izv. VUZ. Radiofiz.* **2**, 450.

Gehrels, T. (ed.) (1976). *Jupiter.* University of Arizona Press: Tucson.

Ginzburg, V.L. & Zheleznyakov, V.V. (1958). On the possible mechanism of sporadic solar radio emission (Radiation in an isotropic plasma). *Soviet Astron.* AJ **2**, 653.

Goldman, N.V. (1983). Progress and problems in the theory of type III solar radio emission. *Solar Phys.* **89**, 403.

Goldman, M.V. (1984). Strong turbulence of plasma waves. *Rev. Mod. Phys.* **56**, 709.

Goldman, M.V. & DuBois, D.F. (1982). Beam-plasma instability in the presence of low-frequency turbulence. *Phys. Fluids* **25**, 1062.

Grognard, R.J.-M. (1975). Deficiencies of the asymptotic solutions commonly found in the quasilinear relaxation theory. *Aust. J. Phys.* **28**, 731.

Gurnett, D.A. (1974). The Earth as a radio source: terrestrial kilometric radiation. *J. Geophys. Res.* **79**, 4227.

Gurnett, D.A., Marsch, E., Pilipp, W., Schwenn, R. & Rosenbauer, H. (1979). Ion acoustic waves and related plasma observations in the solar wind. *J. Geophys. Res.* **24**, 2029.

Harris, E.G. (1959). Unstable plasma oscillations in a magnetic field. *Phys. Rev. Lett.* **2**, 34.

Harris, E.G. (1961). Plasma instabilities associated with anisotropic velocity distributions. *J. Nucl. Energy* C **2**, 138.

Hartree, D.R. (1931). The propagation of electromagnetic waves in a refracting medium in a magnetic field. *Proc. Camb. Phil. Soc.* **27**, 143.

Helliwell, R.A. (1967). A theory of discrete VLF emissions from the magnetosphere. *J. Geophys. Res.* **72**, 4773.

Helliwell, R.A. & Inan, U.S. (1982). VLF wave growth and discrete emission triggering in the magnetosphere: A feedback model. *J. Geophys. Res.* **87**, 3537.

Herlofson, N. (1950). Magnetohydrodynamic waves in a compressible fluid conductor. *Nature* **165**, 1020.

Hewitt, R.G. & Melrose, D.B. (1985). The loss-cone driven instability for Langmuir waves in an unmagnetized plasma. *Solar Phys.* **96**, 157.

Hewitt, R.G., Melrose, D.B. & Rönnmark, K.G. (1982). The loss-cone driven electron-cyclotron maser. *Aust. J. Phys.* **35**, 447.

Huba, J.D. & Ossakow, S.L. (1981). On 11-cm irregularities during equatorial spread F. *J. Geophys. Res.* **86**, 829.

Ionson, J.A. (1978). Resonant absorption of Alfvénic surface waves and the heating of solar coronal loops. *Astrophys. J.* **226**, 650.

Kamilov, K., Khakimov, F. Kh., Stenflo, L. & Tsytovich, V.N. (1974). The enhancement of the interaction between transverse and longitudinal waves by turbulent plasmon condensation. *Physica Scripta* **10**, 191.

Kennel, C.F. (1969). Consequences of a magnetospheric plasma. *Rev. Geophys.* **7**, 379.

Kindel, J.M. & Kennel, C.F. (1971). Topside current instabilities. *J. Geophys. Res.* **76**, 3055.

Kingsep, A.S. Rudakov, L.I. & Sudan, R.N. (1973). Spectra of strong Langmuir turbulence . *Phys. Rev. Lett.* **31**, 1482.

Kintner, P.M. & Gurnett, D.A. (1978). Evidence of drift waves at the plasma pause. *J. Geophys. Res.* **83**, 39.

Krall, N.A. (1968). Drift waves. *Adv. Plasma Phys.* **1**, 153.

Kuijpers, J. (1980a). Turbulent bremsstrahlung of Langmuir waves. *Astron. Astrophys.* **83**, 201.

Kuijpers, J. (1980b). Theory of type IV dm bursts. In *Radio Physics of the Sun* (M.T. Kundu & T.E. Gergely eds) D. Reidel: Dordrecht, p. 341.

Kuijpers, J. & Melrose, D.B. (1985). Nonexistence of two forms of turbulent bremsstrahlung. *Astrophys. J.*, **294**, 28.

Kundu, M.R. (1985). *Solar Radio Astronomy* Interscience: New York.

Kuperus, M., Ionson, J.A. & Spicer, D.S. (1985). On the theory of coronal heating mechanisms. *Ann. Rev. Astron. Astrophys.* **19**, 7.

Kurth, W.S., Craven, J.D., Frank, L.A. Gurnett, D.A. (1979). Intense electrostatic waves near the upper hybrid resonance frequency. *J. Geophys. Res.* **84**, 4145.

Landau, L.D. (1946). On the vibrations of the electronic plasma. *J. Phys.* (USSR) **10**, 25.

Lin, A.P., Kaw, P.K. & Dawson, J.M. (1973). A possible plasma laser. *Phys. Rev. A* **8**, 2618.

Lin, R.P., Anderson, K.A., McCoy, J.E. & Russell, C.T. (1977). Observations of magnetic merging and the formation of the plasma sheet in the Earth's magnetotail. *J. Geophys. Res.* **82**, 2761.

Liu, C.S. & Kaw, P.D. (1976). Parametric instabilities in homogeneous unmagnetized plasmas. *Adv. Plasma Phys.* **6**, 83.

Manchester, R.N. & Taylor, J.H. (1977), *Pulsars*. Feeman: San Francisco.

Matsumoto, H. (1979). Nonlinear whistler-mode interaction and triggered emissions in the magnetosphere: A review. In *Wave Instabilities in Space Plasmas*, (P.J. Palmadesso & K. Papadoulos eds) D. Reidel: Dordrecht, p. 163.

McKenzie, J.F., Ip, W.-H. & Axford, W.I. (1979). The acceleration of minor ion species in the solar wind. *Astrophys. Space Sci.* **64**, 183.

Melrose, D.B. (1978). Amplified linear acceleration emission applied to pulsars. *Astrophys. J.* **225**, 557.

Melrose, D.B. (1980). The emission mechanisms for solar radio bursts. *Space Sci. Rev.* **26**, 3.

Melrose, D.B. (1982). 'Plasma emission' without Langmuir waves. *Aust. J. Phys.* **35**, 67.

Melrose, D.B. & Dulk, G.A. (1982). Electron–cyclotron masers as the source of certain solar and stellar radio bursts. *Astrophy. J.* **259**, 844.

Melrose, D.B. & Kuijpers, J. (1984). Resonant parts of nonlinear response tensors. *J. Plasma Phys.* **32**, 239.

Mikhailovskii, A.B. (1967). Oscillations of an inhomogeneous plasma. *Rev. Plasma Phys.* **3**, 159.

Mikhailovskii, A.B. (1974a,b). *Theory of Plasma Instabilities, Volumes 1 & 2*. Consultants Bureau: New york.

Nishikawa, K. (1968). Parametric excitation of coupled waves 1. General formulation. *J. Phys. Soc. Japan* **24**, 916.

Nunn, D. (1974). A self-consistent theory of triggered VLF emissions. *Planet. Space Sci.* **22**, 349.

Omidi, N. & Gurnett, D.A. (1982). Growth rate calculations of auroral kilometric radiation using the relativistic resonance condition. *J. Geophys. Res.* **87**, 2377.

O'Neil, T.M. & Malmberg, J.H. (1968). Transition of the dispersion roots from beam-type to Landau-type solutions. *Phys. Fluids* **11**, 1754.

O'Neil, T.M., Winfrey, J.H. & Malmberg, J.H. (1971). Nonlinear interaction of a small cold beam and a plasma. *Phys. Fluids* **14**, 1204.

Ott, E., Wersinger, J.-M. & Bonoli, P.T. (1978). Theory of plasma heating by magnetosonic cavity mode absorption . *Phys. Fluids,* **21**, 2306.

Parker, E.N. (1965). Suprathermal hydromagnetic waves. *Astrophys. J.* **142**, 1086.

Penrose, O. (1960). Electrostatic instabilities of a uniform non-Maxwellian plasma. *Phys. Fluids* **3**, 258.

Petschek, H.E. (1964). Magnetic field annihilation. In *The Physics of Solar Flares* (W.N. Hess ed.) NASA Publ. SP-50, p. 425.

Porkolab, M. (1976). High frequency parametric wave phenomena and plasma heating: A review. *Physica* **82C**, 86.

Priest, E.R. (1984). *Solar Magneto-hydrodynamics*. D. Reidel: Dordrecht.

Pringle, J.E. (1981). Accretion discs in astrophysics. *Ann. Rev. Astron. Astrophys.* **19**, 137.

Pritchett, P.L., Lee, Y.C. & Drake, J.F. (1980). Linear analysis of the double-tearing mode. *Phys. Fluids* **23**, 1368.

Puri, S. Leuterer, F. & Tutter, M. (1973). Dispersion curves for generalized Bernstein modes. *J, Plasma Phys.* **9**, 89.

Ramaty, R., McKinley, J.M.& Jones, F.C. (1982). On the theory of gamma-ray amplification through stimulated annihilation radiation. *Astrophys. J.* **256**, 238.

Roberts, B. (1981). Wave propagation in magnetically structured atmosphere 1: surface waves at a magnetic interface. *Solar Phys.* **69**, 27.

Roberts, C.S. (1969). Pitch angle diffusion of electrons in the magnetosphere. *Rev. Geophys.* **7**, 305.

Rudakov, L.I. & Tsytovich, V.N. (1978). Strong Langmuir turbulence. *Phys. Reports* **40**, 1.

Ruderman, M.A. & Sutherland, P.G. (1975). Theory on pulsars: polar gaps, sparks, and coherent microwave radiation. *Astrophys. J.* **196**, 51.

Sagdeev, R.Z. (1979). The 1976 Oppenheimer lectures: Critical problems in plasma astrophysics—I. Turbulence and nonlinear waves. *Rev. Mod, Phys.* **51**, 1.

Sagdeev, R.Z. & Shafranov, V.D. (1961). On the instability of a plasma with an anisotropic distribution of velocities in a magnetic field. *Soviet Phys. JETP* **12**, 130.

Sarris, E.T. & Axford, W.I. (1979). Energetic protons near the plasma sheet boundary. *Nature* **277**, 460.

Scheuer, P.A.G. (1960). The absorption coefficient of a plasma at radio frequencies. *Mon. Not. Roy. Astron. Soc.* **120**, 231.

Sedlácek, Z. (1971). Electron oscillations in cold inhomogeneous plasma I. Differential equation approach. *J. Plasma Phys.* **5.**, 239.

Shapiro, V.D. (1963). Nonlinear theory of the interaction of a monoenergetic beam with a plasma. *Soviet Phys. JETP* **17**, 416.

Sieber, W. & Wielebinski, R. (1980). *Pulsars*. D. Reidel: Dordrecht.

Sitenko, A.G. & Stepanov, K.N. (1957). On the oscillations of an electron plasma in a magnetic field. *Soviet Phys. JETP* **4**, 512.

Sizonenko, V.L. & Stepanov, K.N. (1967). Plasma instability in the electric field of an ion-cyclotron wave. *Nucl. Fusion* **7**, 131.

Spicer, D.S. (1981). Loop models of solar flares: revisions and comparisons. *Solar Phys.* **70**, 149.

Sprangle, P. & Coffey, T. (1984). New sources of high-power coherent radiation. *Physics Today* **37**, *No. 3*, 44.

Sprangle, P. & Drobot, A.T. (1977). The linear and self-consistent nonlinear theory of the electron cyclotron maser instability. *IEE Trans. MTT* **MTT-25**, 528.

Sprangle, P. & Smith, R.A. (1980). The nonlinear theory of efficiency enhancement in the electron cyclotron maser (gyrotron). *J. Appl. Phys.* **51**, 3001.

Stix, T.H. (1962). *The Theory of Plasma Waves.* McGraw Hill: New York.

Stix, T.H. (1975). Fast-wave heating of a two-component plasma. *Nucl. Fusion* **15**, 737.

Sturrock, P.A. (1958). Kinematics of growing waves. *Phys. Rev.* **112**, 1488.

Thornhill, S.G. & ter Haar, D. (1978). Langmuir turbulence and modulational instability. *Phys. Reports* **43**, 43.

Tidman, D.A. & Krall, N.A. (1971). *Shock Waves in Collisionless Plasmas.* Wiley-Interscience: New York.

Tokar, R.L. & Gary, S.P (1984). Electrostatic hiss and the beam driven electron acoustic instability in the dayside polar cusp. *Geophys. Res. Lett.* **11**, 1180.

Tonks, L. & Langmuir, I, (1929). Oscillations in ionized gases. *Phys. Rev.* **33**, 195.

Toptyghin, I.N. (1980). Acceleration of particles by shocks in cosmic plasma. *Space Sci. Rev.* **26**, 157.

Trubnikov, B.A. (1965). Particle interactions in fully ionized plasma. *Rev. Plasma Phys.* **1**, 105.

Tsytovich, V.N. (1972). *An Introduction to the Theory of Plasma Turbulence.* Pergamon: Oxford.

Tsytovich, V.N., Stenflo, L. & Wilhelmsson, H. (1975). Current flow in ion-acoustic and Langmuir turbulence plasma interaction. *Physica Scripta* **11**, 251.

Twiss, R.Q. (1958). Radiation transfer and the possibility of negative absorption in radio astronomy. *Aust. J. Phys.* **11**, 564.

Vasyliunas, V.M. (1975). Theoretical models of magnetic field line merging, 1. *Rev. Geophys. Space Phys.* **13**, 303.

Weibel, E.S. (1959). Spontaneously growing transverse wavé in a plasma due to an anisotropic velocity distribution. *Phys. Rev. Lett.* **2**, 83.

Wentzel, D.G. (1974). Cosmic-ray propagation in the galaxy: collective effects. *Ann. Rev. Astron. Astrophys.*,**12**, 71.

Wentzel, D.G. (1979). Hydromagnetic surface waves. *Astrophys. J.* **227**, 319.

Wild, J.P., Smerd, S.F. & Weiss, A.A. (1963). Solar bursts. *Ann. Rev. Astron. Astrophys.* **1**, 291.

Winglee, R.M. (1985). Enhanced growth of whistlers due to bunching of untrapped electrons. *J. Geophys. Res.* **90**, 5141.

Wu, C.S. & Lee, L.C. (1979). A theory of the terrestrial kilometric radiaton. *Astrophys. J.* **230**, 621.

Young, D.T., Perraut, S., Roux, A., de Villedary, C., Gendrin, R., Korth, A., Kresmer, G. & Jones, D. (1981). Wave-particle interaction near Ω_{He^+} observed on GEOS 1 and 2. 1. Propagation of ion cyclotron waves in He$^+$-rich plasma. *J. Geophys. Res*, **86**, 6755.

Zaitsev, V.V. & Stepanov, A.V. (1983). The plasma radiation of flare kernels. *Solar Phys.* **88**, 297.

Zakharov, V.E. (1972). Collapse of Langmuir waves. *Soviet Phys. JETP* **35**, 908.

Zheleznyakov, V.V. & Zlotnik, E. Ya. (1975). Cyclotron wave instability in the corona and origin of solar radio emission with fine structure—II. Origin of 'tadpoles'. *Solar Phys.* **44**. 447.

List of commonly used symbols

Latin alphabet

a   unit vector (8.25, 10.33)

$a_{ij}(\mathbf{k}, \mathbf{k}', \mathbf{v})$ Compton scattering coefficient (6.31)

A   vector potential

A(superscript) anti-hermitian part

b   unit vector along static magnetic field

B   magnetic field

B $= |\mathbf{B}|$

c speed of light

c_s adiabatic sound speed (8.18)

$d\lambda^{(n)}$ n-fold convolution integral (2.6)

$\hat{D}_s$ differential operator (10.88)

$D_{ij}(\mathbf{p})$ diffusion coefficient in **p**-space (6.43)

e unit charge

$\mathbf{e}_M(\mathbf{k})$ polarization vector for mode M

$f(\mathbf{p}, t, \mathbf{x})$ distribution function

$f(\mathbf{p}, \omega, \mathbf{k})$ Fourier transform of $f(\mathbf{p}, t, \mathbf{x})$

$F(v)$ one-dimensional distribution function (4.2)

$g_{ij}(\mathbf{k}, \mathbf{v})$ (6.29)

$G(t)$ gain factor (5.10)

$H(t)$ step function (1.33)

H(superscript) hermitian part

Im   imaginary part

$I_s(\lambda)$ modified Bessel function

j parameter in DGH distribution (11.19)

J   current density

$J_s(z)$ Bessel function

$\mathbf{k}, k_\perp, k_\parallel$ wave vector and perpendicular and parallel components

$K_{ij}(\omega, \mathbf{k})$ dielectric tensor (1.27)

$K_M(\mathbf{k})$ longitudinal part of polarization (10.32)

L(superscript) longitudinal part

L(subscript)   Langmuir mode

m, m_e, m_p, m_i masses

M(subscript) mode M

M_A Alfvén Mach number (8.74)

$n, n_e, n_i, n_1, n(t, \mathbf{x}), n(\omega, \mathbf{k})$ number densities

n(subscript)   normal component

$N, N_\pm, N_M$ refractive indices

$N_M(\mathbf{k})$ occupation number for mode M (6.12)

NL(superscript)   nonlinear

$\mathbf{p}, \mathbf{p}(\phi), \mathbf{p}(t, \mathbf{x}), \mathbf{p}(\omega, \mathbf{k})$ particle momentum

$P, P^{(\alpha)}$ pressure

P Cauchy principal value (1.18)

q charge

r_0 classical radius of electron

R radius of gyration (10.15)

Re   real part

$R_M(\mathbf{k})$ ratio of electric to total energy (2.66)

s(subscript)   ion sound mode

t time

T temperature, normalization time

T (subscript) transverse waves

T(superscript) transverse part

T_M axial ratio of polarization ellipse for mode M (10.32)

$T_M(\mathbf{k})$ effective temperature for mode M (6.69)

TR(superscript)   transient response

$u_{MPQ}(\mathbf{k}, \mathbf{k}', \mathbf{k}'')$ 3-wave probability (6.25)

$U, \mathbf{U}, U_e, U_i$ streaming speeds or velocities

$\mathbf{v}, v = |\mathbf{v}|$ particle velocity

$\mathbf{v}^{(\alpha)}$ flow velocity for species α

v_A Alfvén speed

$\mathbf{v}_b$ beam velocity

$\mathbf{v}_d$ drift velocity

$\mathbf{v}_{gM}$ group velocity for mode M

v_s ion sound speed

$v_\phi, v_\pm(\theta)$ phase speeds

$V, V_e, V_i, V_\perp, V_\parallel$ thermal speeds (1.2)

V normalization volume

$\mathbf{V}(\mathbf{k}, \mathbf{p}, s)$ velocity function for spiralling charge (10.20)

$w_M(\mathbf{k}, \mathbf{p}), w_M(\mathbf{k}, \mathbf{p}, s)$ emission probabilities (6.14, 10.83)

$w_{MP}(\mathbf{k}, \mathbf{k}', \mathbf{p})$ scattering probability (6.35)

$\mathbf{x}$ position vector

X	magnetoionic parameter $(= \omega_p^2/\omega^2)$
$\mathbf{X}(t)$	orbit of particle (5.1)
Y	magnetoionic parameter $(= \Omega_e/\omega)$
Z_i	charge number $(= q_i/e)$

Greek alphabet

α	pitch angle
α_0	loss cone angle
$\alpha_{ijl}(k,k_1,k_2)$	quadratic response tensor (6.16b)
$\alpha_{ijlm}(k,k_1,k_2,k_3)$	cubic response tensor (6.16c)
β	plasma beta
γ	Lorentz factor $(= (1 - v^2/c^2)^{-1/2})$
$\gamma_M(\mathbf{k})$	absorption coefficient for waves in mode M
δ_{ij}	Kronecker delta (1.23)
$\delta(\omega), \delta^3(\mathbf{k})$	Dirac delta functions
ε	particle energy $(= \gamma m c^2)$
ε_0	permittivity of free space
$\epsilon, \epsilon_\alpha$	sign of charge $(= q/|q|, q_\alpha/|q_\alpha|)$
ε_{ijl}	permutation symbol (10.3)
η	mass density
θ	wave angle (between $\mathbf{k}$ and $\mathbf{b}$)
κ	unit vector along $\mathbf{k}$
λ_α	$k_\perp^2 V_\alpha^2/\Omega_\alpha^2$
$\lambda_{D\alpha}$	Debye length for species α
$\lambda_{ij}(\omega,\mathbf{k})$	cofactor of $\Lambda_{ji}(\omega,\mathbf{k})$
$\Lambda(\omega,\mathbf{k})$	determinant of $\Lambda_{ij}(\omega,\mathbf{k})$
$\Lambda_{ij}(\omega,\mathbf{k})$	coefficient of wave equation (1.29)
$\ln\Lambda$	Coulomb logarithm (9.24)
μ_0	permeability of free space
$\nu_0, \nu_{ee}, \nu_{ei}, \nu_{ii}$	collision frequencies
ξ	fluid displacement (8.13)
ρ	charge density
σ	electric conductivity (9.27)
$\sigma_{ij}(\omega,\mathbf{k})$	conductivity tensor (1.26)
τ_R, τ_A	diffusion and Alfvén times (9.32 & 33)
$\tau_{ij}^{(\alpha)}(\omega)$	response tensor (10.8)
ϕ	azimuthal angle (10.12)
$\phi(z)$	plasma dispersion function (2.30)
Φ	electrostatic potential
χ_{ij}	susceptibility tensor
ψ	azimuthal angle for $\mathbf{k}$ (10.16)
$\psi(t,z), \psi(t,\mathbf{x}), \psi$	phase factors

ω wave frequency

$\omega_p, \omega_{pi}, \omega_{p1}, \omega_{p\alpha}$ plasma frequencies

ω_i $= \mathrm{Im}\,\omega$

ω_{UH}, ω_{LH} hybrid frequencies (10.62)

ω_T bounce frequency (5.33), (11.64)

ω_x, ω_z cutoff frequencies (10.46)

Ω $= \omega_0 - k_0 v_0$ (chapter 5 only)

$\Omega, \Omega_0, \Omega_e, \Omega_i, \Omega_\alpha$ gyrofrequencies

Author index

Subject index

Printed in the United States
By Bookmasters